AF553628

HORMONES OF FISHES

ENCYCLOPAEDIA OF FISH AND FISHERIES-V

HORMONES OF FISHES

By

Dr. Arvind N. Shukla

School of Studies of Zoology & Biotechnology

Vikram University

Ujjain (India)

DISCOVERY PUBLISHING HOUSE PVT. LTD.

NEW DELHI-110 002

Published by:
Namit Wasan
DISCOVERY PUBLISHING HOUSE PVT. LTD.
4383/4B, Ansari Road, Darya Ganj
New Delhi-110 002 (India)
Phone : +91-11-23279245; 23253475; 43596065
E-mail : discoverybooksindia@gmail.com
discoverypublishinghouse@gmail.com
namitwasan9@gmail.com
web : www.discoverypublishinggroup.com

***Edition:* 2020**

ISBN: 978-81-8356-303-1 (Set)

ISBN: 978-81-8356-391-8

Hormones of Fishes

Printed at:
Infinity Imaging Systems
Delhi

Preface

The present title *"Encyclopaedia of Fish and Fisheries"* has been designed for undergraduate and postgraduate students of all Indian Universities. The text of the present title is organized on some major areas like fishing techniques, fish behaviour, and fish physiology, fisheries of important Indian forms. Fishes inhabit every kind of aquatic environment, and their wide distribution resulted in many different designs for their special mode of life. The present title, with a skilful mix of breadth combined with detail, reviews what is known about fishes. Biological inter-relationship are fully discussed, and fishes with features of special interest are highlighted, whether they are economically significant or not. The present text brings together many scattered observations as well as the results of recent investigations. It is hoped that the general biologist, the zoologist and the Icthyologist will find wealth of exciting information in it.

The present title gives a comprehensive overview of the fishery science and also offers diverse information on the subject of fisheries to make the readers up-to-date with the latest in this field where proliferation of scientific information has not only been fast but enormous too in the last decade. Written in scholarly yet easy to understand language. It combines the usefulness of a reference work with the readability of a browsing book, perfect for any one attuned to the splendor of our natural world. It is hoped that the book will prove indispensable to fisheries organizations, lecturers and the young zoologists engaged in teaching and research.

In the preparation of this book large number of books and research papers have been consulted. So no authenticity is claimed.

The author expresses his gratitude to Mr. Wasan and staff of M/s Discovery Publishing House for their whole hearted co-operation in the publication of this book.

The author tried hard to be accurate and upto date in statement and realises the impossibility of completely avoiding errors therefore, the author will greatly appreciate having his attention called to any questionable statement.

Author

Contents

1

INTRODUCTION

Vertebrates respond to environmental stimuli through sense organs, the brain, and/or the spinal cord which relay impulses to muscles or glands. In fishes the muscular responses frequently result in movements of the entire body rather than of the appendages only. Certain cyclic and slow changes in the environment, such as seasonal temperature rises and changes in day length also affect one or more endocrine glands, by way of the nervous system. The hormonal secretions of endocrine glands act on specific target organs, which may be endocrine themselves, or the secretions have a diffuse effect on metabolism in general.

The nervous and endocrine systems are highly interdependent and often act together. Reproductive behaviour of many fishes, for example, is influenced by the senses through their perception of the environmental changes that come with approaching spring; several endocrine glands are affected and in turn act on the ovaries and testes. There is further interaction among sense organs, central nervous system, and endocrine glands at the time of physiological readiness for spawning. At spawning time, more or less complex behavioural patterns are released through visual, chemical, or auditory signals. The spawning act itself, through nervous impulses, readjusts the state of endocrine tissues.

The difficulty of describing briefly the complexity of the interaction of the components of the nervous and endocrine systems in fishes is amplified by the extreme numbers of fishes and their diverse adaptations to the great variety of habitats they occupy. Fishes have more varied habitats and therefore greater anatomical differences than any other group of vertebrates. Extensive reviews of the frequently controversial information on fish nervous and endocrine systems exist; they, as well as selected specific articles, are cited at the end of this chapter.

THE NERVOUS SYSTEM

The basic pattern and progress of cephalisation, the concentration of stimulus receiving and integ rating units in the head, can be followed from the lamprey-hagfish group (Cyclostomata) through the shark-like fishes (Chondrichthyes) to the bony fishes (Osteichthyes). The general organisation of the nervous system in each of these three major fish groups. The general anatomy and location of the organs of special sense and the endocrine glands have been given in the same chapter. We shall now consider certain details of the anatomy and physiology of the nervous system, sense organs, and the endocrine glands.

Connections of the various parts of the central nervous system are summarised in Figure elsewhere in this chapter.

The Telencephalon

The most anterior region of the brain, the telencephalon or forebrain, in fishes is prominently devoted to the reception, elaboration, and conduction of smell impulses. Its relative size varies with the degree to which smell plays a role in the life of the kind of fish examined. The olfactory nerve, or cranial nerve I, enters the brain from the sensory epithelium of the olfactory plate or placode that is situated in the nostril or nasal pit. In sharks and their relatives (Elasmobranchii) and bony fishes (Osteichthyes), the right and left olfactory nerves are accompanied by another pair of nerves, the small terminal nerves (0). They are believed to have vasomotor rather than sensory functions. Although the organ of smell of lampreys has secondarily assumed a median position and is unpaired, the underlying brain structures are paired as in other vertebrates. Anteriorly and on each side, the telencephalon consists of an olfactory bulb followed caudally by an olfactory lobe; its two internal cavities are brain ventricles I and II. Elasmobranchs especially and those bony fishes that depend largely on smell for feeding or social interaction have pronouncedly enlarged olfactory lobes. Ventrolaterally these lobes contain a large ganglion, the corpus striatum, which is a center of correlation mainly to relay smell impulses to posterior sensory or somatic nerve centers. The anteriormost large bundle of nerve fibers that connects the two sides of the brain is the anterior commissure. This commissure is located in the telencephalon and runs, in part, through the so-called terminal lamina, the median anterior wall of the entire neural canal system.

Although olfaction is an obvious function, it is not the sole concern of the rayfin fishes (Actinopterygii) telencephalon which is thought to serve an additional generalised function facilitating activities of lower

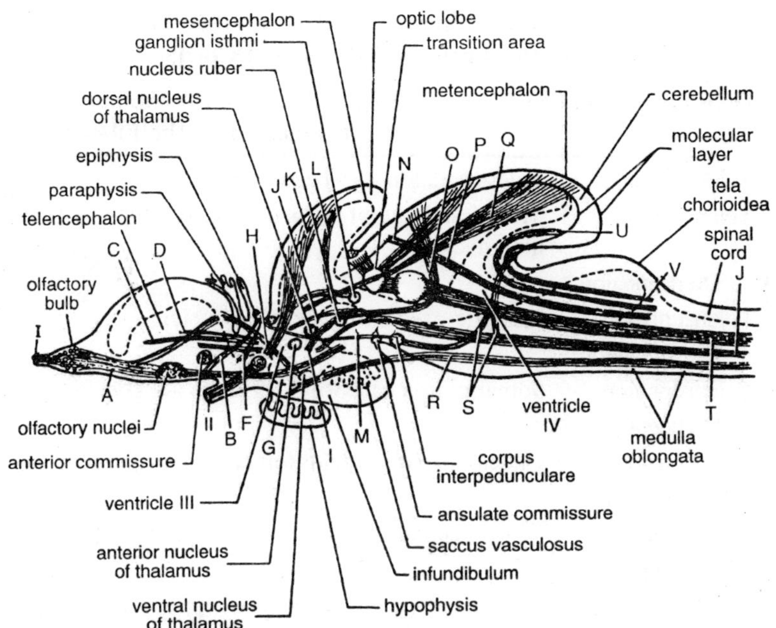

*Figure 1.1 : Diagrammatic section of bony-fish brain with main fiber tracts, as shown in a carp (*Cyprinus*). Nerve tracts: A, olfactory; B, olfactohabenular; C, hippocampothalamic; D, striothalamic; E, striohypothalamic; F, preopticohabenular; G, olfactohypothalamic; H, pallial; I, spinohypothalamic; 1, spinothalamic; K, opticotectal, L, tectobulbar; M, fasciculus retroflexus; N, tegmentocerebellar; 0, diencephalocerebellar; P, spinocerebellar; Q, mesencephalocerebellar; R, lobopedunculospinal; S, cerebellotegmental; T, facialicerebellar; U, lateralicerebellar; V, vestibulocerebellar. Cranial nerves are: I, olfactory; II, optic; III, oculomotor; and IV, trochlear.*

brain centers and mechanisms. The morphogenetic development of the forebrain of the rayfins is entirely different from all other vertebrates in that it forms by eversion of the lateral walls of the dorsal (pallial) area as opposed to evagination which results in lateral ventricles. Consequently, the actinopterygian forebrain retains a single, median, ventral ventricle (subpallium) in contrast to the paired ventral (basal) ventricles in lower fishes and other vertebrates.

The telencephalon of fishes is extensively innervated by secondary olfactory fibers from the olfactory bulb. These fibers project throughout the subpallial portion of each hemisphere, and also project to the posterior basolateral areas of the pallium. The remainder of the pallium of the actinopterygian telencephalon appears not to receive secondary olfactory fibers. There is increasing differentiation and subdivision of

Table 1.1 : Important Nerve Fiber Tracts in the Brain and the Spinal Cord.

Divisions of Central Nervous System	Regions of Brain or Spinal Cord Connected by the Fibers
Telencephalon	Telencephalon $\leftrightarrow$[d] Telencephalon
Telencephalon Diencephalon	Medial and lateral walls of olfactory lobe $\rightarrow$ Habenular nuclei
Telencephalon Diencephalon	Preoptic area $\rightarrow$ Habenular nuclei
Telencephalon Diencephalon	Somatic area of corpus striatum $\rightarrow$ Hypothalamus
Telencephalon Diencephalon	Olfacto-somatic area of forebrain $\rightarrow$ Ventral thalamus
Telencephalon Diencephalon	Olfacto-somatic area of forebrain $\leftrightarrow$ Dorsal thalamus
Telencephalon Diencephalon	Hypothalamus $\leftrightarrow$ Primordial hippocampus and precommissural septal areas of forebrain
Diencephalon Mesencephalon	Lateral geniculate nuclei $\leftrightarrow$ Optic tectum
Diencephalon Mesencephalon	Habenular nuclei $\rightarrow$ Interpeduncular nuclei in base of mid brain
Mesencephalon Diencephalon	Optic tectum $\leftrightarrow$ Thalamus
Mesencephalon Diencephalon	Optic tectum $\rightarrow$ Hypothalamus
Mesencephalon Metencephalon	Optic tectum $\rightarrow$ Cerebellum
Metencephalon Mesencephalon	Cerebellum $\rightarrow$ Tegmentum
Mesencephalon Myelencephalon	Nuclei of nerve V $\rightarrow$ Optic tectum
Mesencephalon Spinal cord	Optic tectum $\rightarrow$ Spinal cord

Mesencephalon Myelencephalon	Lateral line nuclei of medulla → Torus semicircularis and valvular of nuclei in tegmentum
Mesencephalon Spinal cord	Spinal cord → Tegmentum
Myelencephalon Mesencephalon	Gustatory lobes of medulla → Secondary gustatory nucleus of midbrain and also hypothalamus
Spinal cord Diencephalon	Spinal cord → Thalamus
Metencephalon Myelencephalon	Vestibular nuclei and fibers of Nerve VIII → Cerebellum
Metencephalon Myelencephalon	Lateral line nuclei of medulla ↔ Cerebellum
Myelencephalon Spinal cord	Nuclei of nerves VII to X in medulla → Spinal cord
Di-, Mes-, and Met- encephalon Spinal cord	Several parts of brain → Spinal cord of trunk and tail

Name of Tract	Function of Tract
Anterior commissure	Connects the two sides of the forebrain
Median and lateral olfacto-habenular	Project smell impulses to habenular nuclei as way stations for feeding reflexes
Preoptico-habenular	Tertiary olfactory tract, a feeding reflex path
Strio-hypothalamic	Conveys forebrain impulses to hypothalamic centers
Ventral peduncle of lateral forebrain bundle	Evolutionary beginning of regulation of motor responses by forebrain
Dorsal peduncle of lateral forebrain bundle	Primitive discharge path between thalamus and basal ganglia of forebrain

Medial forebrain bundle	Leads visceral, including taste, sensations to and from olfactory forebrain region
Geniculo-tectal Tectogeniculate (branch of superior colliculus)	Relays optic impulses
Fasciculus retroflexus (habenulo-interpeduncular)	Part of discharge paths of olfacto-somatic areas to motor centers (feeding reflexes)
Thalamo-tectal-Tectothalamic	Relate the highly correlative optic tectum with the developing dorsal thalamus
Tecto-hypothalamic	Connects sight impulses with autonomic centers and pituitary region
Tecto-cerebellar	Relates sight and other sensory impulses with muscle tonus for locomotion
Cerebello-tegmental	Portion of relay system to motor centers regulating muscle tonus
Bulbo-tectal	Coordinates tactile sense organs on head with vision
Tecto-spinal	Coordinates sight and other sensory impulses with locomotion
Acoustico-lateral lemniscus	Projects lateral line impulses onto midbrain
Spino-tegmental (probably multisynaptic)	Regulates locomotion
Secondary and tertiary ascending gustatory tracts	Coordinate gustatory and olfactory feeding stimuli, especially in catfishes (Siluroidei)
Spino-thalamic	Projects general body sensations on thalamus (the presence of this tract in fishes has been doubted)
Vestibulo-cerebellar and direct vestibular root fibers	Part of equilibration system
Latero-cerebellar Cerebello-	Coordinate lateral line and locomotory impulses lateral
Vestibulo-spinal	Relays lateral line and vestibular

	impulses to lower medullary and spinal centers
Giant cells of Mauthner	Coordinate trunk and tailbeat in swimming

the dorsal area through the phylogenetic series from Polypteriformes to Teleostei.

Ablation of the forebrain has shown it to be concerned with activities other than integration of olfactory input. Forebrainless goldfish (Carassius *auratus*) show less initiative and spontaneity than intact animals in exploring their environment. Certain tilapias (*Tilapia*) neglect their young when the forebrain is removed and certain minnows (Cyprinidae) so treated are less wary and distrustful of a new situation than their normal experimental running mates. Forebrain ablations result in suppression of aggressive, sexual, and parental behaviour in the threespine stickleback (*Gasterosteus aculeatus*). They also suppress spawning and nest-building behaviour in the paradise fish (*Macropodus opercularis*). Forebrain extirpation results in a selective suppression of colour-vision discrimination in goldfish.

The Diencephalon

A dorsoventral and often also a pronounced lateral constriction marks the transition from fish telencephalon to diencephalon or 'tween brain. Dorsally the parencephalon, or the thin saccus dorsalis, forms the roof of the cavity within the diencephalon, also called the third ventricle. In some fishes such as lampreys (*Lampetra*), gars (*Lepisosteus*), and the bowfin (*Amia*), the saccus dorsalis is enlarged and extends laterally and frontally, even over the telencephalon. There are two landmarks along the diencephalic midline in the lampreys (Petromyzonidae), the parapineal and pineal organs, also known as the epiphysial organs, located slightly left and right of the midline respectively. However, in adult sharks and relatives (Elasmobranchii) and in higher bony fishes (Actinopterygii) only the pineal organ is developed. Embryos of some bony fishes, such as the whitefishes (*Coregonus*), show both organs during early embryonic development, but the parapineal is subsequently lost and only the pineal remains. Wherever both or one of these evaginations persist their structure and function have been linked to the reception of more or less diffuse light stimuli.

In lampreys and hagfishes (Cyclostomata), the pineal organ is connected to the right habenular ganglion, and has a retina, pigment cells, and a lens-like structure. There is an orifice in the cranial roof above it, just behind the nasal opening, and a depigmented spot in the skin so that light stimuli can reach the pineal and parapineal organs. Some sharks (Squaliformes) have an unpigmented spot in the skin of this region of the head and some of them also have a perforation of the cartilaginous cranial roof. However, in the sharks there is generally a less well-developed pineal organ than among cyclostomes and some bony fishes. Particularly well-developed pineal spots are found in vertically migrating, mid-depth fishes such as the hatchetfishes (*Argyropelecus*), in many catfishes (such as *Arius* and *Macrones*), and also in halfbeaks (*Hemiramphus*).

In spite of numerous studies on the pineal body of fishes, no distinct function has been unequivocally demonstrated. The current hypotheses may be divided into two categories-one emphasizing primarily a sensory role and the other- a secretory function. The following hypotheses relate to sensory roles: the pineal body is a photosensory structure; it acts as a baro- or chemoreceptor for cerebrospinal fluid; or it functions as a mediator in the olfactory response to exohormones. In secretory roles the pineal gland may be primarily a gland of external secretion related to the chemical composition of the cerebrospinal fluid or the metabolism of brain tissue or, alternatively, the pineal may be a gland of internal secretion with an endocrine function.

The rest of the diencephalon can be divided into an epithalamic region with its habenular ganglia, a thalamus, and a hypothalamus as its largest and most important component. The optic nerves enter the brain and cross anterior to the diencephalon, and posteriorly this region of the brain is joined to the ventral portion of the mesencephalon through the convoluted, thin-walled saccus vasculosus. Below the hypothalamus lies the hypophysis or pituitary gland. This gland is closely attached to the floor of the brain in lampreys but among sharks and higher bony fishes it is carried on a stalk or infundibulum. The thalamus serves as a relay center for the transfer of olfactory and striate-body impulses to thalamo-medullar and thalamospinal tracts. In its ventral region, toward the brain floor and the hypothalamus there lie the geniculate nuclei or ganglia which receive some optic nerve (II) connections before the optic tectum is reached by the axons of this cranial nerve. These nuclei are well-developed in sharks, even to an extent that warrants their being called geniculate lobes; however, nothing certain is known about their function. In the optic chiasma of all fishes the nerve fibers from the retina of one eye cross to the optic tectum on the other side of the mesencephalon, the decussating or crossing fibers from one eye passing over or even through the bundle of fibers from the other eye.

The diencephalon of fishes seems to be an important correlation center for incoming and outgoing messages relating to internal homeostasis. The hypothalamus especially affects the endocrine system through the pituitary gland. Neurosecretory cells have been described in some diencephalic nuclei and special paths have been found that lead secretions from the diencephalon to the hypophysis.

The Mesencephalon

The mesencephalon or midbrain of fishes is relatively large; it consists of the dorsal optic tectum, appearing in dorsal view as the two optic lobes, and of the ventral tegmentum. The tectum is made up of a number of zones of different-sized nerve cells or neurons. Most of the fibers of the optic nerve end in the tectum in such a manner that the frontal part of the retina is projected onto the contralateral caudal portion of the tectum and the converse applies to the anterior retina, whereas dorsal retinal elements of one side are connected to the ventral portion of the tectum on the other side of the fish. Fishes, like other vertebrates, have convex lenses in their eyes that create an inverted image on the retina; through the above tectal pattern the image may become projected essentially as it is in nature. In the lampreys (Petromyzoniformes), the midbrain is covered with a vascular choroid plexus that helps in nourishing the brain and connects with the fluid-filled ventricles and the central canal. Among higher bony fishes (Actinopterygii), tectal gray matter extends into the ventricle as a paired torus longitudinalis and joins the posterior comthissure between the two hemispheres which lie at the border of the diencephalon. The torus is presumably connected with vision since it is very small in a blind goby (*Trypauchen*).

Electrical stimulation and removal of portions of the midbrain indicate that the tectum contains functional neural units that correlate visual impressions with muscular responses such as would occur, for instance, in facing prey or adjustment of swimming to objects moving in the visual field, in short, to various complex goal-seeking movements. Crucian carp (*Carassius carassius*) were unable to distinguish the positions of light (spatial orientation) after the tectal hemispheres were separated by a longitudinal midline incision. These carp also appeared to have difficulties in locating the position of sound stimulus. Experiments using a puffer (*Sphoeroides*) in which one eyeball was rotated front to back and the other eye was blinded, led to persistent forced circling even when other parts of the brain, except the medulla, were removed. Thus there is strong evidence for the optic tectum as an eyebody coordinating center. Electric stimulation of the tegmentum, however, leads to generalised but relatively uncoordinated locomotory responses. The optic tectum also bears some relation to learning. Responses to smell stimuli have been associated with visual signals; upon midbrain removal the olfactory responses are extinguished.

In view of the complete crossing over of the fish optic nerves it should be noted that the goldfish (*Carassius auratus*) and a goby (*Bathygobius*) could be trained to sensory optic transfer. Somatic and autonomic responses showed that the untrained eye recognised the stimulus learned with the other eye but that motor responses to the transfer are difficult to obtain.

Because of its functional importance in the central nervous system in many fishes and its widespread connections and because it has several layers of nerve cells, the optic tectum has been loosely compared to the cerebral cortex of mammals.

The Metencephalon

In the metencephalon, the cerebellum develops from the underlying enlarged medulla as a dorsal outgrowth of the upper rim of the 4th ventricle of the brain. Its main functions are reported to lie in swimming equilibration, maintenance of muscular tonus, and orientation in space; in some fishes it is by far the largest component of the brain. Several distinct layers or differently shaped nerve cells can be distinguished in the cerebellum and include the molecular or chief receptive cell layer and the so-called Purkinje cells that assume prominence in the cerebellum of the higher vertebrates.

In the lampreys (Petromyzonidae), the cerebellum is only a small dorsal bridge between the acoustic-lateral-line nerve areas. In the hagfishes (Myxiniformes), the cerebellum is represented by a small commissure that consists of eighth nerve and lateral line fibers, the octavo-lateralis system. In sharks and rays (Elasmobranchii) the cerebellum roughly increases with the size of the species. Relatively small sharks, such as the dogfishes (Squalidae), have a simple bilobate cerebellum whereas in the large, swift, and predatory mackerel sharks (Lamnidae) the cerebellum has become the largest part of the brain and is convoluted and has an increased surface area. Prominent lateral or auricular outgrowths may form also on the cerebellum and are mainly connected to lateral-line and vestibular sense organs.

In higher bony fishes (Actinopterygii) the cerebellum acquires a characteristic forward projection, the valvuli cerebelli, which extends under the optic tectum. In the elephantfishes (Mormyridae), many of which have extended and highly motile snouts, the valvuli cerebelli arches far forward beyond the forebrain and becomes the largest component of the brain.

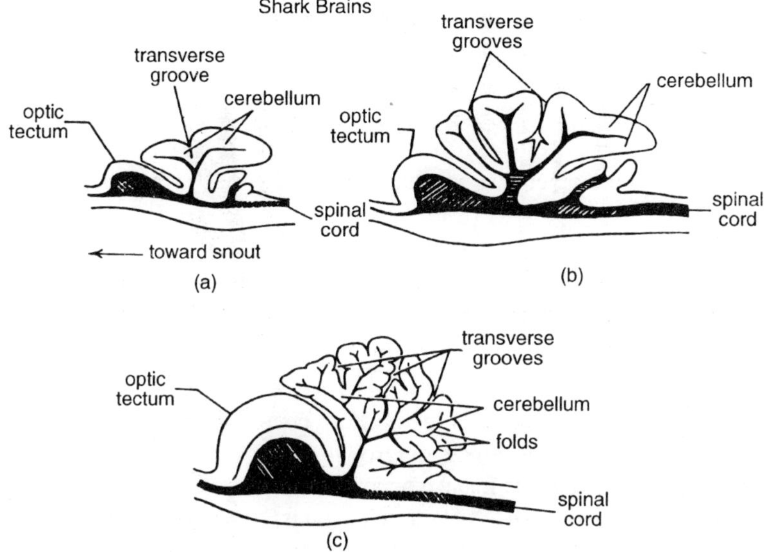

Figure 1.2 : Relative size and development of cerebellum among shark as seen in diagrammatic section : (a) cat shark (Pristiurus); (b) a smooth dogfish (Mustelus); (c) a sand shark (Carcharias).

Among catfishes (Siluroidei), the structure is also well developed. Since the mormyrids emit and respond to weak electrical currents and since there is also at least one catfish with electric discharges (the electrical catfish, *Malapterurus*) it has been suggested that the enlarged cerebellum in these groups deals with electrical impulses.

The ventricle of the cerebellum still prominent in sharks and rays (Elasmobranchii) has almost completely disappeared in the cerebellum of higher bony fishes, as have the sulci or transversal folds so prominent in some large sharks. The lateral auricles and interauricular fibers increase in importance, especially in fishes with well developed lateral-line organs.

A prominent function of the cerebellum apparently is to coordinate muscular tonus and swimming.

The Myelencephalon

The brain divisions from cerebellum to telencephalon are, in their widest meaning, sensory and sensory-coordinating centers that connect

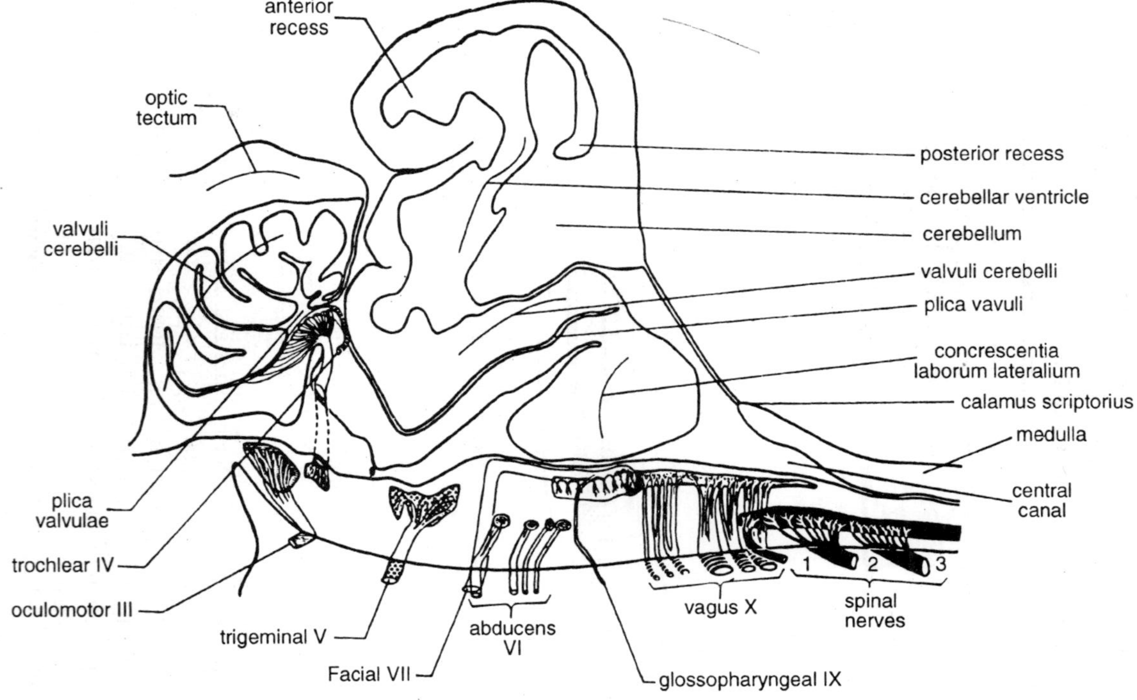

Figure 1.3 : The cerebellum and the arrangement of the motor roots and nuclei in a relatively primitive teleost, a tarpon (Megalops).

with the brainstem, or myelencephalon, by various neural tracts. These anteriormost four divisions have grown in importance as sensory integrators and have become more and more refined in the course of evolution. The myelencephalon, with the medulla oblongata as its main component, is the center to which lead the sensory nerves except those of smell (I) and sight (II). The boundary between medulla oblongata and spinal cord in fish is indistinct. The medulla can be divided into columns of nerve fibers based on the type of information transmitted. Thus, there are somatic and visceral sensory columns and also somatic and visceral motor columns.

In the brainstem, nuclei of cranial nerves III to X are arranged anteroposteriorly; the tracts of nerves III and IV cross to the contralateral side. The afferent nerve components of the brainstem can be divided into somatic and special sensory nervous inflows with cranial nerve VII (facial),VIII (stato-acoustical), IX (glossopharyngeal), and X (vagus) mostly sensory. There are more somatic sensory fibers in the foregoing nerves in lampreys (Petromyzonidae) than in shark-like fishes (Elasmobranchii). The somatic sensory fibers become further reduced in bony fishes (Osteichthyes). Details of the distribution of the cranial nerves which are, in the main, connected to the brainstem.

Depending on the relative importance of the various senses, different portions of the brainstem are enlarged. Such fishes as the Pacific herring (*Clupea pallasi*), the striped mullet (*Mugil cephalus*), and certain carangids such as *Trachurus* have prominent paired swellings designated the cristae cerebelli on the antero-lateral rim of the fourth ventricle. Nerve connections of the cristae are not known but their function may be connected with the strong schooling tendencies of these species. Certain suckers, such as the highfin carpsucker (*Carpiodes velifer*), and minnows, such as the goldfish (*Carassius auratus*), have prominent vagal lobes posterior to the region of the cristae cerebelli from which emerge cranial nerves IX and X. Such fishes also have developed an organ in the roof of the mouth (palatal organ) where food is tested by taste and touch. Another prominent feature, .found among minnowlike fishes (Cypriniformes), is the tuberculum impar, or facial lobe, a central hillock posterior to and sometimes partly covered by the cerebellum. The tuberculum impar arises from a fusion of root elements of the facial (VII) and vagal (X) nerves, where gustatory and tactile impulses are correlated with visceral sensory ones as in an European barbel (*Barbus fluviatilis*) and the carp.

The medulla of the higher bony fishes (Actinopterygii) contains a

Table 1.2 : Cranial Nerves of Fishes and Their Functions

Nerve Number	Name	Peripheral Connection
0	Terminal	Olfactory bulb
I	Olfactory	Mitral cells; olfactory bulb; nucleus olfactorius anterior
II	Optic	Retina of eye
III	Oculomotor	Superior, inferior, anterior, rectus, and inferior oblique muscles of eyeball
IV	Trochlear	Superior oblique muscle of eyeball
V	Trigeminal divided into deep ophthalmic, maxillary, and mandibular branches	Anterior portion of head, upper and lower jaw regions
VI	Abducens	Posterior rectus muscle
VII	Facial and	1. Dorsal group: postorbital lateral line canal; taste buds on body, where present
VIII	Acoustic	2. Orbital group: supra- and infraorbital lateral line canals; taste buds on snout
		3. Acoustic group: inner ear; temporal lateral line canals
		4. Branchial group: operculo-mandibular oral and jugal line canals; taste inside mouth; taste and touch from opercular hyoid and mandibular regions; head muscles antagonistic to those under control of mandibular branch of V
IX	Glossopharyngeal often fused with	1. Dorsal group: part of temporal lateral line canal; dorsolateral skin sensibility and proprioception, first gill slit region
		2. Branchial group: taste organs in dorsal and ventral pharyngeal mucosa, muscles of first gill slit (third gill arch)
X	Vagus	1. Supratemporal branch: skin and lateral

line organs of supratemporal region

2. Dorsal recurrent branch: joins with body taste bud innervation of VII
3. Body lateral line branch: from body lateral line organs
4. Visceral branch: nerves to and from internal organs
5. Branchial branches: to and from four posterior gill slits or equivalent regions

Central Connection	Function
Forebrain; lamina terminalis	Somatic sensory, tip of snout? vasomotor?
Smell centers of forebrain	Special sensory, carries smell impulses
Optic tectum of midbrain	Special sensory, carries visual impulses
Anterior brainstem under optic tectum (Mesencephalon)	Somatic motor, innervates four of the six striated eye muscles and muscles inside eye
Anterior brainstem under cerebellum (Metencephalon)	Innervates one of the six striated muscles of the eye
Lateral column of medulla oblongata region of brainstem, frequently overlaps with centers of VII (Myelencephalon)	Mixed somatic sensory and motor functions. Thermal and tactile sensibility of skin on anterior portion of head; proprioceptors of head musculature
Ventral brainstem behind origin of V	Innervates sixth striated muscle which moves the eyeball
Elongate series of connected nuclei in middle of brainstem (Medulla oblongata; Myelencephalon)	A truly mixed nerve; special and somatic sensory visceral and motor functions. Brings lateral line acoustic and gravity impulses to brain centers, as well as taste and touch impulses from head and mouth; proprioception and innervation of some head muscles
Series of ganglia, or roots, mainly in dorsal and median end region of brainstem (Medulla oblongata; Myelencephalon)	A mixed nerve; taste and lateral line sense in afferent components; innervates ante rior gill muscles (first gill slit)
	A mixed nerve; motor control of bulk of

Series of ganglia posterior to but fused with roots of nerve IX; often overlapping with roots of first occipitospinal nerves at posterior end of brainstem (Medulla oblongata, Myelencephalon)	gill muscles, special fibers from taste and lateral line organs, visceral motor and sensory to and from internal organs; para sympathetic autonomic nervous system

pair of large neurons, the giant cells of Mauthner, at the level of cranial nerve VIII. The lateral dendrites of these giant cells connect with fibers of cranial nerves V, VII, IX, and X and to the cerebellum and the optic tectum. Their axons pass through the spinal cord to the muscles of the tail and are best developed among the good swimmers such as the Atlantic salmon and certain trouts (*Salmo*), whereas they are less prominent in bottom dwellers such as some of the gobies (*Gobius*) and scorpionfishes (*Scorpaena*). In eels (Anguilliformes) and the molas (*Mola*) cells of Mauthner are absent. Apparently the cells are motor coordinators for relaying multiple sensory impulses mainly from lateral-line centers to the caudal and abdominal swimming musculature. In the lampreys (Petromyzonidae), a comparable coordinating system connects the oculomotor levels of the brain with the tail and body musculature through the so-called neurons of Mueller and also cells of Mauthner.

Aside from also being a relaying area between the cord and higher brain areas, the medulla has centers that control certain somatic and visceral functions. Among bony fishes these include respiratory, paling (acting through melanophore aggregation), and osmoregulatory centers. Medullar nuclei are also involved in the maintenance of swimming equilibrium. Tests on medullary control of fin movements in some fishes that use their fins prominently in locomotion, such as wrasses (Labridae), show that the rhythm of fin beats is set by a center in the anterior cord or the brainstem.

THE SPINAL CORD

The central nervous system of fishes loses much of its structural complexity in the transition from brain to cord. Already in lampreys (Petromyzonidae) but more pronouncedly in sharks and rays (Elasmobranchii), a cross section of the cord shows a central region of gray substance consisting primarily of nerve cells and a surrounding area of white matter, the nerve fibers.

These fibers are sheathed, or myelinated, and are collected in

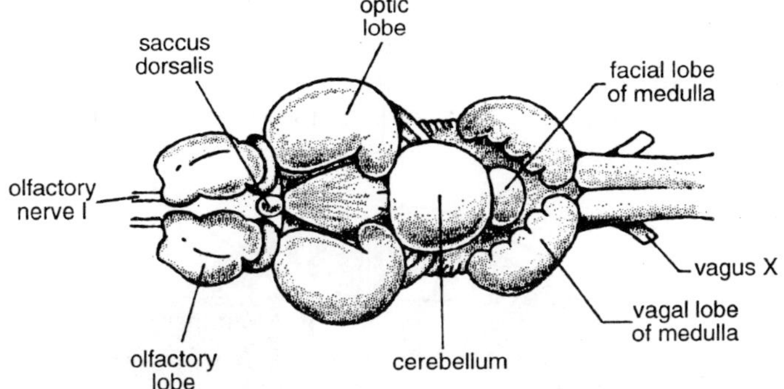

*Figure 1.4 : Dorsal view of brain of the carp (*Cyprinus carpio*) showing development of facial and vagal lobes.*

bundles according to their function and connections. Centered in the gray matter of the cord is the central canal. As in other vertebrates, the gray matter of the cord of bony fishes (Osteichthyes) roughly resembles a letter X with paired dorsal and ventral horns. The dorsal horns receive somatic and visceral sensory fibers and the ventral horns contain motor nuclei, with centers for dorsal or ventral musculature lying in median and lateral positions respectively.

THE SPINAL NERVES

The spinal nerves are segmentally arranged. In the lampreys (*Petromyzonidae*) the ventral roots appear between the levels of the dorsal roots, thus this group of fishes does not have true mixed spinal nerves; only among the hagfishes (*Myxinidae*) there begin to occur true mixed spinal nerves like those present, with few exceptions, throughout the jawed fishes-the sharks and rays (*Chondrichthyes*) and bony fishes (*Osteichthyes*). The jawed fishes in general, have spinal ganglia for the neurons of the sensory or dorsal root nerves although in many families, such as the minnows (*Cyprinidae*), cods (*Gadidae*), perches (Percidae), and drums (Sciaenidae), some afferent fibers come to the cord also from supramedullary and inframedullary ganglia.

From sharks (Squaliformes) to bony fishes (Osteichthyes), there is an increasingly finer differentiation and subdivision of ascending and descending fiber tracts between the brain and the spinal cord. In the searobins (Triglidae), the long and separate anterior rays of the pectoral fins carry special receptors, for tactile and chemical clues; sensory nerves from these rays are marked on the spinal cord by separate swellings corresponding to their dorsal root nerves, and their fiber projections also reach the medulla and the midbrain.

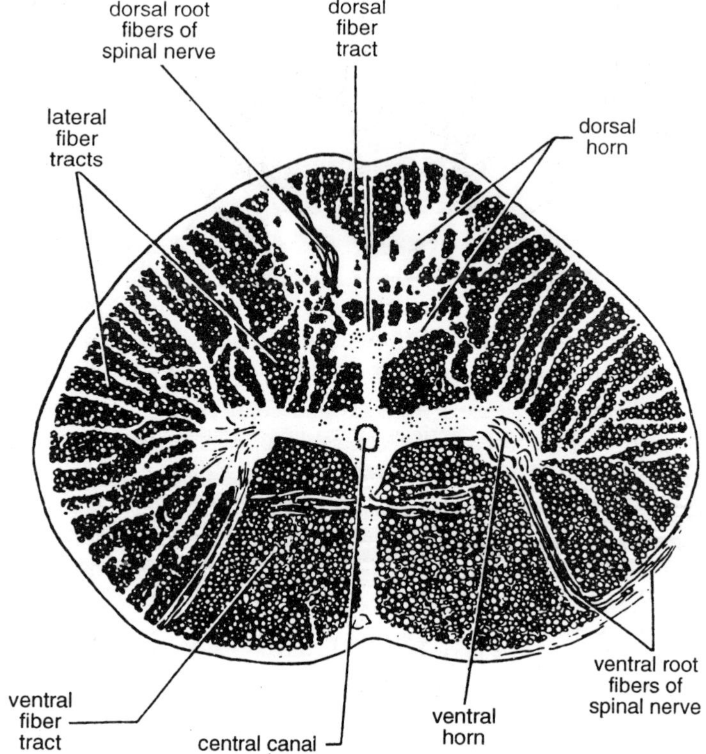

Figure 1.5 : Cross section of spinal cord in the shark group (Elasmobranchii).

Ipsilateral and contralateral reflexes of spinal nerves in one segment of the cord can occur in elasmobranchs; plurisegmental reflexes are common in sharks and teleosts. In view of the important role of the trunk in fish locomotion the question of the nervous control of swimming movements has received considerable attention. There is controversy as to whether a fish requires outside stimulation in order to make coordinated swimming movements or whether swimming is due to the automatic activity of cord and medullar centers, only to be coordinated by peripheral stimuli. Von Holst, after performing many experiments with dogfish sharks (Squalidae), eels (*Anguilla*), loaches (Cobitidae), and minnows (Cyprinidae), suggested that swimming is due to endogenous rhythm from the posterior medulla, in turn passed on to the cord but influenced at all levels by outside "setting" stimuli. Lissmann and Gray, mainly on the basis of experiments with certain cat sharks

(Scyliorhinidae), maintain that there is no automatic swimming center but that swimming movements are of reflex nature and depend on the coordination of afferent impulses in spinal cord and medulla. The movements of young ammocoetes of *Lampetra* (a lamprey) are described as entirely reflexogenic.

THE AUTONOMIC NERVOUS SYSTEM

In the autonomic nervous system of lampreys and hagfishes (Cyclostomata), there is no ganglionated sympathetic chain. Nevertheless, in these fishes a diffuse system of nerves and some plexi of sympathetic function have been described. In the lampreys and hagfishes, both the heart and the organs of digestion are innervated by the vagus. Subcutaneous nerve plexi that are innervated by spinal nerve fibers, apparently of the sympathetic system, are found concentrated in the walls and at the orifices of the slime glands of *Myxine* (hagfish). In the sharks and rays (Elasmobranchii), sympathetic ganglia are found along the sides of the vertebral column; they are irregular and do not extend into the head region. The smooth muscles of the gut and arterial walls are innervated by them. However, there is no connection between them and the nerves which supply the skin such as in mixed communicating rami of bony fishes (Osteichthyes) and higher vertebrates. In all shark-like fishes, the hearts and the internal organs of digestion are abundantly supplied by vagal fibers. In the bony fishes, segmentally arranged ganglia form a sympathetic chain as far forward as the trigeminal nerve (cranial nerve V); most preganglionic fibers for sympathetic ganglia of the head arise from the spinal cord of the trunk region and pass forward. Sympathetic connections to the skin nerves are found in the highest bony fishes (Teleostei).

The parasympathetic system is almost exclusively represented in portions of the widely ramifying branches of the vagus (cranial nerve X). A parasympathetic innervation of eye parts (in cranial nerve III) has already been mentioned. A sacral parasympathetic system similar to that of mammals is suggested in skates by the response of the claspers to stimulation of the posterior spinal nerves. The sympathetic cardio-accelerating fibers of higher vertebrates are absent in all fishes and consequently vagal stimulation has a powerful inhibitory action on the sinus venosus and the auricle but no effect on the ventricle. The stimulation of branches of the vagus leading to other internal organs as well as the administration of cholinergic (parasympatomimetic) and adrenergic (sympatomimetic) drugs has produced conflicting results, especially in action on the gas bladder and certain portions of the alimentary canal.

The nerves from the paling center run in the spinal cord and leave it at various levels in different fishes (in the minnow *Phoxinus,* segments 12 to 18; in the salmon-trout genus *Salmo*, segment 26; and in the searobins, Trigla, segment 3). The melanophore nerves then go to the subvertebral sympathetic chain where they divide into an anterior bundle leading to the colour cells of head and nape, and a posterior bundle, to the trunk , and the tail. Little is known about melanophore control in cyclostomes; however, certain elasmobranchs have at least some sympathetically innervated melanophores. The presence of a darkening center in the midbrain and of parasympathetic melanin disperser nerves has been postulated.

SUPPORTING TISSUES OF THE CENTRAL NERVOUS SYSTEM

Some embryonic ectodermal derivatives and some mesodermal tissues in the region of the brain and spinal cord assume supporting, investing, and nourishing functions of the nerve cells proper. Ectodermajependymal cells line the cerebrospinal canals of all fishes, permeate the body of the cord and the brain, and are especially concentrated along the ventricular walls of the thalamus of bony fishes (Osteichthyes) where they carry cilia. The histology of these cells suggests that they have a secretory function. Neuroglial cells of special shape and function have been reported from the spinal cord of the sea lamprey (*Petromyzon*). In the spinal cords of sharks and rays (Elasmobranchii) occur many branched, astrocyte or spidery neuroglial cells. In the cerebellum and in the spinal cord of bony fishes various neuroglial elements are also differentiated.

The choroid plexus, a highly vascularised brain area, especially prominent in the gars and relatives (Holostei) and elasmobranchs, consists mainly of mesodermal tissue often in several layers and therefore called leptomeninx. Choroid plexi or tela choroidea are filled with cerebrospinal fluid, are richly supplied with blood vessels, and participate in the nutrition of the nervous tissue they surround and cover.

INTELLIGENCE AND BEHAVIOUR

Sensory components of the fish nervous system are highly discriminative, and the sensitivity of the sense cells is often surprisingly great, as, for example, smell, vision, and temperature perception. Most sensory mechanisms work by summation, and full efficiency at the normal level of stimulation may be achieved only with a threshold lower than that which would normally be received. Such low thresholds imply the existence of central processes that allow the animal to disregard random "white noise" in most or in all sensory modalities, but they also imply that the animal can make choices with a high degree of accuracy and confidence of goal-oriented success.

Many acts of discrimination or choice in nature are based on conditioning of the fish to single or multiple stimuli and, even without experiments, there would be no doubt that fishes are easily conditioned. The ability to associate stimulus and response is present even in simple invertebrates; but relative levels of "intelligence" can be distinguished among other criteria, on the basis of the speed of learning simple tasks and on the retention-span of learned responses. Aquatic analogs of the "Skinner box" and the shuttle box are increasingly popular devices used to study instrumental learning in fish.

Electrical shock conditioning of the goldfish (*Carassius auratus*) to visual stimuli and of the fathead minnow (*Pimephales promelas*) among others to olfactory stimuli lead to learning accomplishments. The accomplishments are detected by altered heartbeat (an autonomic response) after thirty-five trials, or by swimming movements (a motor response) after fifty or more trials. Great individual differences of response exist. Some fish in an experimental group are slower than others to learn, and some seem unable to learn even "simple" tasks.

Conditioning by a reward such as food can be established quicker than by punishment; ten to twenty trials are sufficient, depending on the task, to accomplish visual responses in a minnow of the genus *Phoxinus* or to have *Tilapia macrocephala* learn space reversal of food location. Downstream migrating salmons (*Oncorhynchus*) learn quickly to swim a constant course in a circular channel. These responses are probably established speedily because the fish are in physiological readiness for oceanward migration which involves swimming in schools for long distances. One-trial learning has been shown with tagged carp (*Cyprinus carpio*) which became "hook-wise" for months after being caught only once.

Not only higher bony fishes (Actinopterygii) such as the foregoing examples are capable of reflex learning, but also the tiger shark (*Galeocerdo cuvieri*). This shark learns to strike a target and to swim a staked-out path for a food reward. The bull shark (*Carcharhinus leucas*) can be trained to associate sounds of 400 to 600 cycles per second with food.

Second-order conditioned reflexes are also established by fishes. The goldfish is able, after considerable training, to transfer a response learned on the basis of olfactory signals to visual clues. It can also make a similar transfer from auditory to thermal clues.

Experiments of sensory discrimination which first and foremost test sensory acuity rely on generalisation, after a conditioned response has been established; that is, they depend on the ability of the animal to respond to a stimulus similar to, but not identical with, that used in the original training procedure. Generalisation for different values of brightness and hue as components of vision are readily accomplished by various species in several families of fishes; generalisation of olfactory clues by minnows (*Pimephales*) has also been observed.

Retention of learned responses without subsequent reinforcement (that is, long lasting memory) may be another criterion of relative intelligence. Salmons (*Oncorhynchus* and *Salmo*) may retain stream odors, learned during a crucial juvenile period, throughout their entire lives of 2 to 6 years or more. The home-stream odor memory is acquired by an imprinting process. Because the home-stream waters are greatly diluted by the time they reach the river mouth, a multiple imprinting hypothesis has been proposed. A salmon may remember the distinctive odors of a number of points on his downstream migration. The return voyage would then be a series of subvoyages with the salmon reading out the series of imprinted odors in reverse order as "subgoals." While there is strong evidence that imprinted odors are the principal source of guidance in the remarkable homing performance of salmon, the use of additional, perhaps visual, sensory cues is quite possible.

Apprehension and conceptual use of spatial relations are attributes of insight, by laymen credited only to the higher mammals because such abilities are more easily demonstrable in animals with prehensile limbs than they are in other vertebrates. Nevertheless, apprehension of spatial phenomena and insight learning exist in the frillfin goby (*Bathygobius soporator*) and probably in many other fishes. The frillfin goby swims about over submerged tidepools during high tide, but later, when the water has receded, it jumps from pool to pool with an uncanny aim and without seeing the next water area at the take-off point. Transplanted fishes that have not swum over the region and are placed in the pools at low tide either do not jump or, when prodded violently, will eventually jump in a disoriented way. After spending 12 hours, overnight and during high tide in a series of experimental pools, the same individuals jumped directly to the largest pools. Effective orientation was retained up to forty days without additional "high tide" experience. Comparable, though less elaborate performances are reported for the green sunfish (*Lepomis cyanellus*).

Delayed responses, successfully learned, have been taken as indication that the experimental subject in question is endowed with insight. Minnows (*Phoxinus*), after having been allowed to become familiar with a three-doored chamber by swimming freely through it for a time, were later required to reach a previously sighted food morsel by swimming forward and doubling back through one or two possible pathways to reach the reward which was then out of sight. After many hundreds of trials, the performance of the fish did not improve over an average of six out of ten correct solutions. The subjects also developed erratic and vicarious trial-and-error behaviour at the point of choice. They also developed displacement activities, such as picking at stray pieces of sand at the entrance into the channel which led to the point of decision. These observations suggest that stimulus satiation, a negative "attitude" to the task or the experimental setup, after repeated experimental runs, may have bedeviled these experiments rather than their being an unequivocal demonstration that fishes are unable to learn to solve delayed response experiments.

Field observations that corroborate the findings of the delayed-reaction experiments, in spite of the earlier caution of stimulus satiation, have been made on the escape of trapped snappers (*Lutjanus*, family Lutjanidae) and groupers (*Epinephelus*, family Se'ranidae). The fishes that had entered the traps would make a hundred or more trial-and-error attempts to find a way through the wire or netting before chance would help them to locate the exit. An hour of confinement within a trap did not lead to any apparently systematic exploration for an escape route or portal. However, some fishes of the foregoing and other families, such as the sunfish family (Centrarchidae), are known to make such traps their homes, and they will then swim in and out freely.

Another, perhaps oblique, approach to the insight or the conceptualiing ability of fishes comes from a comparison of rats and a fish (*Tilapia macrocephala*) with regard to performance on inconsistent reinforcement of stimuli. On twenty trials per day, demanding visual discrimination between horizontal and vertical bars, responses to one stimulus were reinforced 70 percent and those to the other, 30 percent of the time. After 30 training days the previous 70-percent stimulus was consistently reinforced 100 percent. Records of the responses each day to the 70-percent stimulus showed that the rats "gambled" and that their preference for the more frequently reinforced stimulus rose gradually from a 70:30 level to reach 90 percent at day thirty. The fish, however, after reaching the 70:30 level on about the tenth day, corresponding to the more frequently reinforced stimulus, remained at this matching level of performance for the rest of the experiment.

Position training, demanding a choice between two positions of food placement, was conducted in a fashion similar to the foregoing and gave comparable results. Sensory deficiency was ruled out as a cause for the difference in behaviour between the mammal and the fish and it was suggested that the rat, in such test situations, makes an all-or-none response selection, an indication not only of a high level of versatility in cerebral processes, but also of a certain abstractive capacity. The fish, in contrast, matched continuously the probability of reward or punishment with the choice of its response and indicated, as in the case of the delayed reaction, that it tends to perform in a more stereotyped and rigid manner than the mammal. Another aspect of fish behaviour, prominently involving the nervous system, is the occurrence of pronounced, often spectacular, highly specialised and stereotyped reproductive behaviour that has been the subject of much ethological (animal behaviour) research. The basis of behaviour components are patterns of neuromuscular coordination that lead to movements in the wake of external or internal stimuli. These patterns are genetically determined and the movements that make up the behaviour train during reproductive activities recur with surprising homogeneity in all individuals of a species. It has been possible to study fishes well in this respect, because they are easy to keep and behave quite normally in captivity. Detailed studies of reproductive behaviour have been made of the threespine stickleback (*Gasterosteus aculedttus*), various cichlids, a sculpin (*Cottus*), the fightingfish (*Betta*), and others.

Fishes perform species-typical movements of courtship and/or nestbuilding even when raised to maturity in isolation. Therefore the pattern and sequence of these movements is just as much a part of the genetic makeup of the fish as is its body shape or colouration. Early ethological literature neglected the role of learning in inherited behaviour but it is now indicated that learning plays an important role here also, and that modification, and perhaps refinement, of behavioural components is possible if not the rule (as known, for example, in a cichlid, *Cichlasoma meeki*).

The movements that make up the courtship behaviour of sticklebacks (Gasterosteidae), for instance, can. be analysed down to the activity of single muscles or groups of muscles such as those that move a single fin. Such localised movements are components of larger units and are coordinated into hierarchically arranged groups to build up instincts. The implication is that these movements belong to distinct causal and functional systems that act as units. When in action they reduce the activity of other equivalent but functionally different systems. Instinct also implies a course of patterned rather than haphazard activity. This pattern operates through the common working of sense organs, central nervous system, and effectors in such a manner that we encounter a more or less fixed and obligatory sequence of stimuli and responses for the entire activity to go to completion. The most thoroughly studied such patterned sequence is that of the spawning behaviour of the threespine stickleback. For this fish it has been repeatedly shown that one particular movement in the behaviour train acts as a stimulus for the next. It has also been shown how one may envisage the entire reproductive instinct of this fish to be subdivided into phases and hierarchically organised into components consisting of individual movements.

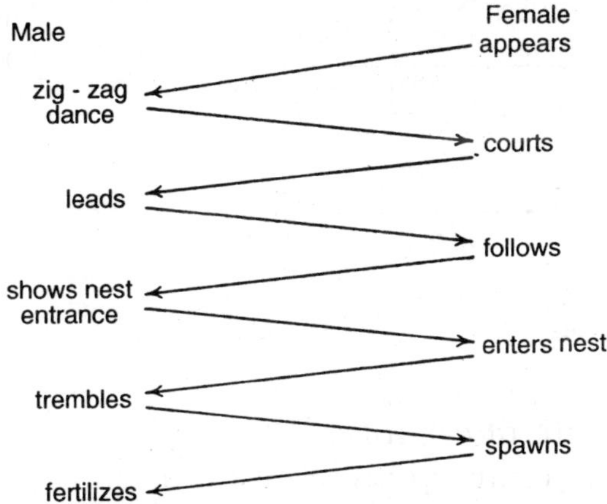

Figure 1.6 : Interlocking patterned sequence of courtship movements of the threespine stickleback (Gasterosteus aculeatus).

The interplay between genetic and environmental forces during evolution has so fashioned the pattern and the responses of the nervous system, including the receptors, that simplified, shorthand, or "part-for-the-whole" stimuli often suffice to "release" further movements in a sequence. For instance, the silver colour or the swollen outline of a ripe stickleback female elicits the zig-zag dance of the male at the initiation of the courtship. It has been shown that many prominent colour patterns of fishes and other animals, especially spots that can be displayed, have such releasing functions as, for example, the black and white markings on tail and flanks of the guppy *Lebistes*. Several releasing signs may be used in one situation, each one adding to the releasing power of the other so that one may fashion a supernormal model with a greater than natural releasing value (principle of heterogeneous summation); even a crudely fashioned model of a female stickleback which is very silvery and has an excessively large abdomen assumes such a function. The principle of stimulus summation also applies to hydrographic and other environmental variables which act as releasers in the spawning behaviour of such fishes as the northern pike (*Esox lucius*), whitefishes (*Coregonus*), and the brown trout (*Salmo trutta*).

Early theory on releaser mechanisms probably overemphasised their central nervous system components. Now we must note that the term "releasing mechanism" is operational rather than indicative of a location in the animal. We must also note that the concept stands for the sense organ-nervous system-effector complex rather than for a central pattern or template device.

The course of stereotyped behaviour is influenced both by external and internal factors. Among the external factors, the releasing by sign stimuli, such as species-specific colours and shape, has been mentioned above. Among the internal factors in reproductive behaviour, it is primarily the hormone levels that determine the intensity and also, to some extent, the course of the behaviour. In nonreproductive activities, such as feeding, hunger as an internal factor may determine the extent and sequence of feeding movements. In the analysis of patterns and motivation of movements there appear certain distinct tendencies, such as sex, flight, and aggression, which influence the course of instinctive behaviour. The zig-zag dance in the courting male stickleback, for instance, varies in duration of its approaching, the "zig," and its leading of the female, the "zag" phase, according to the level of both the sexual and the aggressive tendencies in an individual male.

In the course of an integrated behaviour chain such as courtship, a fish may interpose certain movements such as feeding, chafing, or other comfort movements which belong to entirely different activities or instincts. Thus, for instance, if two male sticklebacks meet at the boundary between their territories they often do not fight but they perform digging movements that belong to the nest-building drive to be activated much later in reproductive behaviour and not during the period when they establish their territories. These "displacement movements"-they are displaced out of the normal sequence of events-result from a conflict between different tendencies, in this example, aggression and flight.

Activities not released by their proper stimuli for a considerable time may be performed without any of the adequate releasing stimuli being present-a phenomenon called "vacuum activity." Single male guppies (Lebistes), raised in isolation, may, when they become sexually mature, perform sigmoid display movements which they normally show at the sight of a ripe female. Cichlids (Cichlidae) which have not been able to rear their own young can begin to guard a flock of water fleas (Daphnia) as an "overflow" reaction.

The nervous system has played a key role in the evolution of speciesspecific behavioural patterns. At the same time, the study of the behaviour of related species reveals considerable plasticity of pattern change and modification so that the behaviour derived from one activity can be taken over into another and acquire new significance (ritualisation). The "eye spots" that commonly occur on the gill covers, tails, or fins of fishes are prominently displayed in fright and release flight behaviour or search for cover by species mates but may also be used in courtship displays. Social releasers, that is, patterns especially adapted to a signal function, play a role in many kinds of behaviour. They help the young to find the parent, they serve to maintain a rank order among individuals, and they are instrumental in ritualizing fighting behaviour where they allow the beaten animal to show signs of inferiority before it is hurt. They are very important in the synchronisation of sexual activities and they may even be operative in interspecific relations. Many territorial fishes such as surgeonfishes (*Acanthurus*) and damselfishes (*Eupomacentrus*) use simple threat displays as well as chasing and ramming behaviour to defend their food supply of benthic algae from the invasion of other herbivores. Reef fishes throughout the world perform more or less ritualised solicitation displays to attract the attention of cleaning organisms. These solicitations and displays show a remarkable degree of variability among families but the species within many families show nearly identical symbiotic behaviour even though they have been separated from their congeners in different oceans for many thousands of years.

Lack of information on the organisation and detailed function of the central nervous system retards the analysis of the seat of coordination of instinctive behaviour. Removal of forebrain of the goldfish or a tilapia resulted in massive defects in the acquisition and retention of avoidance performance. Other experiments indicate that forebrain removal depresses many functions but eliminates few. The fish forebrain may serve primarily as a modulator of lower brain centers. Fishes will remain among the best suited research subjects for the ethologists with their threefold study aims: the explanation of behaviour in physiological terms, the study of its biological significance in conjunction with ecology, and the understanding of the development of behaviour in the course of evolution.

ENDOCRINE ORGANS

The vertebrate endocrine system is present in its essence already in lampreys and hagfishes (Cyclostomata); further similarities with higher vertebrates appear in both the sharks (Elasmobranchii) and in the bony fishes (Osteichthyes). However, the differences are still considerable between the fish and' mammal endocrine glands. These glandular differences can probably be correlated with differences in the evolution of other body systems in the two groups and with the exigencies of an aquatic existence. Although mammalian endocrinology is well-advanced, only certain aspects of hormonally controlled functions in fishes have been investigated. Those studied have been the influence of endocrines on colour cells, the reciprocal action of sex cells and the pituitary, and thyroid function and adjustment to migration, but the finer details of the action and interrelations of fish endocrines still await elucidation.

The endocrine system, in contrast to the nervous system, is primarily concerned with relatively slow processes of a metabolic nature such as nitrogen metabolism (adrenal cortical tissue, thyroid gland), carbohydrate and water metabolism (adrenal cortical tissue), the maturation of sex cells and reproductive behaviour (the pituitary gland and gonadal hormones).

There follows a list of endocrine glands in fishes, with mention of organs of possible endocrine function(s); the individual glands and their action except those of doubtful function will be treated in separate sections; for location of glands in the fish body see Figure elsewhere in this chapter.

The Pituitary Gland (Hypophysis)

This endocrine mastergland is found throughout the vertebrates and develops embryonically from a neural element that grows downward from the diencephalon to meet an epithelial element (Rathke's pouch) which grows upward from the buccal cavity. These two elements are thus of ectodermal origin and enclose mesoderm between them, from which develops the future copious blood supply of the pituitary. This blood supply is derived from the internal carotid artery. Anterior to the pituitary gland lies the optic chiasma and posterior, the saccus vasculosus. The stalk or infundibulum that attaches the gland to the floor of the diencephalon increases from the lamprey group (Cyclostomata) to the bony fishes (Osteichthyes); simultaneously, the pit in the bottom of the brain case where the gland is lodged becomes more pronounced.

Anatomy and Histology of the Gland

The pituitary gland or hypophysis in fishes consists of a tripartite adenohypophysis and a neurohypophysis (with variances in terminology of parts). The division of the adenohypophysis into pro-, meso-, and meta-adenohypophysis is based on their relative anteroposterior positions and on different types of their secretory cells. These divisions of the adenohypophysis are also termed the rostral pars distalis, the proximal pars distalis, and the pars intermedia to emphasise their homology with the gland in tetrapods. The posterior, meta-adenohypophyseal portion underlies the thin neurohypophysis; it receives processes of hypothalamic neurosecretory cells. In sharks and rays (Elasmobranchii), the hypophysis projects farther ventrad and has a forward extension, the rostral lobe, which contains pro-adenohypophyseal components of histologic differentiation similar to that in cyclostomes. A down-growth in elasmobranchs is called the ventral lobe. The most anterior portion of the adenohypophysis is its most vesicular division, and cyclic changes in secretory activity with gestation and growth have been demonstrated in it. In the coelacanth lobefin (*Latimeria*), the rostral division of the pituitary (pro-adenohypophysis) is ventrally derived and independently vascularised, an organisational feature strongly reminiscent of elasmobranchs.

The pituitary gland of higher bony fishes (Actinopterygii) has become more compact than in lampreys and sharks, leading to increased difficulty in establishing a clear-cut division among its parts.

Differential secretory activity can, however, be demonstrated by standard staining techniques. Large ganglion cells occur in the adenohypophysis and neurohypophysis in such fishes as the trouts and salmon (*Salmo*) and the cods (*Gadus*). There is evidence that neurohypohyseal hormones are secreted in the neurosecretory tissues of the hypothalamus and that these accumulate in the neurohypophysis whence they find their way into the bloodstream. The control of the adenohypophysis seems, at least in part, to be based on neurosecretion. The hypophysis of the lobefin (*Latimeria*) differs markedly from that of other fishes. It is an unusually elongate, cord-like structure.

Table 1.3 : Fish Endocrine Organs of Doubtful or Unknown Functions.

Gland	Location	Evidence of Endocrine Function
Thymus	Dorsal (and in some ventral) to gill arches	Growth (?)
Pineal body	Diencephalon	Action on melanophores
Pseudobranchs	Gill cavity on hyoid	Chloride balance
Caudal neurosecretory system (Urohypophysis)	End of spinal cord in tail; cells similar to secretory cells of hypothalamo-hypophyseal neurosecretory system	Water and ion metabolism; sodium exchange (?); gas metabolism (?)

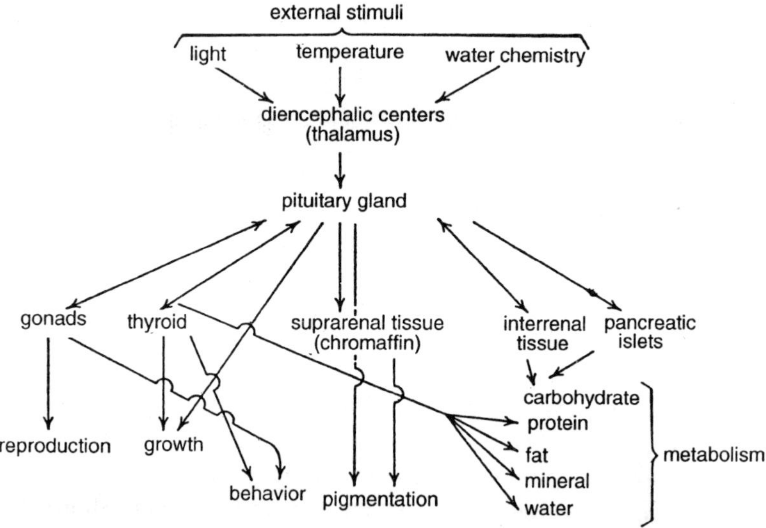

Figure 1.7 Interaction of some endocrine glands of fishes.

The Pituitary Hormones

Many of the hormones from the fish pituitary have the same or analogous effects as those from the incomparably betterstudied mammalian gland. All known pituitary hormones are proteins or polypeptides and in many cases mammalian preparations have been used to establish the role of certain of them in fishes. It is to be noted that minor differences in chemical composition have been established for the same hormone in different families of mammals (pigs and cattle); it is therefore to be expected that similar differences exist between the hormones of different groups of fishes. The specificity of fish pituitary hormones has already been shown in experiments and practical applications with injections and implantation for forcing the spawning of certain fishes of economic importance such as sturgeons (Acipenseridae), trouts (Salmoninae), catfishes (Ictaluridae), and mullets (Mugilidae).

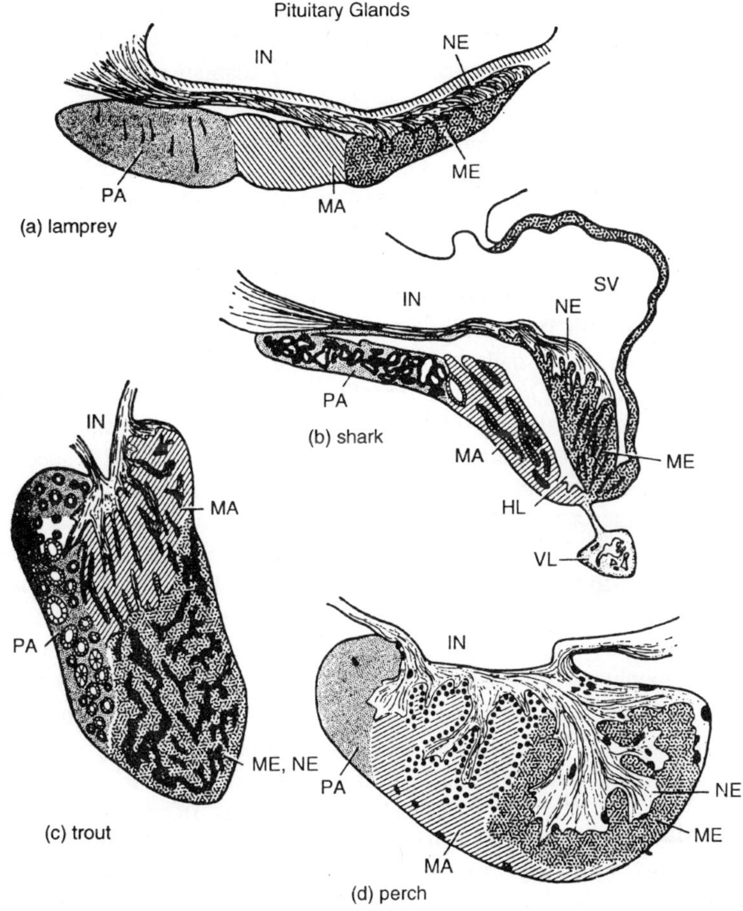

Figure 1.8 : Diagrams of the pituitary gland of fishes: (a) lamprey (Petromyzon); (b) dogfish shark (Squalus); (c) trout (Salmo); (d) perch (Perca). HL, lumen of hypophysis; IN, infundibulum; MA, meso-adenohypophysis; ME, meta-adenohypophysis; NE, neurohypophysis; PA, pro-adenohypophysis; SV, saccus vasculosus; VL, ventral lobe.

Pituitary hormones of fishes can be divided into two groups, one which reciprocally controls the function of other endocrine glands and the other which influences selected enzymatic processes in one or several types of body cells. In the first group are the adrenocorticotropic, the thyrotropic, and the gonadotropic hormones which are secreted in the pro- or meso-adenohypophysis, depending on the species of fish. In the second group are several hormones, described below, that affect melanophores.

The pro-adenohypophysis is the suspected source of a melanin hormone (MAH) acting on the melanin-bearing cells in the skin to bring about a more permanent background adaptation than would be possible through nervous impulses alone. The melanophore-stimulating hormone (MSH or intermedin) from the meta-adenohypophysis acts antagonistically to (MAH) and expands the pigment in the colour cells, thus contributing to long-range background adjustment; melanin synthesis is also furthered by this hormone.

Prolactin, similar to the hormone that influences lactation in mammals, has been demonstrated from the pro-adenohypophysis. In conjunction with intermedin, prolactin promotes the laying down of new melanin in the melanophores of the skin in some fishes such as the mummichog (*Fundulus heteroclitus*). Prolactin is one of several hormones implicated in electrolyte regulation in teleosts, but its importance in maintaining homeostasis appears to vary with species. The adrenocortico-tropic hormone (ACTH, to be discussed later), apart from its target action on the hormone-producing tissue of the adrenal cortex (located diffusely in the kidney of fishes), also has a stimulating effect on melanin production in the xanthic goldfish (Carassius *auratus*). Further work may explain or reconcile these differences in the causation of melanogenesis. A growth hormone from the meso-adenohypophysis promotes increase in body length of fishes. Its mode of action on cell division and protein synthesis has not been explained.

The ventral lobe probably secretes gonadotropins in dogfish shark and skate. A thyrotropic function has also been found in the ventral lobe. The neurohypophysis produces, or stores from hypothalamic neurosecretory cells, endocrine substances that are well known from their effects on mammalian metabolism. Vasopressor and antidiuretic hormones constrict mammalian blood vessels and promote water retention through their action on the kidney. Oxytocin stimulates mammalian uterine muscle, increases the discharge of milk from the lactating mammae, and lowers blood pressure in birds. Fish pituitary material brings about such effects in higher vertebrate test animals or tissues, but it can be presumed that the target organs or the specific sites of action of neurohypophyseal hormones in fishes should differ from those of higher vertebrates where we encounter radically different problems of osmoregulation and therefore of salt and water balance. If fish blood is found to be devoid of neurohypophysial hormones, then it may be that they act as "local hormones" on the adenohypophysis.

The Thyroid Gland

Thyroid hormone contains inorganic iodine, extracted from the blood stream and stored in the follicles of the gland to be released into the circulatory system upon specific metabolic demands. The level of activity of the follicles is under the influence of the thyrotropic hormone of the pituitary which, in turn, is at least in part governed by a combination of genetically determined maturation processes in conjunction with certain environmental variables such as photoperiod, temperature, and salinity. In certain sharks (Elasmobranchii) and most higher bony fishes (Actinopterygii), with exceptions such as the parrotfishes (Scaridae), the thyroid gland consists of diffuse tissue scattered along the base of the ventral aorta; removal or inactivation of the gland, a prerequisite for deficiency studies, is therefore difficult but physiological blocking or radiothyroidectomy using I··· has been accomplished. Little is known of the action of thyroid hormones in cyclostomes or in elasmobranchs. Respiratory stimulation by thyroxine is its best known action in mammals, yet most evidence indicates that thyroxine has no such action in teleosts. Physiologically, small doses of thyroxine and tri-iodothyronine cause thickening of the epidermis and paling of the goldfish (*Carassius auratus*). High thyroid titer retards growth of larvae of salmons (*Oncorhynchus*) but transformation into the juvenile smolt stage in the same genus is accelerated by inducing thyroid hyperactivity. In the mudskipper (*Periophthalmus*), a fish that spends much of its time partly outside of the water, induced thyroid hyperactivity leads to the assumption of an even more terrestrial existence with attendant morphologic and metabolic changes. The thyroid gland seems, in part, to influence osmoregulation in salmons (*Oncorhynchus*) and sticklebacks (*Gasterosteus*) and thyroid hyperactivity has been noted in salmons on their spawning migrations. Last but not least there are indications of a thyroid effect on growth efficiency and nitrogen metabolism in the goldfish, expressed through a rise in ammonia excretion.

It appears that thyroid functions in fishes are tied to a wide variety of processes, that some of them are related to growth and maturation phenomena, and that the thyroid, together with other endocrine glands, plays an important role in diadromous migrations of fishes.

The Ultimobranchial Glands

The small, paired ultimobranchial glands originate embryonically near the fifth gill arch but finally come to lie near the thyroid gland. The hormone calcitonin is found in the ultimobranchial glands of fish and in the thyroid and internal parathyroid glands of mammals, in which it produces hypocalcemia. Experiments with mummichogs (*Fundulus heteroclitus*) and with European and Japanese eels (*Anguilla anguilla* and *A. japonica*) have failed to confirm a hypocalcemic effect of calcitonin. Calcitonin may be related to osmoregulation. Eel calcitonin has been shown to decrease serum osmolality, sodium and chloride in Japanese eels. The ultimobranchial glands may be under pituitary control.

Chromaffin Tissue (Suprarenal Bodies; Medullary Tissue)

The chromaffin tissues receive their name from a characteristic staining reaction. In the lamprey group (Cyclostomata), chromaffin cells are arranged as strands along the dorsal aorta and in the ventricle and the portal vein heart, but in sharks and rays (Elasmobranchii) the tissue lies in association with the sympathetic chain of nerve ganglia. In bony fishes (Actinopterygii), location and concentration of chromaffin tissue shows considerable variation ranging from an elasmobranch-like pattern at one extreme, as in the flounders (*Pleuronectes*), to a true adrenal arrangement, as in the sculpins (*Cottus*) where chromaffin and adrenal cortical tissue are joined into one organ, comparable to the mammalian adrenal gland.

Chromaffin tissue extracts of fishes are high in adrenaline and noradrenaline. Exogenous administration of the catecholamines (adrenaline and noradrenaline) results in the following physiological responses: bradycardia, changes in blood pressure, branchial vasodilation, increased blood glucose levels, diuresis in glomerular teleosts, and hyperventilation. These "fight or flight" reactions in fishes are analogous to those of higher vertebrates and establish the suprarenal bodies as having similar function to that of the mammalian adrenal medulla.

Adrenal Cortical Tissue (Interrenal Tissue)

Endocrine cells producing adrenocortical steroid compounds are arranged throughout the body cavity near the postcardinal veins in the lamprey group (Cyclostomata). Among the sharks (Squaliformes) they lie between the kidneys. In rays (Rajiformes) they assume a more or less close association with posterior kidney tissue with some species having the adrenal cortical tissue aggregated near the left and others near the right central border of that organ. In bony fishes (Actinopterygii) there are one to several layers of interrenal cells to be found along the postcardinal veins as they pass through the head kidney (pronephros).

Secretion of adrenal cortical hormones is under control of the adrenocorticotropic hormone (ACTH) of the hypophysis. In the Mexican tetra (*Astyanax mexicanus*), experimental hyper-function of interrenal cells can be produced as well as a condition that resembles the well-described adrenal stress syndrome in mammals. Cortisol, cortisone, corticosterone, and aldosterone known from mammalian adrenal glands, as well as some unknown compounds, have been isolated from blood plasma of the sockeye salmon (*Oncorhynchus nerka*). These findings suggest, as is true in mammals, that two types of secretions are produced by the fish adrenal cortical tissue: (*a*) mineral-corticoids that influence certain aspects of fish osmoregulation; (b) gluco-corticoids that influence the carbohydrate metabolism, especially the level of blood sugars. Experimental evidence for the first function comes from the rainbow trout (*Salmo gairdneri*) which excretes higher than normal amounts of sodium ions through its gills but retains more than the normal amount of sodium in the kidneys when subjected to internal salt loads under mineral-corticoid treatment. The second function was verified in the oyster toadfish (*Opsanus tau*) where blood sugar levels are raised upon intramuscular injection of corticosteroid compounds.

The concentration of cortisone in the blood plasma of salmons (*Oncorhynchus*) is high at spawning stages of the life history but low during the more sedentary parr stages. In *Oncorhynchus,* 60 percent of the body protein is catabolised during the migration-spawning phase, and this catabolism is associated with a sixfold increase in plasma corticosteroids and an elevation of liver glycogen. Injection of adrenal cortical hormones promotes lymphocyte release in a tetra (*Astyanax*) and antibody release in the European perch (*Perca fluviatilis*), pointing to the similarity in the function of the adrenal cortical tissues in higher and lower vertebrates. Corticosteroids structurally resemble androgens and have androgenic side effects. These side effects suggest some functional equivalence in fishes. Daily rhythms in concentrations of plasma adrenal corticoids have been reported in several teleost fishes, and these rhythms may play a prominent role in the diel rhythm of teleosts.

Corpuscles of Stannius

The corpuscles of Stannius are endocrine organs attached to or imbedded in the kidneys of holostean and teleost fishes. Their number varies with the species and histologically they resemble adrenal cortical cells. The corpuscles of Stannius lower serum calcium levels in mummichogs (*Fundulus heteroclitus*) kept in high calcium environments, such as seawater. It appears that the corpuscles work in conjunction with the pituitary gland, which produces a hypercalcemic effect, to maintain a relatively constant level of serum calcium.

The Sex Glands as Endocrine Organs

In fishes the maturation and possibly also the elaboration of gametes requires the presence of sex hormones; in addition sccondary sex characteristics such as colouration, breeding tubercles, and the maturation of gonopodia depend on the presence of endocrine substances from the sex glands. Sex hormones are produced by specialised cells of the ovaries and testes at levels determined by the output of gonadotropic hormones from part of the pituitary gland (meso-adenohypophysis).

The male hormone testosterone has been measured in the blood plasma of an elasmobranch (*Raja radiata*) and in salmon, in which there is a correlation between plasma levels and the reproductive cycle. Another gonadal steroid, 11-ketotestosterone, is ten times more physiologically androgenic than testosterone in *Oryzias latipes* (medaka) and sockeye salmon (*Oncorhynchus nerka*). Of the female estrogens, estradiol-1713 has been found in many species of fish, often accompanied by variable quantities of estrone and estriol. Progesterone is present in fish, but it is not known if this compound has a hormonal function.

There have been few unequivocal demonstrations for any fish that reproductive behaviour is induced or regulated by gonadal steroids. Many researchers have demonstrated effects of exogenous steroids on sex determination and the development of secondary sexual characteristics in fish, but in most cases there is little mention of the effects of these treatments on reproductive behaviour.

Upon an injection of mammalian testosterone and estrone, a lamprey (*Lampetra fluviatilis*) develops its cloacal lips and coelomic pores, both structures involved in the reproductive process. In sharks and rays (Elasmobranchii) tests for a similarity between the sex hormones from different vertebrate classes have not shown clear results, but among bony fishes (Osteichthyes) such equivalence seems to exist, although experimenters maintain certain reservations. For instance, ethynil testosterone (pregneninolone), which has mild androgenic and progesterone-like effects in mammals and birds, seems to be highly androgenic in fishes. Testerone injections in immature and female parrotfishes (Scaridae) alike evoke appearance of the male colour pattern in the experimental subject. Female sex hormones of fishes are less similar to those of other vertebrates than the testicular hormones which also may strongly affect ovarian development, as shown in a loach, the Japanese weatherfish (*Misgurnus fossilis*). Evidence to date points to less biochemical specificity in both androgenic and estrogenic fish hormones than in mammals and birds and to a greater degree of action on the gonads by sex hormone analogues such as for instance certain corticosteroids.

Prostaglandins, which are potent, hormone-like substances, have recently been discovered in testis and semen of flounder (*Paralichthys olivaceus*), bluefin tuna (*Thunnus thynnus*), and chum salmon (*Oncorhynchus keta*). In mammals, prostaglandins lower arterial blood pressure and stimulate the contraction of smooth muscle, but their function in fish has not been studied.

The Action of Gonadotropins on the Gonads

Removal of the pituitary (hypophysectomy) performed on several species in all three major groups of living fishes reduces or stops gonadal development, both in the transition from juveniles to sexual maturity, and during the seasonal spawning cycle. Dependence on the pituitary gland for the output of sex hormones by the gonads themselves has thus been clearly established in fishes. Present research favors the existence of only one functional gonadotropin in fishes, for which the term Piscine Pituitary Gonadotropin (PPG) has been proposed. The single teleostean gonadotropin possesses properties that in mammals are distributed between two hormones, LH and FSH. On less complete evidence than is available for mammals rests the assumption that the fish mesoadenohypophysis produces two different gonad-stimulating hormones.

Mammalian luteinizing hormone (LH) promotes release of gametes from nearly ripe gonads in fishes and stimulates appearance of secondary sex characteristics. From this it is inferred that there is a similar hormone in the fishes themselves. Gonadotropins purified from salmon pituitaries resemble LH. Furthermore human chorionic gonadotropins and the urine of pregnant mares, which have LH-like effects, promote the release of eggs in female fishes. Evidence for a follicle-stimulating hormone in fishes (FSH), the second gonatropic hormone known from the mammalian pituitary gland, rests, so far, on standard assay for mammalian FSH performed with fish pituitary extracts. Such assays, however, do not furnish proof of a second pituitary gonadotropic hormone in fishes.

The gonadotropins or whole pituitary preparations from mammals, birds, and reptiles are effective in some fish species in hastening spawning or otherwise influencing sex glands and behaviour. When amphibian materials are used success is obtained more often than with mammalian ones. Best results, however, follow the use of glands or gland extracts of the same fish species; however, biochemical interfamilial and interspecific differences of fish gonadotropins seem in many cases to be even greater than those noted between fishes and amphibians.

The Use of Gonadotropins in Fish Culture

In fish-culture operations it may be costly to retain potential spawners over long periods only to have them fail to ripen when desired. Pituitary implants or injections are now widely used to bring about the timely release of eggs and sperm. Among the most extensive users of pituitary-induced forced spawning are Brazil and Russia. In Brazil, the main freshwater-fish crop consists of characins (Characidae). In Russia, the technique is applied to sturgeons (*Acipenser*) for caviar production. Some use is made of the method also for the culture of a variety of fishes, including trouts and salmons (Salmoninae) and carps (Cyprinidae) in Europe, North America, and Asia.

To induce forced spawning, fresh pituitary glands are either injected intraperitoneally after trituration in 0.7 percent solution of salt (NaCI) or they are implanted whole. The glands may be frozen and used within 6 months, or as is frequently done, the glands may be dehydrated in acetone or alcohol, then dried, powdered, and stored in airtight containers for several years. The powder is suspended in acidified saline solution for intramuscular or intraperitoneal injection.

Best results are obtained with glands taken from ripe or nearly ripe donors of the same species although other species can also be used. The dosages depend both on the state of maturity of the donor as well as of the recipient fish. Ripe fish, in general, need only a single dose whereas nearly ripe individuals may require injection of material from a whole gland every second or third day for a month. The use in practical fish culture of the principle of stimulation of gonads by gonadotropins of the pituitary gland is promising. We may expect considerable refinement of procedures as information accumulates on the structural formulas of fish gonadotropins and sex hormones, as already indicated by the effectiveness in inducing spawning in several species of fishes by highly purified salmon gonadotropin.

The Intestinal Mucosa

The intestinal lining (mucosa) produces hormones that join with nervous controls to regulate the pancreatic secretions. The intestinal hormones are secretin and pancreozymin (formerly both subsumed under secretin). Secretin produces flow of enzyme-carrying liquids from the pancreas, and pancreozymin enhances flow of the zymogens. Both hormones are formed mostly in the anterior portion of the small intestine. Their production can be brought about by introducing into the pyloric region of the stomach of carnivorous fishes of an acidified homogenate of fish flesh or through the injection of secretin into the gastric vein which stimulates the secretion of the pancreas. Thus the classic tests for endocrine activity are satisfied for the fish small intestine. In higher vertebrates there exists another intestinal hormone, cholecystokynin, which promotes flow from the gall bladder. A polypeptide with cholecystokinin-pancreozymine-like activity has been found in cells of lamprey intestine.

THE SENSES (SENSE ORGANS) OF FISHES

The sense organs receive physical or chemical stimuli from the environment. Physical changes in heat flow or touch are felt through skin receptors; visual stimuli involve changes in light intensity and quality; and acoustical ones are received through the inner ear or lateral line. Chemical stimuli are those experienced through either smell or taste organs. Pain is probably not experienced as a strong sensation by fishes, though forceful or noxious physical or chemical stimuli evoke violent reactions.

The Cutaneous Senses

The tmperature sense can be used by the fish to remain in a temperature preferendum and may contribute to orientation on long- or short-range movements. However, temperature perception has been thoroughly investigated only among higher bony fishes (Actinopterygii). By conditioned reflex training, several species of both marine and freshwater bony fishes have shown capability of detecting temperature changes of about 0.03 °C, provided the rate of heat flow is fairly rapid (in excess of 0.1 °C per minute), and the detection of small temperature differences relies on a real-summation of impulses from finely branching epidermal nerve endings. Fishes can distinguish a rise in temperature from a fall, but the physiological mechanism for such discrimination is not known. There is evidence that some fishes may be most sensitive to temperature changes on their anterior portions and that temperature sensation is intimately tied to the perception of light (weak) tactile stimuli.

Tampullae of torenzini of sharks and rays (Elasmobranchii), a component of the lateral-line system, respond to cooling by a rise and to warming by a lowering in discharge rate along their nerves. It is not known whether they serve as temperature receptors in the intact animal, or whether or not a diffuse cutaneous temperature sense is also present.

Some fishes are sensitive to touch, as in some ca'tlishes (Siluridae). However, as yet there is little evidence of specialised dermal touch receptors like those in mammals and birds (Vater-Paccinian, Krause's and Herbst's corpuscles). In these higher vertebrates, each touch cdrpuscle has a skein-like nerve net, enclosed in a connective tissue capsule. Comparable structures Have so far been described only in the fins of sharks (Squaliformes), where they act as proprioceptors when the fin is adducted or abducted, and in the snout region of some moray cels (*Gymnothorax*), where they serve for fine touch discrimination in the choice of food. Similar structures also occur on the separated anterior pectoral fin rays of searobins (Triglidae).

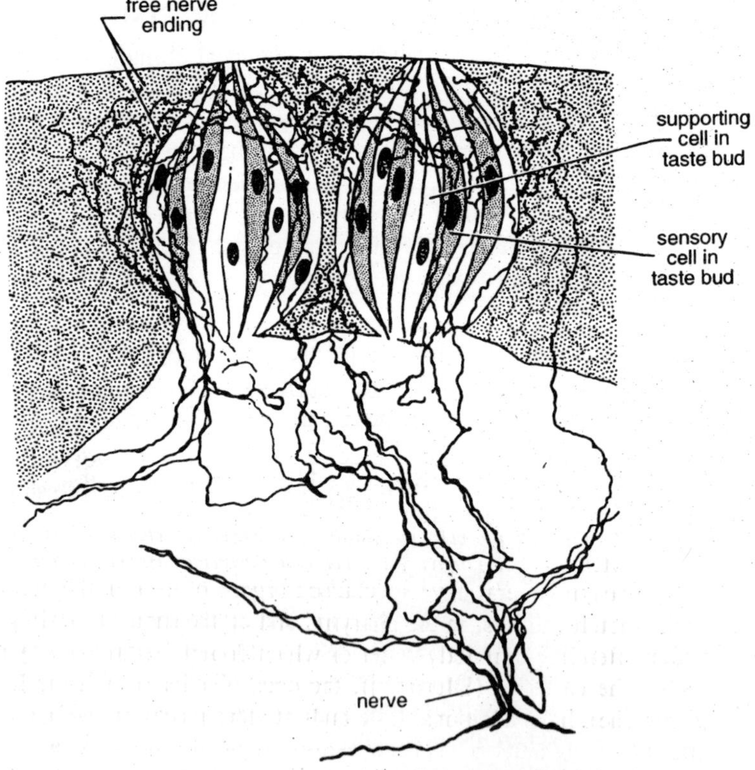

Figure 1.9 : Free nerve endings surrounding taste buds in a barbel of a sturgeon (Acipenser).

Observations of behaviour and electric recordings of nerve discharges have shown bullheads (*Ictalurus*) to be highly sensitive to touch on the head and barbels. These regions carry a net of thin, profusely branching fine nerves with free endings, derived from cranial nerves V and VII; but no specialised touch structures have been found in minnows (Cyprinidae) where they appear to react jointly to tactile and thermal stimuli. On the body surface free endings of the sensory roots of the spinal nerves receive the touch stimuli.

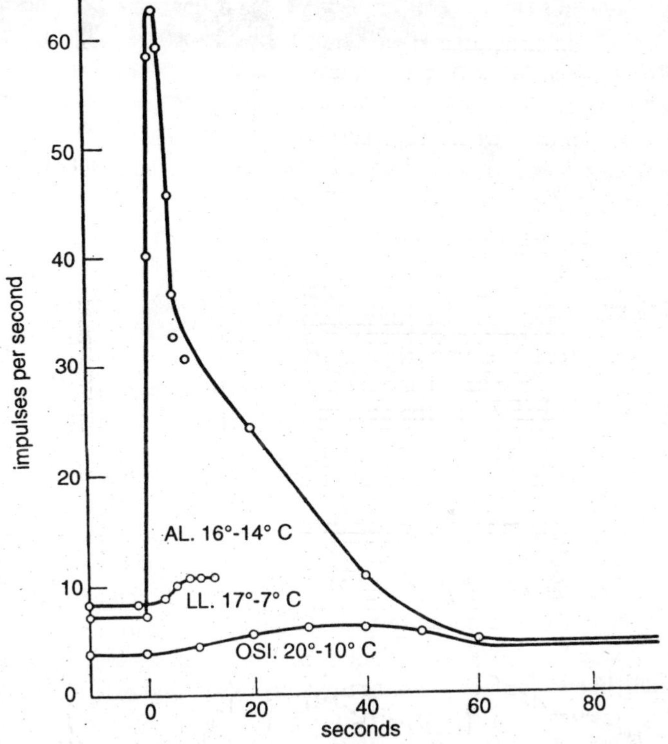

Figure 1.10 : Changes in impulse frequencies upon temperature changes in an ampulla of Lorenzini (AL), lateral-line organ, (LL), and an intramuscular nervous end organ (OSI).

Fishes in general have specialised organs of taste in the epidermis of the mouth and lips, in the pharynx, and on the snout. Certain groups which often live in muddy water or which do not feed mainly by sight, suchas the catfishes (Siluroidei), the cods (Gadidae), and the loaches (Cobitidae), have additional taste buds at other places on the body; over 100,000 such units have been estimated on the body of a bullhead (*Ictalurus*). The buds are mainly innervated by a branch (ramus recurrens)

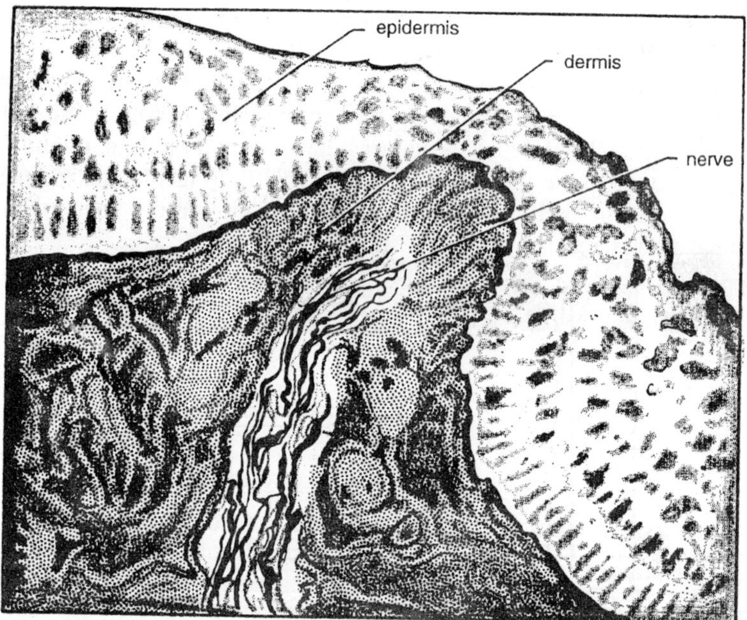

Figure 1.11 : Diagram of presumed touch sensor from the lower lip of a moray eel (Gymnothorax vicinus).

or cranial nerve VII, but elements of cranial nerves IX and X also share in the innervation of taste receptors that occur on outlying parts of the body.

The vertebrate taste bud is typically composed of elongated sensory cells arranged like segments of an orange. The three different cell types in fish taste buds are receptor cells, supporting cells, and basal cells. In addition, there are transitional or intermediate forms of cells that may become either sensory or supporting cells. The sensory cells have short, hair-like extensions in contact with the water and extend from the tip of the taste bud to the underlying dermis where they may have a bottle-shaped expansion and where the nucleus is located. In especially sensitive regions, such as the barbels of bullheads (*Ictalurus*) and sturgeons (*Acipenser*) and the lips of conger eels (Congridae) (and amphibians), nerves that reach the taste buds also give off a net of free nerve endings to the surrounding regions. It appears that taste and touch may thus work in close cooperation in the choice of food before ingestion.

In the lamprey group (Cyclostomata), taste buds have been found in the pharynx, the gill cavity, and on the head. In sharks and rays (Elasmobranchii), taste buds occur in the mouth and the pharynx and taste-bud-like structures, probably also for chemical perception, are present in pits on the body surface.

The carp (*Cyprinus carpio*) has taste sensations similar to mammals; it can sense salty, sweet, bitter, and acid stimuli. It also tastes extracts of fish skin and other, so-called sapid, substances. Amino acids singly or in mixtures are particularly effective taste stimulants in fishes.

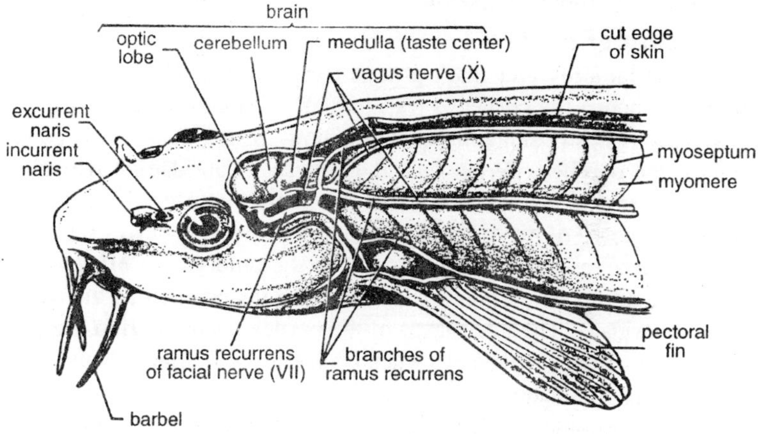

Figure 1.12 : Distribution of the ramus recurrens innervating taste buds on body of a loach (Nomachelius).

Marine fishes (searobins, *Prionotus;* hake, *Urophycis*) with taste- or chemo-receptors on their modified paired fins also sense sapid substances, for example, the extracts of clams or worms. They appear to taste compounds that normally occur in their food but their receptors do not respond to salty, sweet, or bitter substances though they still react to acids. Histologic examination of the skin of a minnow (*Phoxinus*) reveals a subepidermal nerve plexus that sends branches into the epidermis where the nerve fibers divide and taper into fine endings. Some specialised single cells, presumably chemosensory, have also been found in a minnow (*Phoxinus*). Thus, tactile, temperature and chemical sensations are closely integrated in fish skin. The coarser among these divisions are probably tactile whereas the finer ones are believed to serve temperature and chemical sensations.

Vision

Natural waters are rarely pure; therefore light rays in them undergo multiple scattering in all directions. Furthermore, waters are often turbid, almost opaque, and therefore a sharply defined image is difficult to obtain. Nevertheless, it is of great biological significance for fishes to detect slight movements and colour changes that signal prey, predator, companion in a school, or mate. Thus it appears as if the retinal mechanisms of fishes have developed towards contrast enhancement and motion perception besides the formation of a clearly focused image. Basic anatomical features of the eyes of fishes have been described already; here will be added certain details of the functional anatomy and vision of colour and form.

The Cornea of the Eye

In the lamprey and shark groups (respectively, Cyclostomata and Elasmobranchii) the cornea is transparent, without pigment, as it is in most bony fishes (Osteichthyes). Some bony fishes, such as the bowfin (*Amia*), the perches (*Perca*), and the wrasses (Labridae), have yellow or green corneal membranes. The sea lamprey (*Petromyzon*) has a cornea composed of two tissue layers separated by a gelatinous substance. This primitive condition is also found in the embryos of higher fishes and in the adults of certain kinds with poor vision such as the conger eels (Congridae). The inner layer of the cornea continues into a corneal muscle that compresses the eye when contracted and brings about modest distance accommodation. No fine intraocular accommodation mechanism exists in cyclostomes. A distinctly two-layered cornea is found in certain flounders (*Solea*), perhaps to protect the eye of such a bottom-dwelling fish from sand and detritus. The mudskipper (*Periophthalmus*) leaves the water for considerable periods of time; it also has a two-layered cornea and folds of skin beneath the eye which retain water and help to prevent drying of the organ. In the lungfishes (Dipnoi) the cornea is rugged but there are no wetting structures.

The Sclerotic Coat

The sclerotic coat, sclera, or firm layer of the eyeball is variously reinforced in all groups of fishes. In lampreys (Petromyzonidae) the sclera is fibrous and firm but sharks and rays (Elasmobranchii) have cartilaginous elements and also calcified plates in this layer for additional reinforcement. The sclerotic coat in an elasmobranch is highly vascular and has a base of fibrous connective tissue. The sturgeon and its relatives (Chondrostei) carry a cartilaginous support at the corneal border of the sclerotic coat as well as some bony structures. In the sturgeons (*Acipenser*), a halfmoonshaped bone, the conjunctival bone of Mueller, partly surrounds the cornea. The highest bony fishes (Teleostei) have fibrous, flexible sclerotic coats within the orbit and around the optic nerve, but there may be cartilaginous or even bony supporting elements in the outer portion of the eye which can surround the entire cornea.

The living coelacanth lobefin (*Latimeria*) carries cartilaginous reinforcements in the sclera which thicken towards the optic nerve; there are also calcified scleral plates that form a pericorneal ring.

The Choroid Coat

Cyclostomes carry much melanin in the choroid coat, especially toward the posterior pole of the eye. The shiny nature of most shark eyes is due to the reflecting material in a well-developed choroid coat or tapetum lucidum. The tapetum functions to collect a maximum amount of light for the retina by scattering reflection from guanine crystals.

In sturgeons and relatives (Chondrostei) vascular elements become more prominent in the choroid layer of the eye than they were in shark-like fishes (Elasmobranchii) and layers of guanophores are also present.

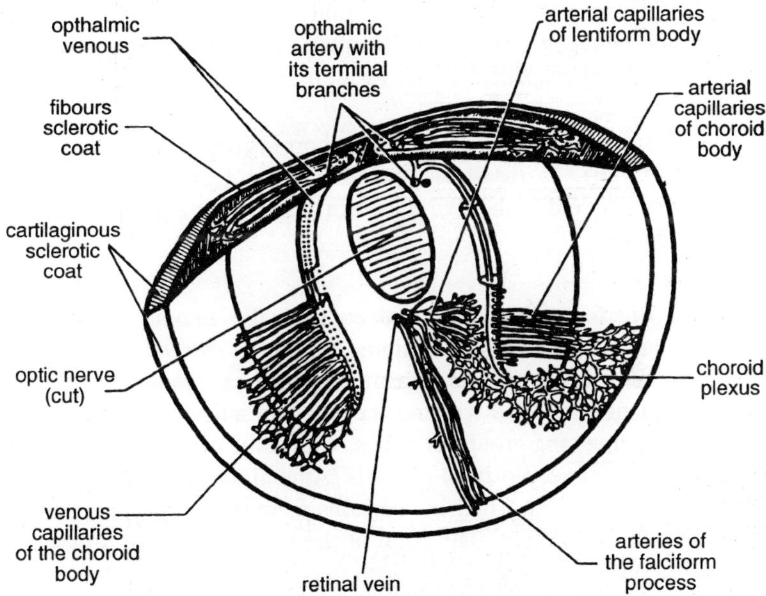

Figure 1.13 : Arrangement of blood vessels in the chorid body in the eye of a trout (Salmon).

Most holostean and teleostean bony fishes, with the exception of the eels (Anguilliformes) and related groups, have the blood vessels of the choroid cost collected into a horseshoe-shaped choroid gland lying between choroid and sclerotic coats of the eye, in a position around the exit of the optic nerve. The gland has its blood vessels arranged like the rete mirabile in the gas bladder, for the secretion of oxygen at higher tensions than those found in the blood in order to provide oxygen in abundance to tissues in the vicinity of the gland. It should be noted that the retina lies close by, that it has the highest oxygen demand of any tissue in the body, and that the retinal circulation proper is not especially well developed in those bony fishes which have a large choroid gland. Measurements with polarographic electrodes in the vitreous humor have shown that fishes with a well developed choroid rete such as the bluefish (*Pomatomus saltatrix*) and remoras (*Remora*) have intraocular oxygen tensions (450 to 780 mm Hg) which are several times that of their arterial blood. Oxygen tensions (210 mm Hg) in the eye lower than arterial blood are found in fishes with a small choroid rete such as the tautog (*Tautoga onitis*) and also eels (*Anguilla, Conger*), as well as sharks and rays (Elasmobranchii), which have no choroid rete and have intraocular oxygen tensions (10 to 20 mm Hg) below that of the surrounding seawater. Thus the choroid gland can provide the retina with a good oxygen supply apart from any other function it may have; it has also been suspected to act as a cushion against compression of the eyeball.

Brilliant guanophores occur in the choroid coat of many bony fishes (Actinopterygii), to maximise the amount of light that reaches the retina. The choroid in the living lobefin (*Latimeria*) also has a brilliant guanine cover.

The Iris

The iris forms the pupil and controls the amount of light that reaches the retina. In the cyclostome *Mordacia mordax* the iris lacks muscles, but sharks and rays (Elasmobranchii) have muscular elements in their guanine-studded iris and thereby can adjust the shape of the pupil. The remaining groups of fishes, with some exceptions among the flounders (Pleuronectiformes) and eels (Anguillidae) have a fixed pupil, either circular or oval. Their iris contains no muscles, but some guanine and melanin are present. The pupil of the living lobefin (*Latimeria*) *is* large and noncontractile. Some deep-sea teleosts lack an iris.

The Lens and Accommodation

The lens of the fish eye is a firm, transparent ball made up of noncollagenous protein. Since the index of refraction of the cornea and the ocular humors are about the same as water (1.33), refraction and image formation depend almost entirely on the lens, which has a very high *effective* index of refraction (1.67). The fish lens is not homogeneous but has an *actual* refractive index from 1.53 at the center to 1.33 near the outside. An almost constant feature of fish eyes is the distance from the center of the lens to the retina divided by the radius of the lens. This ratio is about 2.55. The lens is roughly spherical in most bony fishes (Osteichthyes) and in lampreys (Petromyzonidae). However, sharks and rays (Elasmobranchii) have a horizontally compressed lens and in fishes that leave the water at times the lens face shows modifications; in the foureyed fishes (Anablepidae), for instance, the lens is pyriform to provide for aquatic and aerial vision. The lens is attached from above to the suspensory ligament and from below to the falciform process with its retractor lentis muscle.

In most fishes, accommodation, which is the adjustment of vision from near to far, results from changing the distance between the lens and the retina rather than altering lenticular shape. Some sharks and rays (Elasmobranchii) accommodate by small changes in lens convexity. Others are capable of moving the lens by means of a muscular papilla on the ciliary body and thus achieve a little adjustment. Rayfin fishes (Actinopterygii) show varying degrees of lens accommodation, depending on the development of the retractor lentis muscle. According to Pumphrey, sight feeders such as the brown trout (*Salmo trutta*) have this muscle well developed and possess other remarkable features of lens and retina which suggest that the vision of this fish is acute over a wide range. Trouts (Salmoninae), in fact, have the ability to focus simultaneously on different portions of the retina objects which are distant and other objects which are nearby. The lens in these fishes is slightly ellipsoid with the longest lens diameter parallel to the long axis of the fish. The lens thus has two focal lengths. One is for light rays reflected from distant objects that are lateral to the fish and come to focus on the central retina. The other, shorter one, is for close objects ahead that are projected on the posterior (temporal) retina. Furthermore, in accommodation, the lens is moved in a plane parallel to the long axis of the fish rather than inwards. Consequently, the distance between lens and central retina remains unchanged, and the focus remains adjusted for distance vision. At the same time the contraction of the retractor lentis muscle brings the lens closer to the posterior retina that represents the frontal field of vision of the fish. In this portion of the visual field the fish is thus capable of obtaining sharp images within distances ranging from a few inches to infinity, or the limit of transparency of the water.

Retinoscopy has shown that many species of bony fishes (Osteichthyes) are farsighted. Refractive errors change by several diopters during such measurements on schooling fishes, such as the silversides (*Menidia*), certain jacks (Carangidae) and others, suggesting that fishes can indeed move their lenses toward and away from the retina or deform their ocular globe for accommodation. Some groundfishes such as flounders (*Paralichthys*) and reef fishes such as the squirrelfish (*Holocentrus*) and butterflyfishes (Chaetodontidae) are emmetropic, that is the images focus on the retinal surface. No retinoscopic tests have been made on small stream and lake fishes such as certain sunfishes (Centrarchidae) but analogies with reef fishes in behaviour knd habitat suggest that they also might be emmetropic. In the alewife (*Alosa pseudoharengus*) the focal plane of the eye lies an average distance of 0.2 millimeter behind the retina, suggesting that static images which this fish sees may not be as sharp as those of animals with aerial vision.

In deepsea fishes especially large lenses and tubular, upward-viewing eyes are found, as in the pearleyes (*Scopelarchus*), possibly to aid in gathering on the lower retina signals from the ventral photophores of species mates, of prey, or of predators.

In the front of the lens there is a small space with clear saline aqueous humor. The main cavity of the eyeball is filled by the transparent vitreous humor which is secreted by the ciliary body and contains transparent tendinous reinforcements.

The Retina

Fishes have typical vertebrate retinas made up of several partially interlocking cell layers. As light hits the retina after traveling through the lens and the humors, the retinal layers encountered in order are: (a) fibers leading to the optic nerve, (b) their ganglion cells, (c) bipolar cells which, in turn, are in contact with (d) the photoreceptor cells, proper, namely the rod and cone cells of the retina. In the retina of lampreys (Petromyzonidae) no distinction can be made between the rods and cones but in some sharks and rays (Elasmobranchii) and in all the highest bony fishes (Teleostei) the rods and cones are well differentiated. In certain bony fishes twin cones occur and triple or even quadruple cones have been reported; it is not known whether these cone aggregations, peculiar to the fishes, have any special function. In teleost, but not in elasmobranch retinas there is a fifth layer (e) of retinal cells, the epithelium, which contains melanin and borders on the choroid coat.

In bright light the melanin pigment of the epithelial layer of the retina shades the sensitive rods; in dim light, however, the melanin becomes concentrated close to the choroid border so that photosensitive cells are fully exposed to whatever light is available. At the same time the contractile myoid elements in the base of the rods and cones move the sensitive cell tips in such a fashion that rods are extended away from the lens in bright light to be engulfed by the epithelial melanin, whereas in darkness the rod myoids contract and approach the cells towards the lumen of the eyeball. Cones move in the opposite direction toward the lens in bright and toward the outer epithelium in dim light. Fish eyes show these photomechanical or retinomotor reactions more pronouncedly than the eyes of other vertebrate groups. The level of light reaching the retina' can also be controlled by contraction of the pupil. With the exception of a few groups such as eels and flatfishes, this process is relatively poorly developed in fish. In sharks and relatives (Elasmobranchii) the photoreceptor layer may be shielded by melanin which moves in and out in the choroid layer inasmuch as the retina lacks a pigmented epithelium.

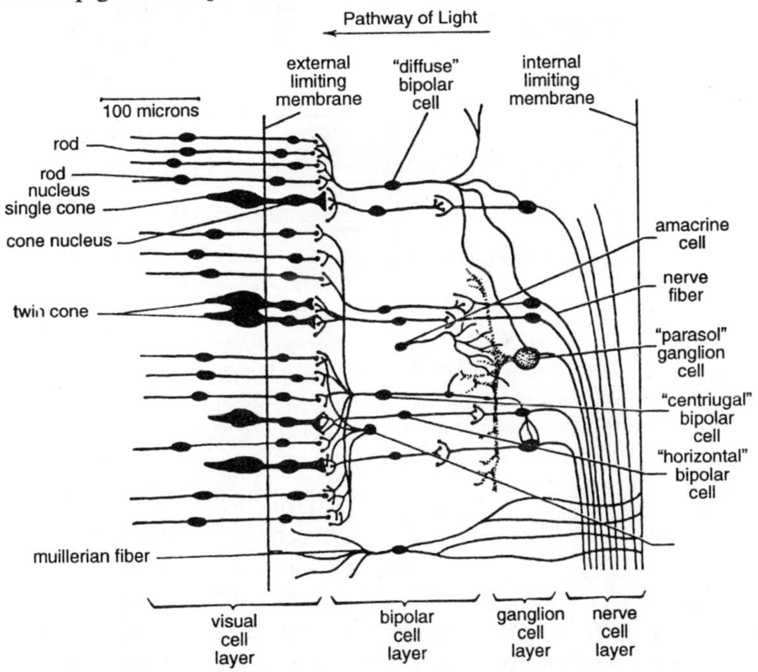

Figure 1.14 : Diagram of the arrangement of neurological components of the retina of a salmon (Oncorhynchus).

The number of rods and cones connected to one ganglion and thus to one fiber of the optic nerve varies greatly in different kinds of fishes. Crepuscular sight feeders and some deepsea fishes have many rods converging on one optic fiber, as an adaptation for vision in subdued light (summation of impulses); in other groups, such as the pikes (Esocidae), that seem to rely on good light for feeding, far fewer retinal receptors are connected to one optic neuron. Relative numbers of rods and cones also vary considerably in different species. The highest cone-to-rod ratios are in sight feeders that are active during the day whereas many crepuscular species have apparently increased the number of rods as compared to the cones. Certain midwater fishes (for example, *Winteria* and *Opisthoproctus*), besides having large eyes with luminous organs adjacent to them, also have very large numbers of rods in the retinas.

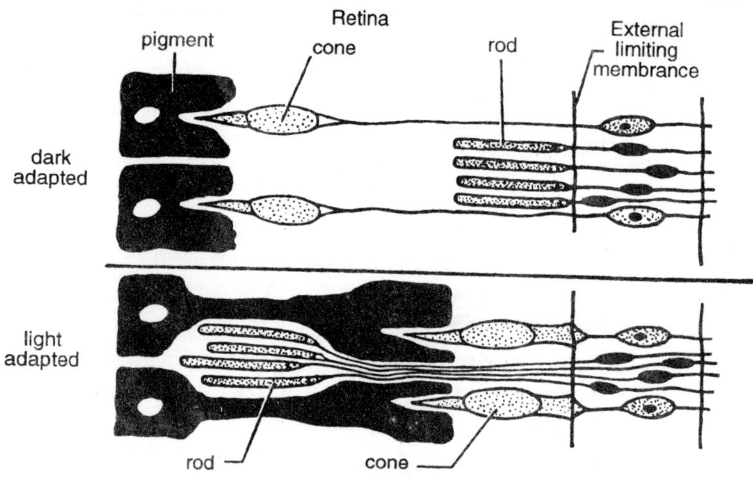

Figure 1.15 : Diagrams of the adaptions of the visual elements of the bony-fish retina to light and dark.

Few bony fishes (Osteichthyes) have a fovea (retinal area of clear vision) where cones and ganglion cells are more numerous than in other regions of the retina. The fovea appears to be restricted to fishes that live in clear waters and especially to those which can move their eyes independently or converge them to obtain a wide angle within which they may have binocular vision; examples include the seahorses (*Hippocampus*) and many members of the sea bass family (Serranidae).

The retinas of fishes yield two kinds of light-sensitive pigments, rhodopsin and porphyropsin. Purple-coloured rhodopsin has been isolated from the eyes of many marine rayfin fishes (Actinopterygii), whereas the retinas of freshwater rayfins and the marine wrasses (Labridae) contain a rose-coloured, light-sensitive substance, porphyropsin. Diadromous fishes, such as the sea lamprey (*Petromyzon marinus*), eels (*Anguilla*), and the Atlantic salmon (*Salmo salar*), possess both rhodopsin and porphyropsin; lamprey (*Petromyzon*) retinas from seaward migrants yield more rhodopsin, whereas porphyropsin predominates in the eyes of spawners in fresh water; converse changes occur in the eel as it travels from lakes and rivers to the Sargasso Sea. Rhodopsin and porphyropsin are synthesised in the dark from Vitamin A. On exposure to light there appears in the rods the yellow pigment retinine which can, however, be used to reconstitute the light-sensitive purple or rose pigments. The light-induced chemical change from purple to yellow pigment in the rods becomes translated in the retina to nerve impulses which travel along the optic nerve. Biochemical and physiological processes occurring in cones and thus connected with colour vision are not well known. Some pigments with different sensitivity maxima have been isolated from vertebrate retinas and electrical recordings have been made from the bipolar cell layer of several families of marine rayfin fishes. These findings suggest that the cones in the retinas of shallow-water species have paired responses to stimuli of different wave lengths of light. There is a family of elements sensitive to the combination of yellow and blue and another to red and green stimuli. When the rods are removed from such a retina, the cones respond to white light thus suggesting that colour vision in fishes is based on a threepigment system. Snappers (Lutjanidae) only respond to white light. They may be colour blind, but it is to be noted that they live in water depths of 30 to 70 meters. (about 100 to 200 feet), where the light consists largely of the blue-green portion of the spectrum and where even to man, with his well-developed colour vision, things appear to be of different shades rather than of different hues.

The initial step in vision is photochemical and involves reactions in the light-sensitive pigments of the rods and cones, as already suggested. It is not known how these chemical changes become transformed into the electrical impulses that can be picked up in the retina, nor how these finally come to be the signals which leave the eye in the optic nerve and travel to the brain. Signals in the optic, as in other afferent nerves are of three kinds: "on," "off, " and "on-off." They are characterised by changes in discharge frequency along the nerve. The retina of the goldfish (*Carassius*) has its lowest threshold for "on" responses in the blue, and for "off" responses in the red portion of the spectrum, indicating that this type of coding of receptor function permits acute contrast vision. Good discrimination of movement and contrast is also implied for this fish by the following observations. When a pointed monochromatic light of 153 microns in diameter is used to test the receptive field of one optic nerve ganglion that covers a field of about 4 square millimeters, the above three types of responses convey highly detailed messages about direction and progress of the moving light stimulus. Progressing diagonally from the lower left to the upper right "on" responses as an increase in volleys of nerve impulses are recorded as the light appears. Then as the point-light proceeds its progress is announced in an "on-off" fashion by volleys both on appearance as well as on disappearance on a certain area of the field. In the center the retinal receptors cause an increase in discharge rate ("off" response) along the nerve only as the light disappears, that is as it moves upwards. The upper half of the diagonal path shows the reverse sequence of coded messages from those observed in the lower half.

The eyes of lungfishes (Dipnoi) resemble those of amphibians, and in their retinas they have no (or poor) distance accommodation, few rods, and many cones, as well as some twin cones. The living lobefin (*Latimeria*) has large eyes, retinas without melanin, and receptor layers which consist almost exclusively of rods. Furthermore, relatively few rods are connected to each optic nerve fiber, and the eyes altogether resemble those of other mid-depth fishes that live in an oceanic twilight-zone.

The coded message carried by the optic nerve is further elaborated in the brain. These processes, of which little is known from any vertebrate, including man, must finally make the visual stimulus meaningful to thg organism.

Sight in the Lives of Fishes

Many fishes rely on sight for capturing food, for receiving signals that bring on or complete mating behaviour, and for finding and relocating shelter, which are only some of the aspects of fish behaviour that are sight dependent or sight influenced.

Experimental training has shown that shallow-water fishes have colour vision, very much like man; they can distinguish twenty-four different narrow-band spectral hues. Some fishes even exceed man's spectral range into the violet. White is to fishes a sensation basically different from that of hues and they see complementary colours. More importantly still, these trained - fishes, mostly minnows (Cyprinidae, including the elritze, *Phoxinus phoxinus*), are capable of experiencing simultaneously brightness and colour contrasts. There is also good discrimination for different shades in a series of greys. The difference in electrical responses in the retinas of snappers (Lutjanidae) from deep waters compared to those of shallow-water fishes has been mentioned, and it may well be that not all but only shallow-water fishes have developed colour vision. Such a difference among fishes would not be surprising, considering that the mammals, which have fewer species and inhabit an equally varied but different medium, also have some groups with and others without colour vision.

Sight-feeding fishes, such as the elritze (*Phoxinus phoxinus*) and the black basses (*Micropterus*), when given conditioned reflex training on different shapes without inherent biological meaning but coupled in training with either punishment or reward, show fair to good form discrimination. The fishes learn quickest and make fewest mistakes if the objects are moved rather than held stationary.

In this connection we should regard closely the feeding behaviour of certain sight-feeding fishes such as the Pacific salmons (Oncorhynchus). In these fishes as well as in those of the training situation above, good use is made of the peculiar retinal mechanism that is geared to detect contrast and movement. Salmon, like trout (Salmo; Salvelinus) hunt with their eyes and switch from cone to rod vision as the light fades. Between 0.1 and 100 or more footcandles of light, the fish relies on the cones to give it maximal information on the hues and shades of its moving prey; when prey are located, they are stalked and taken head first into the mouth. Below 0.1 and close to 0.0001 footcandles (the light equivalent to that of a clear new moon at night), the rods take over, feeding stops, and schools break up. Reaction to light stimuli at these low intensities further suggests that fishes may use vision still at considerable depth-about 700 meters or more in clear ocean waters. In experiments where water fleas (Daphnia) are fed to young salmon in low light intensities, the fishes assume an oblique position below their prey so as to see them in maximal contrast against the screen afforded by the water surface, watch them move for a few moments, and then snap them up one by one.

Binocular vision that exists over varying angles in different species is important in the judgment of distance and depth. Some fishes such as the pikes (Esox) have veritable sighting grooves on their snouts in front of their eyes. Two imaginary lines projected from the grooves to cross the head in front of the eyes will include the angle within which a particular species has overlapping visual field, and thus presumably the field in which the animal has three dimensional vision. Many fishes can move their eyes independently and adjust this portion of their visual area. Stream fishes, such as most trout (Salmoninae) and many minnows (Cyprinidae), fix their position in the current by visually marking a stone, log, or other object.

Incident and background-reflected light falling on the upper and lower portions of the retina initiate chromatophoric adjustment of body colour to background. Light falling on the eye from above in conjunction with the gravity receptors of the inner ear keeps the fish in the proper position for swimming or for resting. In addition to the foregoing treatment of vision, mention must be made of the light-sensitive skin areas of lampreys and hagfishes (Cyclostomata). In the skin of the dorsal tail region of the ammocete larvae of lampreys (Petromyzonidae) and of the cloacal region of hagfishes (*Myxine*) there are round sensory cells with photosensitive pigments of maximal reactivity at light wave lengths of 500 to 530 millimicrons. In the lampreys these cells are innervated by branches of the lateral-line nerves and they function to enhance the negatively phototropic reactions of the animal. The epiphysis and the dorsal region of the brain are also light sensitive in various fishes.

The Sense of Smell

Most odoriferous substances that play a meaningful biological role in the lives of fishes are of organic nature. They are detected through the olfactory epithelium connected to the olfactory nerve in the nasal sac, although some of them are perceived through the taste buds. However, the sensitivity spectra of taste and smell are different for the same substances and some organic molecules are only smelled while others are only tasted. The sensitivity of the sense of smell also serves for orientation when smell gradients are followed.

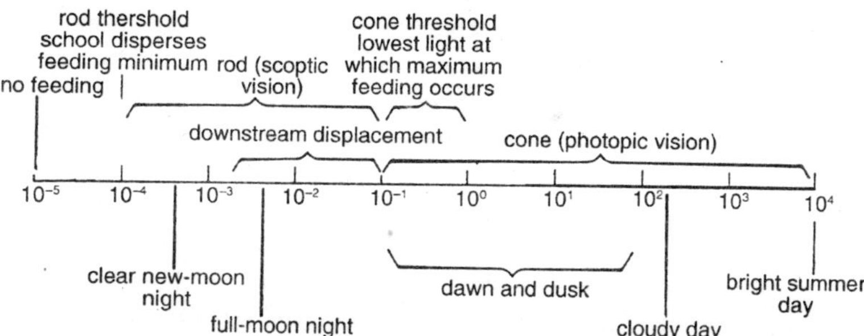

Figure 1.16 : Diagram summarising rod and cone vision under various light intensities together with some other responses of a salmon (Onchorhynchus) to different light intenities.

The nasal orifices as well as the entire nasal structure show wide varieties of adaptations among fishes. The single median nostril of lampreys and hagfishes (Cyclostomata) leads into a short nasal tube that bends posteriorly and ends in a nasopharyngeal pouch at the level of the second internal gill slit. The pouch is subject to the rhythmic respiratory movements that cause an exchange of water in it and in its tributary nasal sac. In the hagfishes (Myxiniformes), the nasopharyngeal pouch connects with the gut, but, in the lampreys (Petromyzonidae), it ends blindly. The nasal sac, branching from the nasal tube, is protected by a valve that directs a stream of water inward at inspiration but allows a reversal of flow at the emptying of the nasal pouch in response to the expiratory movement of the gills. The nasal chamber proper is thrown into twenty-five folds in the sea lamprey (*Petromyzon marinus*) and covered with olfactory epithelium where one fold faces another; the remaining part of the surface of each fold is covered with indifferent nonsensory epithelium. As in all other vertebrates, sensory cells of the olfactory epithelium give rise to the fibers of the olfactory nerve.

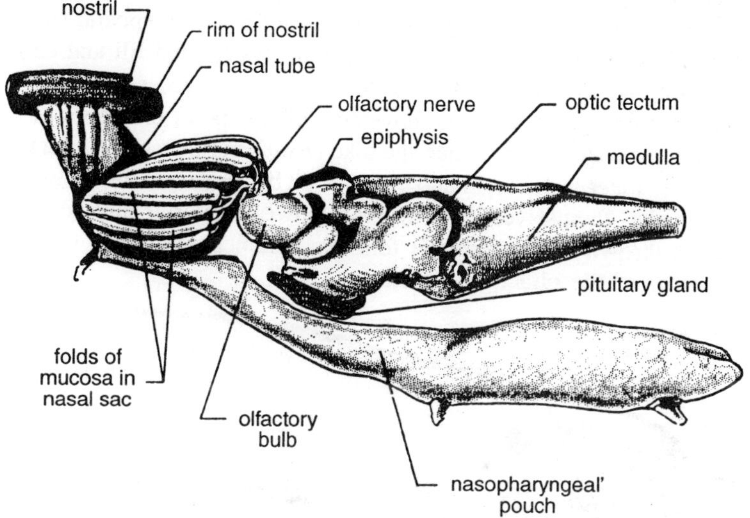

Figure 1.17 : Sketch of a wax restoration of the olfactory organ of a sea lamprey (Petromyzon marinus).

In sharks and rays (Elasmobranchii), the paired pits that lead into the internally folded nasal sac are usually on the ventral side of the snout and divided by a fold of skin into an inlet and an outlet; the latter sometimes leads into the mouth by means of a skin-covered nasal furrow and, as the fish swims and breathes, water reaches the olfactory epithelium. Certain sharks depend largely on their sense of smell for food-finding and follow odor gradients by swimming in circles or following a figure-eight course.

The paired olfactory organs of rayfin fishes (Actinopterygii) are relatively smaller than those of elasmobranchs but have a wider variety of adaptations for intake and exchange of water. The anterior nasal apertures are usually ahead of and somewhat medial to the eyes. The apertures lead directly or through a short tube into the nasal pit with its rosette of folded olfactory epithelium. In most rayfins, the posterior nasal apertures are normally close to the anterior ones; however, in the eels and morays (*Gymnothorax*), the apertures are relatively far apart. Some fishes, such as the sculpins (*Cottus*) and the sticklebacks (*Gasterosteus*), have only one nasal aperture and the nasal pouch is alternately filled and emptied by the breathing movements. An eel (*Anguilla*) has a projecting ciliated tube that leads the water to the anterior nasal opening, over the folds in the egg-shaped nasal sac, and out of the posterior nasal openings near the eyes. In the herrings (Clupeidae) and the catfishes (Siluroidei), there are lymphatic sinuses near the nasal sac which act as a hydraulic system and fill and empty the nasal sac as pressure is transferred onto the accessory pouches during breathing, through the movement of several of the bones of the head. In many fishes, of which the pike (*Esox*) and the cod (*Gadus*) are examples, water is deflected into the nasal pit by the swimming movement of the fishes themselves. The division between anterior and posterior nares is thrown into an erectile fold that conducts in-current water to the center of the olfactory rosette.

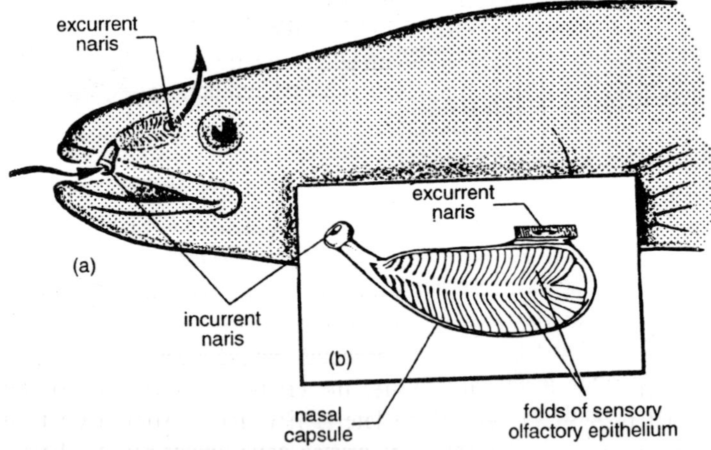

Figure 1.18 (a) Position and extent, and (b) section of olfactory organ of an eel (Anguilla). Arrows indicate direction of water flow. The stalk of the incurrent naris is erectile.

The physiological mechanism of odor reception is incompletely understood. Electrode recordings from single units of the fish olfactory tract of *Abramis brama* (European bream) and *Carassius auratus* (goldfish) show the following: a fiber exhibits different patterns of activity when different odors are tested, different fibers respond to the same odor in different ways (spatial coding), and most fibers exhibit adaptation following continuous stimulation with the same odor.

Many experiments have shown the sensitivity and great importance of the sense of smell to fishes in getting food and in orientation. Bluntnose and fathead minnows (*Pimephales*) can distinguish by smell between the rinses of related species of invertebrates and of aquatic plants. Fishes that feed largely by odor, such as bullheads (*Ictalurus*), some sharks (e.g., *Squalus*), and morays (*Gymnothorax*), cannot find their food when their nares are plugged. Pheromones, "special substances," from prey that are tasted by fish include glycerophospholipids and also nucleotide mono and diphosphates. The bluntnose and fathead minnows can be trained to detect chlorophenols in the water much below a concentration that can be tested chemically. An eel (*Anguilla*) is reported to detect a pure organic substance at a dilution of 2×10^{-18} when only 3 or 4 molecules of the material could have found their way into the nasal sac of the fish. Accordingly, it is no surprise that sunfishes (*Lepomis*) displaced from their home range in a stream, can find it again through their sense of smell, and that stream odors appear to play a decisive role in guiding returning salmons (*Oncorhynchus*) to their home streams. Minnows (Cyprinidae) give off a pheromone alarm substance (Schreckstoff) when frightened or injured. When this substance is smelled by the other members of a school, it causes them to disperse. Fishes with acute olfaction such as the catfishes (Ictaluridae) use their sense of smell in social interactions, differentiating between individuals in a group or social hierarchy by means of the specific odor of the slime of any one fish.

The Inner Ear (Equilibrium and Hearing)

The organs of equilibrium and hearing are divided into a pars superior and a pars inferior. The pars superior is composed of the semicircular canals and their ampullae, and a sac-like vesicle, the utriculus. The pars inferior, the structure of sound reception proper, consists of two more vesicles, the sacculus and the lagena. The ampullae contain patches of receptor tissue, the cristae staticae, with sensory cells that resemble those of the lateral line and to some extent also those of taste buds. However, the cristae have longer sensory hairs at the apices of their sensory cells than taste buds and they are covered by a gelatinous cupula. Displacement of the body in any direction sets up shear forces to which the sensory cells of the ampullae react. The cells have spontaneous, rhythmic, nervous discharge. The bending of a cupula in one direction brings about a decrease in the rhythmic activity of the receptor, whereas bending of the cupula in the other direction accelerates the discharge rate. This is the mechanism for the detection of angular displacement.

In bony fishes (Osteichthyes), the utriculus contains an otolith, or earstone, called the lapillus. This and the otoliths of the sacculus and lagena are limy and grow discontinuously, at least in the Temperate zones, and can therefore serve for the assessment of age of the fish. The lapillus rests horizontally on the hairs of the sensory cells of the crista utriculi. The lapillus responds to the force of gravity and activates the sensory cells of the cristae. For the maintainance of balance, they work in conjunction with the lower portion of the retina. Thus with both the eyes and utriculi intact, light from above and gravity from below push and pull, as it were, to keep the fish in an upright position. Upon extirpation of one utriculus, or upon shining a strong beam of light at the eye of a fish at right angles instead of from above, the fish can be made to lean to one side or the other. When a fish has its pars superior removed from both inner ears, it can be made to swim at a 90° angle or even upside down if it is illuminated from the side or from below.

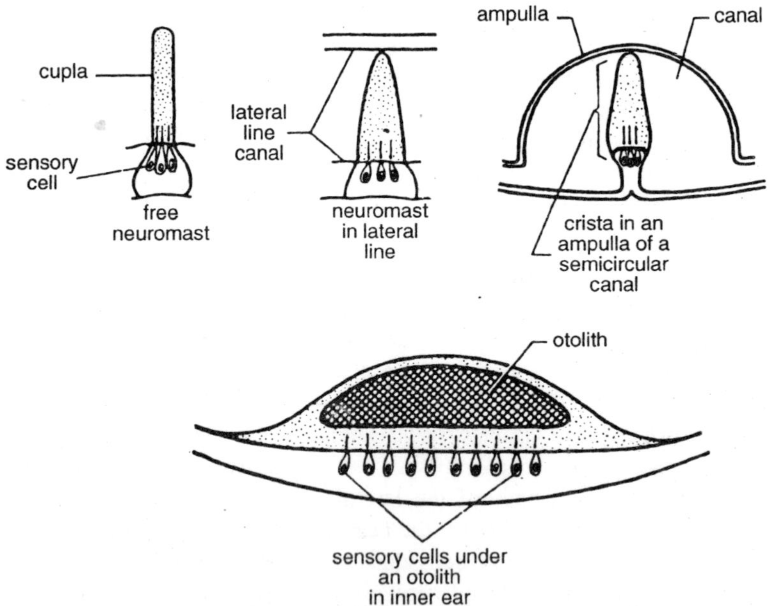

Figure 1.19 : Diagrammatic comparison of various sensory epithelia of the acousticolateralis system.

The adequate utricular stimulus for swimming upright is the full weight of the lapillus resting on the sensory hair cells with no shear. A displacement of the fish by 90° will render the lapillus weightless but will produce maximum shear on the hairs of the sensory cells under the otolith. The lagenar otolith, and the underlying sensory tissue are, to a small degree, also involved in gravistatic or righting reflexes, mainly through their influence on muscle tonus.

In sharks and rays (Elasmobranchii) an endolymphatic duct leads from the inner ear to the outside. Through this duct sand reaches the gelatinous cupulae over the sensory cells of the inner ear. However, it is not known to what extent such particles from outside assist in the stimulation of the sense cells inasmuch as there are in the inner ear of sharks also otolith-like calcareous concretions of endogenous origin.

The size and arrangement of the semicircular canals and the vesicles are considerably different among various fishes. The canals are relatively large in the elephantfishes (Mormyridae) and in certain blind cavefishes (Amblyopsidae). The sacculus and the lagena are more broadly joined to one another in the sharks (Squaliformes) than in bony fishes (Osteichthyes).

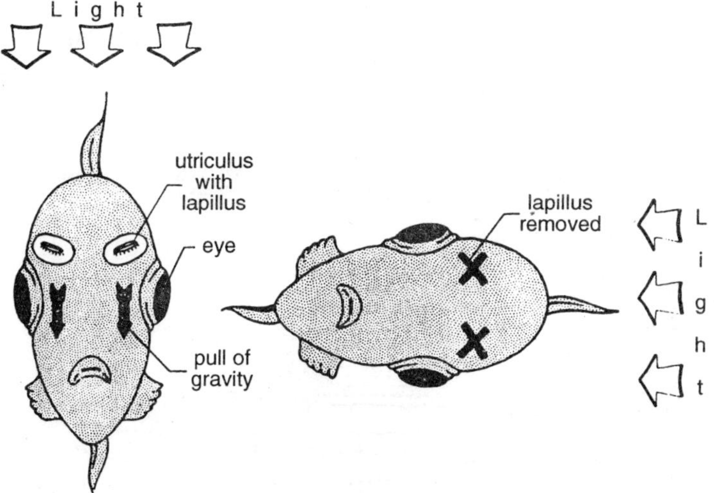

Figure 1.20 : Effects of retina of the eye and lapillus of the utriculus of the inner ear in equilibration of fishes. When the lapilli are removed, the fish no longer orients to gravity and light but orients to light only.

The relative development of superior and inferior portions of the inner ear also differs in rayfin fishes (Actinopterygii) of different habits. Flying fishes and pelagic fast swimmers that require superior three-dimensional orientation have a better developed pars superior than demersal species where in turn the inferior portion reaches the highest degree of development.

The pars inferior of the inner ear, lagena and sacculus with their otolithsnamed asteriscus and sagitta, respectively-is the seat of sound reception. These two structures, as well as the posterior semicircular canal, are innervated by the posterior branch of the auditory nerve (cranial nerve VIII) whereas the superior part of the inner ear and the two remaining semicircular canals are connected to the anterior branch of the same nerve. The otoliths of lagena and sacculus are held vertically over the acoustic sensory maculae by tendinous attachments to the inner walls of the vesicles.

The physiological mechanism of discrimination of different sound frequencies is not known but it is seated mainly in the utriculus and one must distinguish between two types of hearing structures among bony fishes, the common and the cypriniform types. Among the minnows, catfishes, and suckers (Cypriniformes), the inner ear is connected uniquely to the gas bladder through a chain of bones, the Weberian ossicles. The Cypriniformes have a finer and wider range of hearing than fishes of the normal type, their lagena and sacculus are divided by a narrow constriction, and their sagitta carries a large wing into the endolymph, apparently better to pick up vibrations that are transferred to it from the resonating gas bladder.

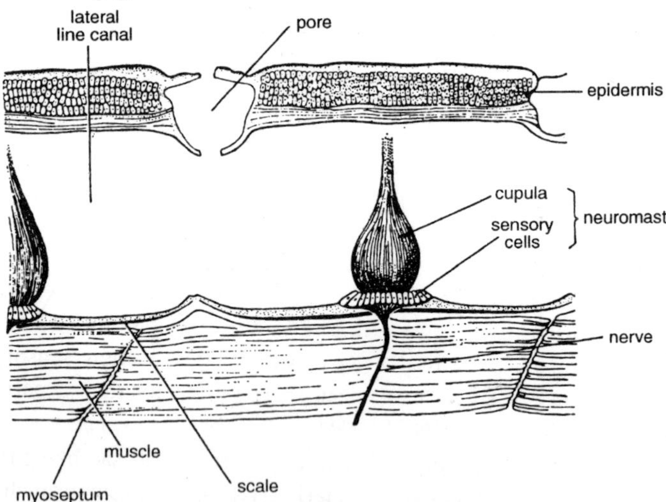

Figure 1.21 : Neuromasts in the lateral-line system of a japanese eel (Rhynchocymba).

Nothing is known of the range and acuity of sound reception in lampreys and hagfishes (Cyclostomata). Sharks (Squaliformes) respond to various noises and sounds, some of pure frequencies ranging from 400 to 600 cycles per second. Rayfin fishes (Actinopterygii) can perceive both intensity differences and a range of vibrations from 16 to 13,000 cycles per second (cps) as in the bullheads (*Ictalurus*). Low frequencies of 16 to 100 cycles are probably sensed through skin and/or lateral-line receptors. The most acute level of sound perception in the creek chub (Semotilus *atromaculatus*), a minnow, lies around 280 cps; this fish is capable also of distinguishing quarter-tone intervals between 400 and 800 cps. Still, true directional hearing seems to be limited to a few dozen meters. The upper range of sound perception of fishes without accessory devices lies near 5000 cps, as in the labyrinthfishes (Anabantidae), and they have less capacity to distinguish between tone intervals.

The Lateral Line

The lateral-line system is a set of sense organs restricted to the fishes and aquatic stages of the amphibians. Although it has connections in the brain with the afferent nerve pathways from the auditory system of the inner ear, it is a different system. It was already present in fossil ostracoderms, and is thought to have originally evolved from primitive cutaneous sense cells. Fossil evidence from *Pteraspis* (Agnatha: Pteraspidomorphi) onwards permits some tracing of literal-line evolution, but embryological evidence is less clear because in some fishes the lateral line arises from the otic or auditory placode, whereas in others, as for example, an European minnow (*Leuciscus*), several separate lateral-line placodes form on head and body.

Apart from the typical lateral-line canals with their pores in~ skin or scales on the head and body of bony fishes (Osteichthyes), the following lowing widely differentiated sensory structures are, at present, tentatively ranged within the lateral-line complex:

a. Sensory crypts or pits and papillae between or under the placoid scales of sharks and rays (Elasmobranchii), arranged in groups or lines, usually in the anterior and dorsal body regions; certain histologic peculiarities of their sensory cells suggest that they are taste receptors rather than a part of the lateral-line system proper.

b. The ampullae of Lorenzini are more or less sac-like structures in the head region of sharks and rays (Elasmobranchii) and also of *Plotosus, a* southeast Asian catfish, to which various groups of pores (supraorbital, buccal, hyoidean and mandibular) give access through a canal. The structures are jelly-filled and may have few to several diverticula lined with sensory epithelium. The sensory cells are innervated by branches of

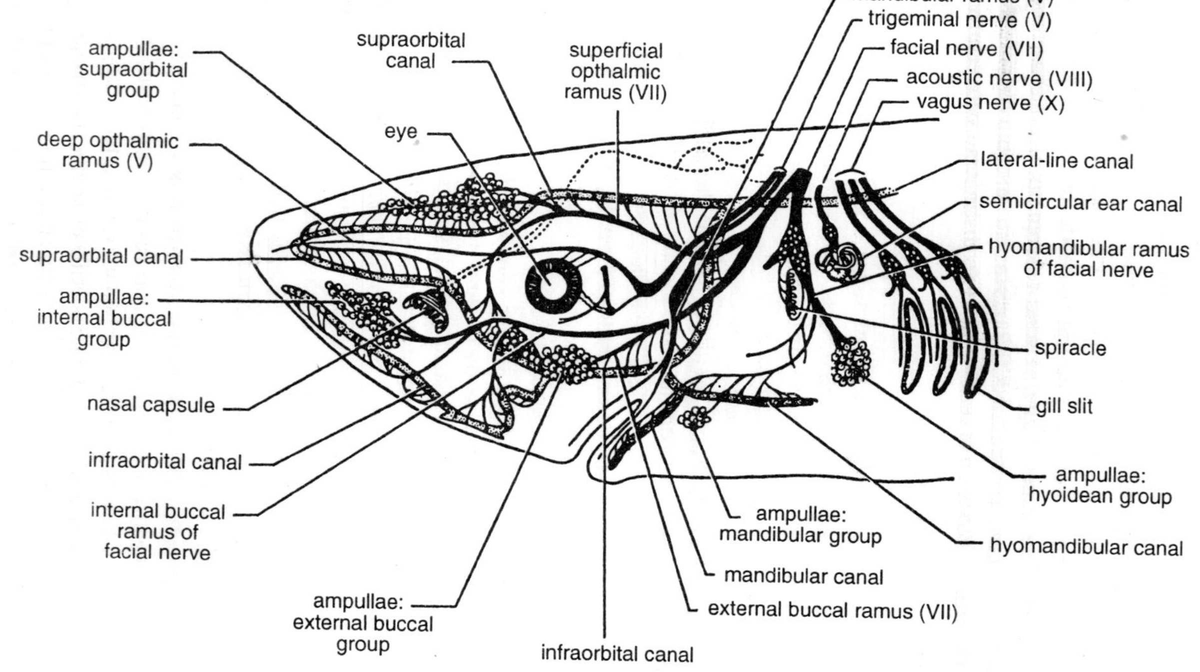

Figure 1.22 : Distribution of the ampullae of Lorenzini and sensory canals and their innervation on the head of a shark.

the facial nerve (VII). Isolated preparations of such ampullae give rhythmic nervous discharges which are accelerated on cooling and slowed on warming. They are also sensitive to mechanical and weak electrical stimuli as well as to small changes in salinity. Their highest sensitivity, however, is to differences in electric potential such as would be generated by muscular movements, for instance, the opening and closing of the gill covers. The ampullae of Lorenzini are now believed to be mainly electro-receptors assisting in the location of prey as well as serving proprioceptive functions in swimming. Their sensitivity is extremely high, reacting to a change of 0.01 microvolts/cm of seawater as shown in the European thornback ray (*Raja clavata*) in which the rate of heartbeat changes when stimulated with such small changes in electric potential. The cat sharks (*Scyliorhinus*) hold similar sensitivity indicating that they and probably other elasmobranchs are capable of detecting the presence of potential prey animals in their vicinity by changes in muscle potential of the prey.

c. The vesicles of Savi, found in several genera of rays (Rajiformes) including certain electric rays (*Torpedo*), mainly on the underside of the snout as linear series of gelatin-filled follicles which enclose gray, granular, amorphous material. These vesicles are innervated by a branch of the trigeminal cranial nerve (V) but their function is unknown.

d. The organs of Fahrenholz are pits in the head region of larval lungfishes (Dipnoi). These pits have hair-bearing sensory cells th1t resemble those of the lateral-line organs; theif function, however, is unknown.

e. The spiracular organs, sensory structures of a basic lateral-line type are of unknown function in the spiracles of the Chondrichthyes, the Holostei, and the Chondrostei.

f. The cutaneous sense cells of the elephantfishes and gymnotid eels (Mormyridae and Gymnotidae, respectively), which are presumably electrosensitive.

Only some fishes among all animals are now known to have organs which generate an electric discharge. The voltage of this discharge varies considerably in different groups of fishes as does the nature and origin of the currentgenerating structure. Two groups can be distinguished: (1) those which produce a strong stunning current such as the electric eel (*Electrophorus*), the electric rays (*Torpedo*), and the electric catfish

(*Malepterurus*); and (2) those which produce currents of low voltage (Mormyridae and Gymnotidae with the exception of *Electrophorus*). The second group emits a continuous series of pulses to set up an electric field around the animal. The field becomes distorted when prey enters it or when inanimate obstacles, such as may be encountered in swimming, obstruct it. The device makes possible the location of objects in the turbid habitat of these fishes. The number of discharges per second varies with the species, but in both the South American gymnotids (Gymnotidae) and the African elephantfishes (Mormyridae) certain species discharge with relatively low frequencies of up to 200 cycles per second, whereas others emit up to 1600 low-voltage impulses per second. Speciesspecific discharge rates may play a role in the recognition of mates.

Generation of an electric field for the sensing of surrounding objects implies the existence of electric receptors. In the elephantfishes there are pores, concentrated in the head region, with narrow canals leading into single pits. Each pit has a patch of tissue, resembling lateral-line receptors. These structures are renewed throughout life, like other skin cells. They may respond to chemical and pressure stimuli aside from being sensitive to electrical impulses. The sensitivity of the mormyrid *Gymnarchus niloticus* to static electric charges is 0.15 microvolt per square centimeter. Gymnotids have comparable skin organs composed of several sense cells that lie in a vesicle which opens to the outside through a jelly-filled canal. Nonsensory cells may also occur in these structures which belong to the lateral-line system. The structures are presumed to act as electrical receptors. Catfishes (*Ictalurus*) have also been shown to respond to weak electrical stimulation with this sensitivity being located in the small pit organs of the lateral-line system. In the remaining groups of fishes the lateral-line organs proper may be arranged structurally as follows:

1. Free, arranged in groups on the head and in lines along the body (the elritze, *Phoxinus;* the loach, *Cobitis*).
2. Sunk in pits, in groups, or in several lines on the head or along the body (many elasmobranchs, cyclostomes) or in bands crossing the ventral surface (elasmobranchs, cyclostomes).
3. In partially open or closed, variously developed canals located

a. on the head where there are typically three main canals, the supraorbital, the infraorbital, and the hyomandibular, with branches, extensions, or connections such as the supratemporal commissural canal that connects the canal system of the two sides of the body;

b. on the body where there is typically one canal extending from the cleithrum of the shoulder girdle to the base of the caudal fin.

Both the body lateral-line canal and that of the head exhibit several modifications. For example, the body canal may be extremely short as in the European bitterling (*Rhodeus amarus*), intermediate in length as in the fathead minnow (*Pimephales promelas*), interrupted as in a parrotfish (*Scarus*), strongly arched towards the back or decurved towards the belly. There may also be several roughly parallel lateral lines as in the greenling (*Hexagammus*) and mullet (*Mugil*) or ones scattered in short segments over most of the side of the body, as in the pikes (Esox).

The different parts of the lateral line are innervated by cranial nerves VII, IX, and X and have close central associations with the auditory centers served by nerve VIII. Centers in the medulla oblongata of the brain are divided into a dorsal group for the head lateral lines and into ventral ganglia for the body lateral-line system. Secondary connections in the brain relay impulses from the lateral-line system to the thalamus, the hypothalamus, the optic tectum (especially in sharks, Squaliformes), and the cerebellum (particularly in the Mormyridae). The giant cells of Mauthner, with their final common path as wholesale movers of trunk and tail musculature, have axon tracts that bring them in direct contact with the lateral-line branches of nerves VII, IX, and X as well as with the optic tectum, the sacculus, and the cerebellomotor centers.

The receptor unit of the lateral-line system is the neuromast, an area of sensory tissue made up of pyriform receptor cells each with a hair-like extension at its apex that reaches into a gelatinous cupula. In contrast to the sensory cells of taste buds, which they resemble, they do not reach the base of the neuromast but are cushioned from below by nonsensory, supporting cells.

Like the end organs of equilibrium in the semicircular canals, the neuromasts have a continuous rhythmic discharge. In the skates (*Raja*) two types of receptors can be isolated upon perfusion of a few functional units with a stream of water. One group responds with a strong increase in discharge frequency to a headward slow jet of water whereas tailward perfusion inhibits the discharge activity. The end of headward perfusion is marked by a silent period but the cessation of water flow toward the tail brings about a vigorous resumption of nervous activity; the second group of receptors behaves in the opposite manner. The

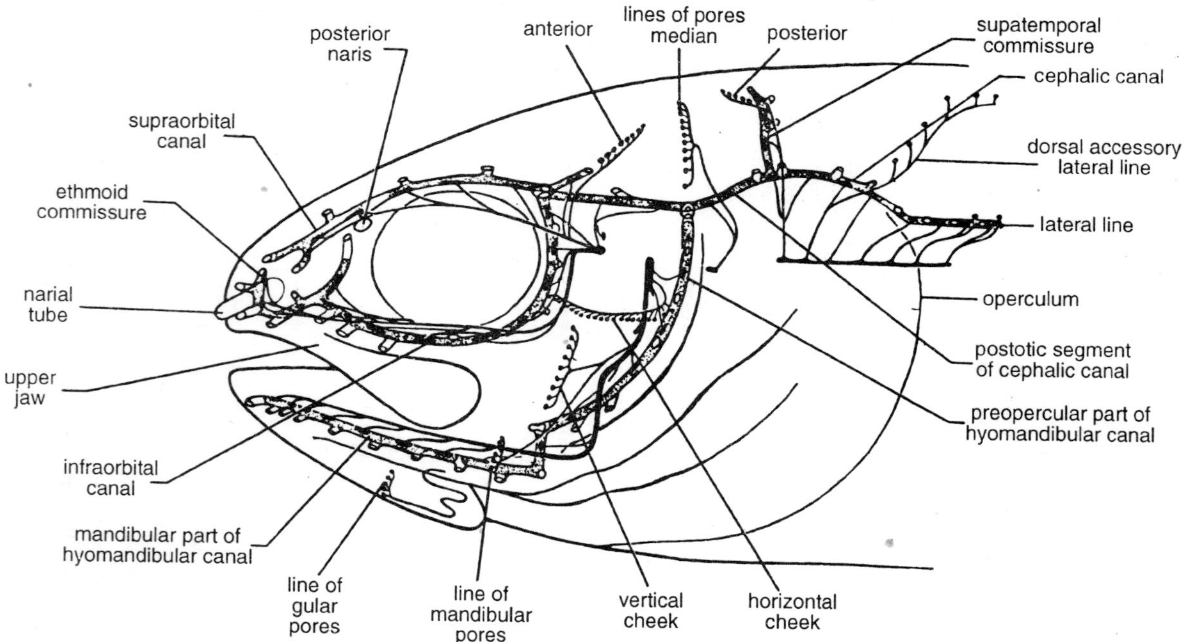

*Figure 1.23 : Electric discharges in a lateral-line nerve of a skate (*Raja clavata*) with arrows indicating the beginning of stimulation by a jet of water: (a) stream of water toward the head; (b) stream of water toward the tail.*

antagonistic, rhythmically discharging receptors are mixed randomly in each group of sensory cells, but there have also been found, again in the skates, "spontaneously silent" units in the lateral-line system that can be made to emit impulses during perfusion in one direction and at the end of perfusion in the other. In some Japanese eels (*Anguilla japonica* and *Rhynchocymba nystrani*), the sense cells in the center of neuromasts are without resting discharge and adapt rapidly in contrast to the peripheral cells which have a slow adaptation rate and a spontaneous discharge. The central cells are connected to substantial nerve fibers of 15 microns in diameter whereas the peripheral cells are connected to small fibers of 4 microns in diameter.

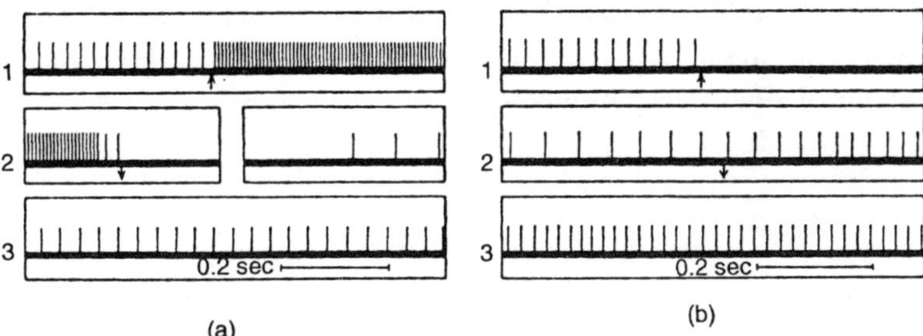

Figure 1.24: Electric discharges in a lateral-line nerve of a skate (Raja clavata) with arrows indicating the beginning of stimulation by a jet of water; (a) stream of water toward the head; (b) stream of water toward the tail.

The anatomical and physiological mechanism underlying detection of hydrodynamic displacements by prey, predators or social partners is as follows. The sensory cells of the neuromasts have hair-like cilia consisting of one elastic kinocilium and numerous softer stereocilia (40-50 in *Lota,* the burbot). Adjacent sensory or hair cells have their kinocilia oriented in opposite directions, alternately towards the head or the tail of the fish, but always along the axis of the lateral-line canal. As head- or tail-ward bending of the cupula would affect the kinocilia of different hair cells differentially one can envisage the pattern of lateral-line nerve discharges to convey to the brain exact information on direction and/or provenience of water displacement in the vicinity of the fish.

Training experiments, operations to eliminate lateral-line reception, and physiological investigations show that the system responds to a variety of stimuli. Irregular pressure waves and low-frequency vibrations of 50 to 100 cycles per second (cps) excite the receptors of the larger lateral-line fibers of the bullheads (*Ictalurus*), whereas a weak but regular water flow or vibrations of 20 to 50 cps act on the receptors connected to the thinner nerves. Several fishes have been trained to locate small vibrating rods or discs close to their heads and sides as well as to react to small localised water currents. After the difficult operation of cutting only nerves of the lateral line on the head and trunk, and not nerves of the touch sense, the capacity of fine discrimination of small vibrations or current disturbances is lost.

It can be concluded that the lateral-line system informs the fish of localised disturbances, caused by small currents, by mechanical vibrations below 100 cps, or by irregular nonviolent displacement of the surrounding water. The system is also involved in "distant touch" location of moving objects such as predators or prey or in the sensing of fixed objects that reflect the water movements brought about by the swimming fish itself. Lateral-line organs also react to monovalent cations (e.g., Na^+) but the functional significance of this sensitivity has not been evaluated.

2

Hormones of Pineal Gland

Despite the vast number of publications on the pineal body of fishes, no distinct function has been unequivocally demonstrated. It is still impossible to present a unified concept of pineal function; rather an attempt will be made to review some of the evidence for the various current hypotheses concerning its function. Several monographs on the pineal body are available: Studnicka, Tilney and Warren, Bargmann, and Kitay and Altschule; in addition, the recently published proceedings of an international round-table conference on its structure and function, edited by Ariens Kappers and Schade, contains many facts pertinent to the pineal body of fishes.

THE PINEAL DILEMMA

The most persistent question concerning the pineal gland of fishes is whether it is the vestigial remnant of an ancestral third eye or whether it is a specialised functional organ. Its location on the top of the head orients the pineal organ directly toward the primary source of light and thereby effectively eliminates the pineal body as a receptor of a visual field. But this orientation renders the pineal well suited for detecting changes in light intensity or for acting as a dosimeter of incident radiation. In other words, the pineal of fishes probably never was, in the true sense, an eye. Nevertheless, numerous investigators have shown that the pineal organ of the fishes does have a light receptive function. The problem then is not the presence or absence of a function, but rather, a question of the precise function or functions.

Although several of the current hypotheses are not mutually exclusive, they can, for present purposes, be divided into two categories-one emphasizing primarily a sensory role and the other a secretory function. Of the sensory theories, the following are the most credible: (1) The pineal body is a photosensory structure, (2) it acts as a baro- or chemoreceptor for the cerebrospinal fluid (CSF), or, (3) it performs the function of a mediator in the olfactory induced response to exohormones. Conversely, (4) the pineal may be primarily a gland of external secretion related to the chemical composition of the CSF or the metabolism of the brain tissue, or (5) it may be a gland of internal secretion with an endocrine function.

It is not proposed here to review all of the literature concerning these various hypotheses but to provide a framework into which the accumulated knowledge of pineal function in the fishes can be channeled.

ONTOGENY AND STRUCTURE OF THE PINEAL ORGAN

The description of pineal ontogeny, morphology, and histology given here may be amplified from detailed accounts and bibliographies by Hill, Terry, Tilney and Warren, Holmgren, Oksche, Van de Kamer, and Rudeberg.

Ontogeny

Much of the polemic surrounding the pineal organ of the fishes originates from the prevailing ignorance of its embryological origin. This has given rise to inconsistent nomenclature and a failure to distinguish clearly the pineal body from other epithalamic outgrowths.

In vertebrates, two approximately middorsal evaginations develop from the roof of the diencephalon. The rostral evagination represents the anlage of the parapineal body which is homologous with the parietal eye of certain reptiles. The pineal organ develops from the more posterior of the diencephalic evaginations. Therefore, the parapineal and pineal organs are two independent structures and are not homologous. The paraphysis, which has occasionally been confused with the parapineal, is morphologically part of the telencephalon and is not part of the pineal area.

In fishes, the parapineal anlage usually shows some early ontogenetic development but is generally absent or rudimentary in adults, including the primitive ganoid *Amia calva* and the dipnoan *Protopterus annectens*. However, a parapineal body resembling a parietal eye of lizards was described in at least one adult teleost. In other groups of fishes, a well-developed parapineal body with distinguishable photosensory-like cells is found only among the nonmyxinoid cyclostomes. However, the same author notes that the problem of the parapineal organ is not resolved and should be completely reexamined.

Structure

Topography and General Morphology

The pineal body, *sensu stricto, is* a conspicuous structure in most adult fishes ; it is, however, rudimentary in *Syngnathus acus* and *Hippocampus spinosa* and absent in *Torpedo ocellata, T. marmorata* (Selachii), and the adult *Myxine glutinosa*. Despite contradictory opinions, a pineal organ is present in the tuna *Thynnus thynnus*.

In basic form, the fish pineal organ consists of a hollow, variously invaginated and well-vascularised structure lying dorsal to the diencephalon to which it is connected by a hollow narrow stalk. The pineal body consists of an expanded distal vesicular area containing a central lumen which connects with the third ventricle via the narrow proximal stalk. Posteriomedially the stalk connects with the subcommissural organ and the commissura posterior and makes contact with the commissura habenulae on its anterior side before joining the dorsal sac epithelium which comes to lie directly ventral to the distal vesicle of the pineal body. With advancing age, the epithelium of the distal vesicle under-goes a gradual infolding with the result that the central lumen, although retaining its connection with the third ventricle, becomes divided into narrow sinuous spaces. Many of the blood vessels which surround the pineal body in embryonic and larval stages, are, with the increased development of the animal, carried into the infoldings of the pineal epithelium and come to lie in the intralobular septa. This centripetal invasion by the blood vessels results in the extensive vascularisation characteristic of the adult pineal. Generally, the distal vesicle is closely apposed to the overlying cranial roof but is not, in teleosts, associated with a pineal foramen in the skull. However, as in many deep-sea fishes, the pineal *body* often lies beneath a more or less depigmented and translucent area of the cranial roof. In other fish, the pineal body is located in a depression in the roof of the skull or may be associated with a very complex structure of skull bones which may allow the passage of light directly to the pineal body. Some authors, in questioning a photosensory role of the pineal body, have attached special significance to the fact that some fish possess thick skull bones and have no special structure to allow light to pass though to the pineal region. It should be noted, however, that measurable quantities of light have been demonstrated to penetrate the skulls of sheep, dogs, rabbits, and rats.

Histology

Many of the references on the pineal histology are found in papers concerned with the secretory activity of this organ and will be discussed in a later section. Histologically, the pineal body of fishes most closely resembles a sensory structure. The structure of the vesicular epithelium resembles the retina except that the pineal "retina" is of a reversed type with the sensory cells in the inner retinal layer; thus, they are turned in the direction from which the light enters. In cyclostomes the dorsal wall of the pineal forms a lenslike pellucida, but a lenslike differentiation is never found among other fishes where the pineal seems to have acquired some characteristics of a secretory organ. Despite this histological progression from an eyelike structure in some fish to a glandular organ in others, there is now general agreement that the pineal body of fish is largely composed of three cell types: sensory cells, supporting cells, and ganglion or nerve cells.

Most of the pineal cells are sensory cells. These are most abundant in the vesicular area on the side of the lumen but occur throughout the pineal epithelium. These cells are large, ovoid, or club-shaped with well-defined nuclei surrounded by clear cytoplasm containing many mitochondria and capped with a membranous structure-an outer segment. Cytoplasmic outgrowths arising from the sensory cells and extending into the pineal lumen have also been described . On the pole to the cytoplasmic outgrowths, the sensory cells give off axons, which, together with the dendrites from the ganglion cells, form a network of axons and dendrites within the pineal organ.

Supporting cells of various types show a marked affinity for nuclear stains and lack the typical structure of the sensory cells. It is most likely that these cells are of an ependymal glia type.

Ganglion or nerve cells together with the dendrites and Nissl substance indicative of their nervous nature, have been described in the pineal of some but not all fishes examined. Although few in number compared with the other cell types, ganglion cells may be readily distinguished by their large size and by the presence of two or three wellstained nucleoli. Recently, Riideberg has reported that within the pineal of the sardine, *Sardina pilchardus sardina,* the ganglion cells are limited to three small groups. If this localisation of ganglion cells proves to be the case in other fishes, it may account for the failure of some workers to find pineal ganglion or nerve cells.

Innervation

Apart from the sensory cells proper, afferent nerve fibers connecting the pineal body with other parts of the brain have been described. These nerves, which appear to represent the axons of ganglion cells, arise from the extensive nerve net formed within the pineal by the dendrites of the ganglion cells and neuraxis of the primary sensory cells. In the stalk region, these fibers form variable numbers of nerve bundles before reaching the roof of the diencephalon. Efferent fibers originating within the central nervous system and ending on pineal cells have rarely been described. These efferent fibers are probably the same type of aberrant commissural fibers described in mammals.

Aside from a well-documented nervous connection between the pineal body and the posterior commissure relatively little is known of the central terminations of the pineal nerve. Possible connections have been suggested with the superior commissure, and with the right habenular nucleus as well as other habenular structures. In one case only, fibers have been described as running to the optic tectum. Some workers have suggested a link between the pineal body and the subcommi ~ suial organ, but this is apparently not present in all fishes. Le Gros Clarke believes that the pineal organ of cyclostomes is heavily supplied with olfactory input channels and this possibility should be investigated. Notwithstanding the paucity of information concerning the exact site (s) of the pineal nerve terminations, connections already described could permit a link between the pineal body and the efferent centers of the brain through which the pineal body could influence various integrative afferent centers. Further information on the innervation of the fish pineal gland together with the phylogenetic implications and earlier bibliographies may be consulted in the excellent papers of U. Holmgren and Ariens Kappers.

PHYSIOLOGY OF THE PINEAL BODY

Although neurophysiological studies have proved that the pineal organ of *Salmo gairdneiri irideus is* responsive to illumination there is only scanty literature on the actual physiology of the pineal organ in fishes. Active cellular metabolism has been indicated by at least two authors who studied the uptake of radioactive phosphorus by the fish pineal.

Recently, interest has been directed toward the possible presence of melatonin and other tryptophan derivatives within the fish pineal. Quay localised the enzyme hydroxyindole-O-methyl transferase

(HIOMT) which is respcvisible for the formation of melatonin within the pineal of *Salmo irideus*. Working with the same species, Oguri *et al.* found that a melatonin precursor, 5-hydroxytryptophan is taken up by the pineal organ in greater quantities than it is by other parts of the brain studied; they infer from this finding that the pineal of fish may also be concerned with melatonin synthesis. These studies have now been substantiated by Fenwick who isolated melatonin from the pineal organs of the Pacific salmon *Oncorhyncus tshawytscha*. Thus, the pineal organ of fish is responsive to light, has a high cellular metabolism, and shows an active tryptophan metabolism which is capable of producing the mammalian hormone metatonin.

The Pineal Body as a Sensory Organ

There remains little doubt that the pineal body of some, if not all fish, functions as a sensory organ. Krabbe and Walter and more recently Van de Kamer as well as Hafeez and Ford have suggested that the fish pineal is associated with the detection of the pressure or chemical composition of the CSF. Evidence favoring this hypothesis includes the reported open connection between the pineal lumen and the ventricles of the brain, the changing shape of the pineal body of *Esox lucius* in relation to the pressure of the CSF, and the cytoplasmic extensions of the primary sensory cells toward the lumen of the body thus suggesting a functional relationship with the CSF. Hafeez and Ford go on to suggest that the similarities between the pineal body and the subcommissural organ with regard to the chemical nature of their secretions and the common direction of release into the ventricles support the view of a CSF-pineal relationship. There have been no reported studies of the effect of pinealectomy on either the pressure or composition of the CSF.

Although the hypothesis that the pineal gland is associated with the limbic system or visceral brain suffers from the same paucity of experimental evidence, some suggestive findings have been reported. Both Le Gros Clarke and Boon noted a close anatomical relationship between the pineal body and the olfactory system of fishes; the latter has been shown to have at least some connection with hypothalamic nuclei responsible for pituitary hormone release. Moreover, Hoffman and Reiter reported that olfactory stimuli are capable of interfering with the primary inhibitory effects which the active pineal gland has on the reproductive system of male hamsters. Such studies have not yet been extended to fishes.

In contrast to the scanty evidence for these pineal functions, many publications suggest a photosensory role as the primary sensory function of the pineal organ. The evidence, based on a number of fishes, comes from direct neurophysiological studies as well as indirect evidence from histology or ultrastructure, colour changes, and behavioural tests. Only the more pertinent evidence will be reviewed here.

By light as well as electron microscopic investigation, it has been clearly demonstrated that morphologically the neurosensory cells present in the epithelium of the pineal body are very similar to the ciliary type of photosensory cells present in the retina of the lateral eyes and possess characteristics which are indisputably conelike. It has been suggested by De la Motte that the light receptive role of the pineal body in *Phoxinus is* based on porphyropsin, although Thines and Kahling had previously reported a different visual pigment in *Anoptichthys*. More directly, Grunewald-Lowenstein observed histological and histochemical changes in the pineal organ of *Astanax* following prolonged exposure to continuous darkness or illumination. Similar changes, however, were not found in young sockeye salmon held under different light regimes. Recently, Dodt and Morita have demonstrated the photosensory potential of the fish pineal by electrophysiologically detecting alterations in nervous activity in the pineal of *Salmo irideus* during and after illumination of the pineal even though this species lacks any specialised tissue overlying the pineal organ. This finding is important when attempting to define a light receptive role for the pineal in those fishes which possess apparently light impermeable skulls. It questions the theory that such a skull precludes a photosensory role.

A light receptive role of the pineal organ is also indicated by lightinduced alterations of pigment distribution within the chromatophores. Although several studies suggested that the pineal had no effect on pigment distribution, von Frisch did find that some part of the diencephalon was involved in chromatophore responses. The pineal was more directly implicated by Young who found that pineal extirpation abolished the marked diurnal rhythm of colour change in ammococtes and distributed the rhythm in adult *Lampetra*. Subsequently, Breder and Rasquin, Hoar , and Schonherr reported some degree of pigmentary dispersion after pineal occlusion or destruction-a finding similar to that reported by Young in pinealectomised ammocoetes. These results are supported by the observation that administration of beef pineal extracts in embryonic and larval *Fundulus* caused marked pallor although similar results could not be obtained in adult *Fundulus* or adult *Phoxinus*. The difference in responsiveness between the young and adults was presumed by Fain and Hadley to result from the acquisition, in the adults, of nervous innervation mediated by catecholamines and a loss of sensitivity to the pineal extracts. Whether the pallor following pineal extract administration is the result of pituitary inhibition or a direct action on the pigment cells is not known. A direct action is, however, suggested by the ability of the pineal extract to cause pigmer concentration *in vitro*.

Photosens ovity of the pineal body has been indirectly demonstrated in several species of fish by the use of behavioural tests. Breder and Rasquin, Hoar, and Fenwick report that phototaxis is abolished following pinealectomy. Further, Breder and Rasquin and Fenwick found that phototaxis, whether positive or negative, depended on the presence of intact optic cysts or intact eyes; they concluded that although the sign of phototaxis was governed by the pineal organ, the phenomenon itself depended on the presence of lateral photic receptors. Hoar, however, working with young sockeye salmon smolts, *Oncorhynchus nerka,* found that the negative phototaxis of otherwise intact animals is not disturbed following damage to the pineal region; this difference may result from species or age differences of the animals or differences in the intensity of light employed in the experiments. The pineal organ of fishes becomes increasingly invaginated with age and undergoes a decrease in sensory cell number with continued development together with a marked degeneration of nervous elements. Thus, the pineal body of fishes may change from a primary photosensory structure in the young fish to a secondary photosensory structure in the older fish where it can no longer autonomously produce a phototactic response. On the other hand, Hoar performed his experiments outside where the light intensities were much greater than those employed in other studies. Previously, von Frisch and Young suggested a general light sensitivity of the diencephalc roof; thus, since the smolts used by Hoar had thin skulls, the phototaxis demon-strated in "pinealectomised" fish could have resulted from the effect of high light intensity on the brain itself.

Recent studies by Fenwick have shown that intact goldfish in a light gradient were unevenly distributed and spent most of their time in the darker half of the tank. Conversely, pinealectomised, bilaterally enucleated, or pinealectomised plus bilaterally enucleated goldfish were distributed uniformly throughout the gradient. From this evidence it was concluded that the phototactic response of goldfish depends upon the presence of the pineal organ as well as the eyes. Furthermore, in a conditioning situation, pinealectomised animals with intact vision showed significantly more responses to the conditioned stimulus than did the controls when the conditioned stimulus was light, but not when the conditioned stimulus was sound. Blind goldfish, with or without an intact pineal, could not be effectively conditioned to light although they did become conditioned to sound. From this data it was concluded that although the pineal organ of goldfish is a photosensory organ, in the absence of the eyes, the photic information received by the pineal cannot be translated into a directional response or be used as a sensory mechanism for the initiation of active behaviour. It appears, therefore, that the photosensory role of the goldfish pineal organ is to modulate the response elicited by photic information received by the eyes.

Although these studies do not provide conclusive evidence of the mode of action or the exact role of the pineal body in light reception, they do demonstrate the importance of the pineal body in phototactic behaviour and suggest that the pineal organ and the eyes function as a unit in phototaxis.

The Pineal Body as a Secretory Organ

External Secretion

In addition to its photosensory role, many investigators have shown that fish pineal produces an apocrine secretion which contains glycogen. The secretion is generally considered to enter the CSF. Although Van de Kamer suggests that the secretory droplets may be experimental artifacts, the high cellular metabolism demonstrated by the cytological studies of Palayer and U. Holmgren and the presence of a well-developed Golgi complex in the supporting cells and the sensory cells indicate secretory activity.

Internal Secretion

There is also the possibility of an internal secretory or endocrine role. Although both Friedrich-Freksa and U. Holmgren suggest that some pineal secretion is taken up by the blood vessels in teleosts, Hafeez and Ford did not find any secretory granules within the nerve fibers or in the proximity of the blood vessels of sockeye salmon, *Oncorhynchus nerka*. However, since the demonstration of secretory granules depends upon the nature of the secretion and on the specificity of the techniques employed, an internal secretory role of the pineal organ was not definitely ruled out and few investigators have applied techniques suitable for detecting the one known mammalian pineal hormone melatonin.

Melatonin and its precursor serotonin, together with the enzyme HIOMT necessary for the formation of melatonin, have now been demonstrated in fish pineal organs. Also, Oguri *et al.* report that the pineal takes up more ^{14}C-5-hydroxytryptophan than any other tissues studied. Since this substance is a precursor of serotonin, an active tryptophan metabolism within the fish pineal organ is indicated. Thus, an endocrine role of the fish pineal, possibly based on the hormone melatonin, is riot improbable.

Despite earlier reports by Krockert that the growth rate of young *Lebistes* could be decreased by feeding them desiccated bull pineal glands, Pflugfelder clearly implicated the pineal as an endocrine organ of fishes and suggested a pineal-pituitary relationship in teleosts. His observations on pinealectomised guppies indicated hypertrophy of the adenohypophysis, hyperthyroidism, decreased growth rate, a slight acceleration in the appearance of secondary sex characteristics in young males, an increase in the activity of the interrenal cells, and a disturbed calcium metabolism resulting in spinal curvature. The thyroidal effects could be partially offset by the injection of epiphysan or thyroxin. Pflugfelder has since reported alterations in the pars distalis and thyroid in pinealectomised goldfish. U. Holmgren reported similar findings with regard to skeletal abnormalities and ,vas able to reduce these by administering beef pineal injections. Further, the latter investigator reported a decreased radiocalcium uptake in pinealectomised fish and an increased uptake in those fish receiving injections of pineal extract. These findings, however, could not be confirmed by Weisbart and Fenwick who report an undisturbed blood calcium level in pinealectomised goldfish.

Few other studies have examined the endocrine nature of the fish pineal, and these have been conflicting. Contrary to the report of thyroidal hypertrophy by Pflugfelder, Pang found a decreased thyroidal cell height in pinealectomised *Fundulus* and Rasquin, U. Holmgren, Fenwick, and Peter report no alteration in this tissue following pinealectomy. Further, neither Peter nor Fenwick could find any apparent increase in the interrenal cells as previously reported by Pflugfelder. While the significance of these findings is not clear, the possibility that the pineal body of mammals is related to electrolyte balance has never been wholly repudiated and should provide the inipefus for further investigation in this area.

Although a pineal-gonadal relationship seems well established in the mammals, little research has been carried out on the possibility of such an axis in the fishes. Krockert was able to delay the appearance of secondary sex characteristics in young *Lebistes by* feeding desiccated bull pineal glands. In the same species, Pflugfelder reports a slight acceleration in the sexual development of males following pinealectomy. In contrast to Pflugfelders results, Schonherr and Rasquin reported unsuccessful attempts to hasten reproductive maturation by pinealectomy. Pang described a delay in the appearance of nuptial colouration in pinealectomised *Fundulus,* but noted that the controls had smaller gonad sizes. It is not clear whether this means that his controls underwent gonadal atrophy or that his pinealectomised fish showed gonadal hypertrophy. Recently, Peter and Fenwick, both working with goldfish, have reported contradictory (although not totally incompatible) results. Peter, in one experiment of 9-honths duration, could find no signficant differences between the gonosomatic indices (GSI) (gonad weight/ body weight $\times$ 100 = GSI) of pinealectomised, sham-pinealectomised, and unoperated control goldfish. However, when experiments were carried out at different stages of the yearly reproductive cycle, Fenwick found that although pinealectomy had no effect on the size of the gonads during most of the year, the operation, when carried out just prior to the onset of the final maturation phase preceding the normal spawning period, did result in a highly significant increase in the GSI relative to that found in the groups with intact pineal. The time at which pinealectomy was found to have this effect corresponded with that period of the reproductive cycle during which an increasing day length had its greatest stimulatory effect on the reproductive system. Further, the increase in gonad size in goldfish subjected to increasing periods of daily light exposure could be prevented by the daily injection of melatonin.

These results may help to explain the earlier contradictory reports. In the first place it seems probable that the pineal gland may not function to a similar extent, or even in an identical manner, during different periods of the life cycle. As seen previously, the invagination of the pineal gland increases with age and there is a decrease in sensory cell number together with a marked degeneration of nervous structures. Further, the pineal gland becomes less accessible to light with increase in body size; even the effect of pinealectomy on pigment distribution differed in young and older lampreys. Second, if the endocrine function of the pineal organ is dependent upon melatonin as suggested by Fenwick, and if the level of melatonin is depressed by exposure to long periods of bright light as reported in the mammals, then the experimental design must be considered before interpreting the results of pinealectomy. If such controls are not carefully enforced, the intensity and duration of the experimental photoperiod may be sufficient to lower the melatonin level to a point where even animals with a pineal are effectively "physiologically pinealectomised."

The Pineal Body as a Sensory cum Secretory Organ

The apparent disagreement between workers supporting a sensory function as opposed to those who argue for a secretory function does not seem justified. Rather, it seems likely that both functions are linked and that the pineal organ should be viewed neither as an entirely sensory structure nor as a wholly secretory gland; both sensory and secretory functions seem likely. This hypothesis is supported by recent observations made on the chelonian epiphysis. Vivien and Roels noted that certain cells within the chelonian epiphysis could be seen to alternate between characteristics of reception and secretion; it is not difficult to develop a similar theory for the pineal organ of the fishes.

SYNOPIS

The pineal body of fishes should be reexamined in the light of recent knowledge concerning the physiology of the pineal in higher vertebrates. The presence of photosensory cells with central connections, the extensive vascularity, high metabolic activity, evidence for both external and internal secretory activity, the pineal involvement in phototaxis, and the diverse effects of pinealectomy on the pigmentary response and the endocrine system together with the evidence for an active tryptophan metabolism favor the view that the fish pineal body, far from being an evolutionary vestige, is an important functional organ. While its small size and simplicity do not indicate it is a visual receptor, it is sufficiently sensory in appearance to suggest a role as a dosimeter of variations in incident radiation. The pineal's location and its sensory *cum* secretory activity suggest that it is well adapted to play a part in the mediation of light influences on the fish pituitary gland; furthermore, it could probably control the pressure and/or composition of the cerebrospinal fluid. As pointed out by Roth, the phylogenetic trend toward a more glandular structure in the higher vertebrates may reflect only a change in the way the information reaches the pineal rather than a completely unrelated change in function.

3

Hormones of Master Gland

In all vertebrates, the pituitary gland or hypophysis consists of two parts, separable on the bases of embryology, structure, and function. These are the neurohypophysis, a downgrowth from the floor of the diencephalon, and the adenohypophysis originating as an ectodermal upgrowth (Rathke's pouch) from the roof of the embryonic buccal cavity. The two parts meet and enclose between them a mesodermal rudiment which gives rise to their intrinsic blood vessels. Thus the gland is a composite organ, and it has many different endocrine functions.

The adenohypophysis is the site of synthesis, storage, and release into the blood of several different peptide hormones; and the greater part of pituitary histophysiology is concerned with the allocation of each of these hormones to the type of pituitary cell that secretes it. The adenohypophysis is divided into the pars distalis, site of secretion of most adenohypophysial hormones, and the pars intermedia. The neurohypophysis in fishes is rather simpler than in land vertebrates and consists essentially of a hypophysial stalk, suspending the gland from the ventral region of the diencephalon and containing an extension of the third ventricle (infundibular recess), and at the distal end of the stalk an enlargement, the neurohypophysial lobe or core, which forms the middle of the gland. The stalk contains the axonal fibers of neurosecretory cells, the cell bodies being located in the hypothalamus. The neurohypophysial core consists largely of the endings of these fibers interspersed with cells termed "pituicytes." The neurohypophysis seems to be in general a storage-release center for materials which are actually synthesised in the hypothalamus and then transported to the neurohypophysial core along the neurosecretory axons. In many fishes, the neurohypophysial stalk is virtually absent, the pituitary then being pressed close to the ventral surface of the hypothalamus, while in a few teleosts (e.g., *Lophius)* the neurohypophysial stalk is extremely long.

Adenohypophysial Histophysiology and Cytophysiology

Purves has usefully divided the cytological criteria used in the study of the adenohypophysis into two groups. The first category consists of features that are indicators of the specific nature of the functions of individual cell types such as granule size, staining reactions and chemical nature, cell morphology, and reactions to specific physiological alterations; these features are the data of special cytology, which is particularly concerned with allocation of function to each cell type. The second category includes those features which are indicators of the functional state of the cell, indicating high or low rates of metabolic or secretory activity such as nuclear size, nucleolar size, amount of cytoplasmic RNA, state of the Golgi apparatus, and accumulation or loss of secretory granules. These features constitute the field of general cytology. In the study of fishes, as in other vertebrate groups, workers on the pituitary have been concerned with both kinds of criteria. However, more than in the highly worked field of mammalian pituitary histophysiology, most investigations on fishes are still primarily concerned with special cytology. This implies that there is, as yet, no complete general agreement about the functions of the various types of cell distinguishable in the fish pituitary; although perhaps with the greater technical standardisations that have come about in recent years, especially the use of methods developed in mammalian studies by Herlant, and first applied to the fish pituitary by Olivereau and Herlant, the actual morphological and tinctorial characteristics of the teleostean adenohypophysial cell types would now be agreed upon by most workers. However, there is still disagreement about the functions of these cell types since the experimental allocation of function to the cells has been attempted systematically in only a few species. It follows that not all workers in this field would agree on a generally applicable functional nomenclature of cell types. In this review, the nomenclature used is the one established partly on functional and partly on tinctorial grounds following experimental studies in the eel, *Anguilla,* and the molly, *Poecilia*. It is a mixed nomenclature, in Purves' terms, based partly on similarities in staining properties of the secretory granules to those in mammalian cells of demonstrated function but mainly on the characteristic responses of the cells in fishes to experimental situations designed to alter the secretion rates of the different adenohypophysial hormones. Logically, one is on equivocal ground in applying this nomenclature to species in which such experimentation has not been performed. But in the teleosts, which include the vast majority of fishes, the distribution of cell types within the adenohypophysis is extremely regular so that to a far greater extent than in other vertebrate groups the topographical location of a cell type can support its identification on tinctorial grounds. Nevertheless, it must be recognised that the extension of Olivereau's mixed nomenclature to all teleosts would not yet be accepted by all workers on the fish pituitary, and the reader new to this field should bear this in mind. It seems to us that the alternative, to perpetuate yet another purely numerical or Greek letter morphological system, would present further opportunities for the extension to fishes of the nomenclatural confusions that are rife in pituitary studies on higher vertebrates. We think that tinctorial and locational grounds alone are usually sufficiently certain criteria for the extension of the Olivereau nomenclature from those fishes in which it has an experimentally defined functional basis *(Anguilla, Poecilia,* and *Mugil* and for certain cell types a few other species) to the vast majority of teleosts in which the desirable experimental backing is lacking. Unfortunately, there are almost no certain grounds for extending the system to nonteleostean groups.

Frequent reference will be made to the staining properties of the cells. By this is always meant the staining properties of the specific secretory granules elaborated and stored within the cytoplasm. The background colourations of the cytoplasm, or of other inclusions such as lysosomes, are not relevant to the tinctorial classification of pituitary cell types. Since it is obvious and readily demonstrable that the method of fixation can greatly influence the reactions of every type of cell to standard staining techniques, and since the affinity of the secretory granules for the various dyes in common use will be affected by the staining procedure applied, wherever possible the cell types will be described from material fixed in sublimated Bouin-Hollande and stained by the various procedures introduced by Herlant, in particular Herlant's Alizarin blue tetrachrome (Aliz B), periodic acid-Schiff with orange G (PAS-OG), and the combination oxidation-Alcian bluePAS-orange

G (Ox-AB-PAS-OG). The central role played by the PAS procedure in pituitary studies has often been emphasised: In essence, this technique distinguishes cells with glycoprotein-containing granules (PAS positive) from cells with nonglycoprotein granulation (PAS negative), a distinction approximately corresponding to the older division between basophils and acidophils and often marked by the terms "mucoid cells" (PAS +ve) and "serous cells" (PAS -ve). The importance of this distinction for pituitary studies stems from the chemical information we have about the various hormones extracted from mammalian pituitary glands, some of which are glycoproteins [gonadotropins, follicle-stimulating hormone (FSH), luteinizing hormone (LH), and thyrotropin or thyroid-stimulating hormone (TSH)] while others are peptides or proteins with no carbohydrate moiety [prolactin, somatotropin (STH) or growth hormone (GH);corticotropin, adenocorticotropin or adrenocorticotropic hormone (ACTH);and intermedin or melanophore-stimulating hormone (MSH)]. Another important technique is aldehyde fuchsin (AF), usually preceded by oxidation, which has a less secure histochemical basis.

THE PITUITARY GLAND IN TELEOSTS

General Organisation

The organisation of the teleostean gland has been the subject of many reviews, and we shall concern ourselves mainly with the more recent information. The anatomy of the gland at first sight appears almost as varied as the teleosts themselves, but closer examination shows that this superficial variety can be reduced to a common anatomical and histological pattern. A species in which the histophysiology of the pituitary has been particularly fully investigated is the European eel, *Anguilla anguilla,* in a long series of fundamentally important studies by Olivereau. The gland has also been investigated in the cyprinodont *Poecilia latipinna* and *P. formosa* by the senior author and his collaborators. A description of the gland in these two teleosts will serve as a basis for a general account of the teleostean pituitary, and notable departures from these basic types will then be considered.

The Eel

Olivereau has recently summarised her extensive experimental investigations of the eel pituitary The two primary divisions of the gland are obvious, with the central neurohypophysis interdigitating with the shell-like adenohypophysis. The adenohypophysis may be

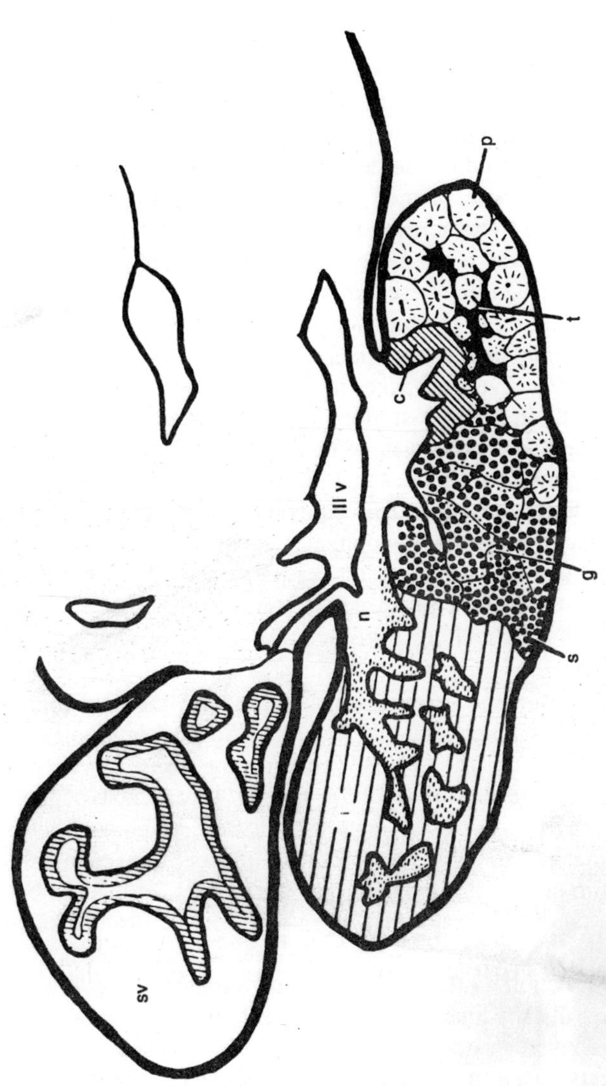

Figure 3.1 : Anguilla anguilla. Diagram of midsagittal section through the eel pituitary, anterior to the right. Follicles of prolactin cells (p), mixed with TSH cells (t), and bordered posteriorly by ACTH cells (c) form the rostral pars distalis. The proximal pars distalis comprises cords of cells below the neurohypophysis (n), mainly composed of growth hormone cells or somatotrops (s) in the sexually immature fish, with scattered immature gonadotrops (g). Posteriorly is the pars intermedia (i), deeply invaded by processes of the neurohypophysis, which in this region displays masses of AF + ve neurosecretory material (horizontal shading). The saccus vasculosus (sv) projects behind the pituitary; lily indicates third ventricle, the floor and sides being the hypothalamus, the floor being also termed the "infundibular floor."

further divided into two main parts: anteriorly the pars distalis, subdivided into rostral and proximal pars distalis, on the basis of cell types; and posteriorly a pars intermedia. As usual in teleosts, the interdigitations of the neurohypophysis with the pars intermedia are particularly deep and elaborate, while the neurohypophysial-pars distalis interdigitations are less pronounced.

The terminology applied to the fish pituitary was for many years in a state of confusion, and a new terminology introduced by Pickford and Atz helped to introduce some uniformity. On this scheme the rostral pars distalis was termed the "pro-adenohypophysis," the proximal pars distalis was termed the "meso-adenohypophysis," and the pars intermedia was termed the "meta-adenohypophysis." At the time they were introduced these terms were valuable in preventing the too easy assumption of the functions of the various regions implicit in applying the mammalian terms "pars distalis" and "pars intermedia" to the fish gland. However, advances in our knowledge of the functions of the various cells in the teleost gland during the past 10 years or so have rendered this terminology in some respects misleading. Current evidence, reviewed below, implies that the pro- and meso-adenohypophysis together form the functional equivalent of the tetrapod pars distalis, and the meta-adenohypophysis is equivalent to the pars intermedia; this being so, there is no reason to accord each of these three regions in the teleost gland equal nomenclature status, nor is there any point in withholding from them the names of their tetrapod equivalents.

Thus, we prefer the terminology adopted by the Division of Comparative Endocrinology of the American Society of Zoologists, and in general use for the teleost pituitary by many workers on histophysiology. A further reason for uniting pro-adenohypophysis and meso-adenohypophysis as a single pars distalis, as pointed out by Olivereau, is that certain cell types occur in different species in either the rostral or the proximal regions of the pars distalis, or even in both regions; there can be no doubt that taken together the pro-adenohypophysis and meso-adenohypophyses have the same complement of cell types and function as the pars distalis of tetrapods. Any division into rostral pars distalis and proximal pars distalis is merely a matter of descriptive convenience, and no longer should be taken to indicate some fundamental and constant distinction equivalent to the separation of pars intermedia and pars distalis. The dangers inherent in accepting the fundamental nature of the rostral-proximal division of the pars distalis are illustrated by the idea current a few years ago that the pro-adenohypophysis was in some way peculiar to fishes, with no homolog in the tetrapod gland, and even that the meso-adenohypophysis was the homolog of the entire pars distalis of tetrapods, although at that time the cells secreting prolactin and ACTH (and which in fact both occur in the rostral pars distalis) had not been identified.

In the eel the rostral pars distalis includes three cell types. The q cells (prolactin cells) are arranged in follicles, a feature typical of the more primitive teleosts. Centrally, between these follicles, occur masses and cords of 8 cells (TSH cells), while at the posterior border of the rostral region with the neurohypophysis is a layer of E cells (ACTH cells). The proximal pars distalis in the eel includes two cell types arranged in cell cords running more or less vertically. Most of the cells are a cells (growth hormone cells), and they are mixed with gonadotrops, of which there are two types. The pars intermedia includes two cell types, the precise functions of which are uncertain. The neurohypophysial fibers contain typical masses and grains of neurosecretory material; even when, as Herring bodies, these accumulations are very large, they are in fact contained within the swollen nerve fibers. This stainable neurosecretory material is usually particularly abundant in the posterior part of the neurohypophysis, where it interdigitates with the pars intermedia. The neurohypophysis also contains scattered cells (pituicytes) of uncertain function.

The Molly

The pituitary gland of this small viviparous cyprinodont is being studied experimentally, and a general account of the gland has been published. The molly pituitary resembles that of the eel, although it is proportionately shorter and deeper. The rostral pars distalis contains only two cell types, a mass of η (prolactin) cells, showing no trace of the follicular arrangement seen in the eel, and a posterior border of ε (ACTH) cells. In contrast to the eel, the δ (T'SH) cells occur in the proximal pars distalis, in a dorsal zone where they lie intimately mixed with α cells. Below is a ventral zone of gonadotrops. The pars intermedia is not penetrated by the neurohypophysis so deeply as in the eel and contains two cell types corresponding to those of the eel.

Other Teleosts

Departures from the general morphology and cell distribution seen in the above-mentioned two basic species occur, the most obvious being the wide variations in shape and proportions of the gland. A common type has a much shorter antero-posterior axis, and is much deeper dorsoventrally (e.g., cyprinids and salmonids). In such cases, the gland in the juvenile fish may resemble in general that of the eel or *Poecilia,* its shape and proportions changing as the fish grows larger. In these deeper glands, the neurohypophysial core commonly forms an elongated central axis with the three regions of the adenohypophysis arranged around this axis, the pars intermedia ventral below the hoop of the proximal pars distalis with the rostral pars distalis embracing the neurohypophysial core amteriorly. In these cases, the extensive ramification of the ventral neurohypophysial core into the pars intermedia is very elaborate.

One of the most important variables in the teleostean gland is the location of the d (TSH) cells, which may lie in the proximal pars distalis, as in *Poecilia (e.g., Astyanax, Caecobarbus, Phoxinus,* and cyprinodonts), or may be rostrally placed, as in *Anguilla (e.g.,* clupeoids and cyprinids) ; or the cells may lie in an intermediate position between the two regions (e.g., cichlids, salmonids, and *Mugil*), this being one of the reasons for uniting the rostral and proximal regions as a single pars distalis. Again, the position of the gonadotrops is variable, being usually in the proximal pars distalis as in *Poecilia* and the eel, but in the trout found in both rostral (few) and proximal regions; and some gonadotrops invade the rostral region even in the eel at full sexual maturity. Another variation concerns the general structure of the rostral pars distalis. In the more primitive teleosts (isospondylous forms, salmonids, clupeoids, and apodes), the prolactin cells of this region are arranged around follicles, as in the eel. This follicular arrangement of the prolactin cells in these primitive teleosts recalls the similar structure of this region in ganoid fishes. In certain clupeoid and salmonid fry, and in the adult *Hilsa ilisha,* the lumina of the rostral follicles communicate with a persistent orohypophysial duct which probably represents the cavity of the embryonic Rathke's pouch. In most teleosts the orohypophysial duct is never present; or if formed it disappears in the adult, and the rostral follicles in the pars distalis of adult isospondylous teleosts should probably be regarded as a primitive feature, as suggested long ago by earlier workers on the gland.

Despite these and other variations in the morphology of the gland, the teleost pituitary when studied in detail usually presents the principal parts described for the eel and *Poecilia,* and the pars distalis can generally be seen to present rostral and proximal regions. In certain cases the distinctions may be less clear, and in *Lepidogobius lepidus* Kobayashi *et al.* have described four distinct tinctorial zones in the pars distalis. Probably the experimental analysis of this species would reveal a functional agreement with the more usual roughly bipartite pars distalis.

Lepidogobius demonstrates further unusual features: the whole pituitary is pressed up into the hypothalamus, almost obliterating the third ventricle, the neurosecretory fibers from the hypothalamus take an unusual course to the neurohypophysis, and the neurohypophysial core does not penetrate deeply into the adenohypophysis, not even into the pars intermedia. Another specialised arrangement is seen in *Hippocampus* where the neurohypophysis extends to enclose the pars intermedia laterally, ventrally, dorsally, and posteriorly, so that the posterior region of the gland consists of a central core of pars intermedia enclosed in a sleeve of neurohypophysial tissue.

Histophysiology of the Adenohypophysis

The teleostean adenohypophysis has been shown to secrete the usual complement of hormones, prolactin, growth hormone, gonadotropins, TSH, ACTH, and MSH. Each of these factors has been allocated by experimentation to the cell type that secretes it, and the cell types will now be treated in turn.

Prolactin Cells η Cells, Erythrosinophilic Cells or Paralactin Cells)

The prolactin cells lie mainly in the rostral region of the gland, where they form a compact mass that is the main defining character of the rostral pars distalis. In some species, perhaps in most, the prolactin cells extend ventrally and laterally for a greater or lesser distance around the proximal pars distalis, in winglike projections. Their specific secretory granules (η granules) stain red with the erythrosin in Aliz B tetrachrome and the Cleveland Wolfe trichrome, and with the azocarmine or acid fuchsin in other methods such as Azan, Mallory, and Masson, but they are negative to PAS, Alcian blue, and aldehyde fuchsin.

In fresh glands of *Poecilia* and *Fundulus kansae* the η granules are opaque so that the rostral part of the pituitary is often a dense white. In the isopondylous forms in which the prolactin cells are arranged in follicles, the material commonly found in the follicular lumen exhibits variable staining, some with orange G and some with PAS, Aniline blue, or light green; with the electron microscope, at least some of this material in the eel consists of a mass of membrane-bound vesicles, probably derived from degenerating cells. The prolactin cells in such fish (e.g., eel, trout, and salmon) are generally columnar, their apices bearing cilia projecting into the follicular lumen, and often with the η granules concentrated toward the outer cell base. The cytoplasmic RNA (i.e., endoplasmic reticulum) usually lies between the nucleus and the outer base of the cell. Ultrastructural studies on the eel revealed electron dense granules in these cells, about 280 mμ diam in freshwater *Anguilla* but about 350 mμ in the marine *Conger;* the endoplasmic reticulum lies basally, while the Golgi apparatus lies on the other side of the nucleus, toward the follicular lumen. The Golgi apparatus in immature freshwater eels consists of six to eight parallel cisternae forming a bowl- or cup-shaped body; ultrastructural appearances suggest that the η granules are formed here, the outer convex surface of the Golgi perhaps receiving newly synthesised material from the closely adjacent rough endoplasmic reticulum, the concentrated materials then being released as membrane-bound granules at the inner concave surface of the Golgi. In *Anguilla* elvers, and in adult *Anguilla* caught in the sea during the spawning migration, the η granules are smaller than in the freshwater stage, about 200 mμ diam.

The prolactin cells of nonisospondylous teleosts are not grouped in follicles and are not generally columnar (although the cells may be elongated in some species such as the cichlid, *Heterichthys cyanognathus)*. Usually they form a compact mass of rounded cells as in *Poecilia,* generally evenly granulated. In cyprinodonts, the nucleus is frequently indented or kidney-shaped, a feature particularly marked in *Poecilia* and *Fundulus heteroclitus*; in these cyprinodonts, too, the Golgi image is easily visible in active cells as one or more clear tubules among the granules, curved in a C or U shape, and the endoplasmic reticulum, visualised by staining for RNA,.forms a cap or halo on the nucleus. In the related guppy *(Poecilia = Lebistes reticulatus)* and platy *(Xiphophorus maculatus),* the η granules are revealed by the electron microscope as membrane-bound osmophilic vesicles, between 200 and 300 mμ diam, and Weiss has described ultrastructural appearances that suggest that the η granules in the platy are probably released from buds projecting from the cell surface, by a process in which the granule membrane fuses with the cell membrane and the granule escapes through the resulting opening; this resembles one of the modes of granule extrusion described in higher vertebrates.

At the light microscope level, η granules appear to vary in size even within individual cells, and commonly vary between individuals and species; in *Poecilia latipinna,* when the granules are rather sparse they appear to be distinct and large, but if they are densely packed they seem smaller, while in the eel and trout the granules are usually coarse and distinct no matter what their density. Here, and in other places in this review, we should emphasize that granule size, even if determined with the electron microscope, is not an exact criterion of cell type in comparisons between species. Fluorescent antibody to ovine prolactin located specifically on the η granules in *Fundulus heteroclitus*, confirming that the granules themselves do contain fish prolactin.

Histochemically the prolactin cells have been shown to contain SH/ SS groups in the eel, *Mugil*, and *Xiphophorus*; but as in other species they do not react with Alcian blue even after oxidation, indicating that they are not particularly rich in cysteine. Nevertheless, *Mugil* prolactin cells will incorporate cysteine-35S. In goldfish in freshwater these cells displayed intense incorporation of acetate-'H, indicating intense protein synthesis. The η granules in fresh *Mugil* glands were precipitated only by 7.5% or stronger trichloroacetic acid, in contrast to the a granules (growth hormone cells) which were precipitated by 2.5% TCA. In ultrastructural studies of eel prolactin cells, acid phosphatase was shown to be distributed in particles in the apical cytoplasm, and also within the Golgi region and in some of the developing *q* granules, possibly concerned in the destruction of excess secretory material as suggested for the lysosomelike bodies in the gonadotrops.

Evidence for the Secretion of Fish Prolactin by the η Cells

Olivereau and Herlant pointed to the similarity in staining properties of the teleostean η cell and the prolactin cell of mammals. However, too little was known at that time about the physiological role of prolactin in fishes to suggest an experimental approach to defining the function of the η cells.

More recently, following the demonstrations by Burden, Pickford and Phillips, and Pickford *et al.* that prolactin is the only mammalian pituitary hormone that will promote tolerance of freshwater in hypophysectomised *Fundulus heteroclitus,* the η cells have been investigated experimentally, especially in *Poecilia latipinna.* Details of the physiological background will be found in the chapter by Ball, this volume. For the present, the important point is that evidence indicates the secretion by the pituitary in both *F. heteroclitus* and *P. latipinna* of a prolactinlike hormone (fish prolactin, paralactin) that specifically promotes survival in freshwater by limiting the outflux of sodium from the body. Fish prolactin is not essential to *Fundulus* in seawater nor to *Poecilia* in dilute seawater, but it is secreted, presumably at a low rate, by *Poecilia* in dilute seawater, with the effect of reducing the rate of sodium exchanges.

Studies on the pituitary in these two fishes have shown that only the η cell displays cytological evidence of changes in secretory activity such as would be predicted of the cells secreting fish prolactin. Thus, in *Poecilia* these cells are always more active in freshwater than in dilute or full-strength seawater; and these cells, but no others, are rapidly activated when *Poecilia* enters freshwater from dilute seawater, in correlation with reversal of plasma sodium loss, a marked curtailment of sodium outflux from the body, and an increase in pituitary prolactin content. Similarly, in *F. heteroclitus* the η cells are consistently more active and numerous in freshwater than in seawater, and a regenerated pituitary remnant in an incompletely hypophysectomised individual, experimentally shown to secrete fish prolactin, consisted almost entirely of active η cells. In more direct experimental approaches, it was shown that removal of part of the zone of η cells impaired freshwater tolerance of *P. latipinna*, and that ectopic pituitary transplants in *P. formosa* and *P. latipinna* are able to secrete fish prolactin, the transplants always containing active η cells The identification was clinched when Ball demonstrated that ectopic pituitary transplants of the rostral part of *P. latipinna* pituitary, composed mainly of η cells, secreted fish prolactin in response to entering freshwater, but that transplants of the posterior part of the gland, containing few η cells, did not. More recently, Emmart *et al.* showed that fluorescent rabbit antiserum to ovine prolactin located only on the η granules in *F. heteroclitus* and in no other cell type. Thus, the functional identity of the η cells in these cyprinodonts is as well established as that of any pituitary cells in any vertebrate, and there can be no doubt that the η cells secrete fish prolactin.

In other teleosts, these cells behave in the same way as in the cyprinodonts. They are more active in freshwater than in seawater in salmonids and in the eel, in *Mugil*, in *Fundulus kansae*, and in *Tilapia mossambica*. They occur in all teleosts that have been examined, as far as can be judged from the published description sometimes based on material not suitably fixed and stained.

In some species, usually marine, the prolactin cells may appear chromophobic, especially after poor fixation; and when the eel enters seawater the η granules first become smaller and then virtually disappear, leaving apparently inactive chromophobic cells. Conversely, the granulation is sparse in elvers newly arrived in freshwater from the sea and the cells may appear totally chromophobic in young eels, but in the freshwater phase of larger eels they are always well granulated.

There are a few observations on changes in the η cells in relation to life history. They are strongly granulated and erythrosinophilic in newly born *Poecilia latipinna* and are already differentiated in the gland of the embryo within the maternal ovarian follicle; they differentiate early in gestation in the guppy embryo *(Poecilia reticulata)* at first with granules smaller than in the adult, and with an active-type endoplasmic reticulum The prolactin cells do not appear to undergo marked activity changes during the monthly cycle of oocyte growth and pregnancy in viviparous cyprinodonts; but in *Zoarces viviparus,* an unrelated viviparous form with a totally different mode of gestation, ultrastructural studies indicated that these cells are hyperactive during pregnancy. In the male *Hippocampus,* which incubates the eggs in a brood pouch or marsupium, the prolactin cells undergo an annual cycle in correlation with the development and functions of the marsupium, being especially active during the first half of the incubation period. These changes correlate with experimental evidence that prolactin is concerned in maintaining the marsupium. The η cells of salmon, *Salmo salar,* are fairly active in the freshwater parr and smolt and appeared to be reorganizing after degeneration in adult fish ascending the river from the sea in the spring, eventually becoming stimulated and hyperplastic. Their activity seemed to be depressed in spawning fish in freshwater, particularly in the female. They cells in hybrid *Xiphophorus* bearing melanomas were extremely large and hyperactive, an interesting correlation in view of evidence that prolactin can promote melanogenesis in *Fundulus heteroclitus*.

In view of all the evidence for the part played by fish prolactin in sodium conservation in freshwater, one might expect the η cells to be maximally active in deionised water; but this is not so, at least in the case of the eel, in which a sojourn in deionised water reduced the cells to a state of inactivity comparable to that in seawater, with marked concomitant changes in other cell types described elsewhere. Deionised water is obviously a highly artificial medium, and in its effects on electrolyte metabolism it does not act simply as a highly dilute freshwater.

There are indications of interactions between the prolactin cells and other endocrine glands. The cells are activated after radiothyroidectomy in the eel and goldfish, but not distinctly so in the trout. However, thyroxine had no very clear effect on the cells in the eel apart from inducing a retention of η granules. Doses of thiourea or thyroxine, both of which inhibited thyroidal ^{131}I uptake, had no obvious effects on the prolactin cells in *P. latipinna;* nor did propyl thiouracil affect the prolactin cells of *Mugil,* although inducing pronounced alterations in the TSH-thyroid axis. Surgical interrenalectomy did not alter the prolactin cells in the eel, although the adrenal inhibitor SU 4885 leads to slight and irregular stimulation of these cells. Given over three days, ACTH inactivated the prolactin cells of *Hippocampus*. The only indications of gonadal influences on these cells are Olivereau's observation that in freshwater the prolactin cells are more active in female than in the male eels (immature fish), and Sokol's report that the η cells of *F. heteroclitus* undergo a transient chromophobia (= partial degranulation) coincident with spawning in the coastal seawater.

Like the mammalian prolactin cell, the η cell in *Poecilia* remains active in ectopic pituitary transplants, an activity manifested both functionally (tolerance of freshwater; reduction of sodium turnover in dilute seawater) and cytologically. In such transplants, as in the normal gland, the η cells are activated rapidly when the fish enters freshwater from dilute seawater. Thus hypothalamic connections are not essential for activation of the η cells in response to salinity reduction, nor for the maintenance of their activity in freshwater. However, we are not in a position to postulate a hypothalamic prolactin-inhibiting factor (PIF) such as exists in mammals, since we do not yet have quantitative data on secretory rates of fish prolactin from transplants and from the *in situ* gland. It is difficult, because of this, to interpret Olivereau's observations that prolonged treatment with ovine prolactin leads to regression of the η cells of *Anguilla*, and Boisseau's finding that a three-day treatment with prolactin led to marked involution of *Hippocampus* η cells. Comparable physiological observations in the rat are suggestive of an increase in secretion of PIF induced by the exogenous prolactin, leading to inhibition of endogenous prolactin output, but obviously this concept may not be extended to teleosts in the present state of our knowledge.

In vitro observations agree with the results from pituitary transplants, the η cells remaining active in cultured trout glands, with partial or total degranulation and cellular enlargement. In cultured glands from *Fundulus heteroclitus,* Emmart and Mossakowski found that new colonies of η cells, arising by mitoses in the outgrowing layers of the explant, contained granules that bound a fluorescent antibody to sheep prolactin, indicating the continued ability of the η cells to synthesize fish prolactin in the complete absence of hypothalamic influences. Fixed material from these cultures showed evidence of release of luorescent-labeled granules across the cell surfaces, and in addition these workers observed with phase contrast the emission of granules from the constantly rippling and undulating membrane of the living cell. Sage, working with *Xiphophorus* glands, used elaborate techniques to show that the η cells lost material *in vitro,* and that this loss was greater on more dilute media than on a medium with approximately the same sodium content as the fish serum. He did not specifically determine whether dilution of the medium increased the synthetic rate of the cells, in addition to enhancing their rate of discharge (although this is implied by their degranulation in the presence of cyanide) ; certainly, in *P. latipinna* transferred to freshwater from dilute seawater, plasma sodium falls through the range of concentrations employed by Sage concomitantly with the onset of increased discharge and synthesis in the prolactin cells.

ACTH Cells (Corticotrops ε Cells, or X Cells)

The ACTH cells, characteristically disposed in a sheet at the interface between the rostral pars distalis and the neurohypophysis, have been recognised morphologically in teleosts for many years, but their significance was not known. Some workers thought that they were aberrant or immature forms of the η cells, while Bugnon suggested they might represent the pars tuberalis of higher vertebrates. In recognition of their enigmatic nature, Olivereau termed them the "X cells."

In *Poecilia,* the ε cells are usually truly chromophobic (that is, without granules that can be stained by any of the standard dyes or histochemical techniques) ; but sometimes they display a very fine, powderlike granulation, faintly staining with erythrosin in Cleveland Wolfe or Aliz B preparations, or with azocarmine in the Azan technique. They may also stain a faint purple after Aliz B or gray with lead hematoxylin (PbH). They are elongated cells, usually arranged with the long axis perpendicular to the adeno-neurohypophysial interface. The cells are small, and the absence of distinct granulation makes it difficult to discern the cell outlines. The most characteristic feature in this species is the elongated nucleus with a distinctive scattering of chromatin which gives the nucleus an easily recognizable spotted appearance. The nucleolus in most normal cells is insignificant or undetectable.

In the eel the ε cells are much better granulated than in *Poecilia,* and the granulation stains weakly with erythrosin and quite strongly with Alizarin blue. The granulation is also strongly stained by PbH which picks out the cells easily, although it is not a specific stain for the ACTH cells; PbH also stains a gonadotrop in the eel and a pars intermedia cell type in most teleosts. The eel cells are usually oval rather than fusiform as in *Poecilia.* The nucleus may be rounded, elliptical, or pear-shaped, and the nucleolus is generally indistinct. In all teleosts in which they have been described, these cells display no trace of an affinity for PAS or Alcian blue; that is, they are nonmucoid or serous cells. In *Anguilla,* they are rich in SH/ SS groups.

ε Cells were described in *Xiphophorus* as polymorphic chromophobes, forming a wedge-shaped area posterior to the y cells; a similar arrangement was found in the surfperch, *Embiotoca.* In the Pacific salmon they border the neurohypophysis behind the mass of η cells, and are rather insensitive to dyes, with a fine granular cytoplasm that stains brown after Masson. Baker described trout ε cells as forming a layer of amphiphils interspersed between η cells and neurohypophysis and coloured by Alizarin blue and PbH. A similar account of the salmonid ε cells was given by Olivereau, who emphasised that their affinity for Alizarin blue and PbH is much less than in the eel and also that they contain fewer SH groups than in the eel. Typical ε cells have also been described in *Perca* and in *Mugil*, and in the goldfish they are small and apparently chromophobic elements among the η cells, but they will stain with Alizarin blue and PbH. Oztan described ε cells in *Zoarces* as columnar acidophils, facing the rostral pars distalis and staining with light green; however, she was not able to separate them from the η cells with the electron microscope. *Fundulus heteroclitus* also has typical ε cells, in a palisade behind the η cells. Sokol described them as blue-green basophils (after aldehyde fuchsia-light green-orange G) ; and they also display the characteristic affinities for erythrosin, Alizarin blue, and PbH. Sokol also described the ε cells of the guppy, *Poecilia reticulata,* as weakly staining green basophils.

At the ultrastructural level, the eel ε cell reveals a moderately electron dense cytoplasm containing numerous fine vesicles or microtubules. The endoplasmic reticulum is diffuse and not prominent, and the Golgi area is rarely seen in normal fish. The cytoplasm is evenly packed with electron dense granules measuring 200-250 mμ in diameter. Follenius described the cells as forming a chromophobic band in several teleosts, and his ultrastructural studies revealed a sparse cytoplasm with ergastoplasmic vesicles in some cells; other cells contained a well-developed Golgi apparatus surrounded by osmiophilic secretory granules.

The most general point of agreement in the various accounts is that these cells tend toward chromophobia, especially marked in cyprinodonts. This tendency may be exaggerated by stress at sacrifice, when the animal may struggle, or by prolonged anesthesia. A similar tendency toward chromophobia is characteristic of the ACTH cell in higher vertebrates.

Evidence for the Secretion of ACTH by the ε Cells. Allocation of ACTH secretion to these cells was accomplished simultaneously and independently by treatment of *P. latipinna* and *Anguilla* with Metopirone (SU 4885 Ciba, metyrapone), a drug which inhibits the biosynthesis of adrenocorticosteroids. Although other actions have been described, its main effect in mammals, following short-term treatment with low doses, is inhibition of 11β-hydroxylase, thus preventing the formation of the normal 11-oxycorticosteroids. As a result, because of the negative feedback relationship between these steroids and ACTH, the use of SU 4885 has led to identification of the ACTH cells of mammals and birds.

In *Poecilia* and the eel, short-term treatment with SU 4885 resulted in hypertrophy and hyperplasia of the interrenal cells, indicating increased ACTH secretion, and this was accompanied specifically by activation of only the ε cells in the pituitary. This morphological stimulation of the interrenal by the drug is partly (eel) or wholly *(Poecilia)* suppressed by hypophysectomy, which supports the interpretation that SU 4885 elicits elevated ACTH output and hence identifies the c cells as corticotrops. With longer treatment in both species the TSH cells are stimulated, along with thyroidal activation, and other effects of the drug have been observed with long treatment in the eel. However, there can be no doubt that the initial and most marked hypophysial effect of S U 4885 is to provoke enhanced secretion of ACTH, and that this originates in the ε cells. Surgical removal of the interrenal in *Anguilla* has confirmed this earlier work, resulting in activation of the ε cells specifically. However, interrenalectomy, unlike SU 4885 treatment, did not lead to stimulation of the TSH cells, suggesting that the action of the drug on these cells may be direct rather than mediated via the ACTH-interrenal axis. Unpublished work on *P. latipinna* has shown that if cortisol treatment is combined with SU 4885, the interrenal stimulation (i.e., elevated ACTH secretion) is prevented, and correspondingly the ε cells are not activated.

This identification has been further strengthened by studies in the eel of the responses of the ε cells to a variety of other treatments designed to alter ACTH-interrenal function; the cells were activated by treatment with reserpine, aldactone, op'DDD, and inactivated by treatment with cortisol or ACTH. Aldactone also activates these cells in goldfish, and a heavy infection of the kidney, which partly destroyed the interrenal, was associated with ε cell activation in a Pacific salmon. In *Hippocampus,* as in the eel, ACTH inactivates the ε cells.

The degranulation of the ACTH cells following conditions of stress such as anesthesia and bleeding was emphasised by Olivereau, and such a stress response could perhaps explain the slight stimulation of these cells when *Mugil* was transferred from seawater to freshwater. The stress response of the ACTH cells was even more marked in response to surgery in the eel, with correlated activation changes in the interrenal.

A lengthy sojourn in deionised water, which had profound effects on other cell types, produced only slight stimulation of the ε cells and interrenal in *Anguilla,* perhaps indicating that the ACTH-interrenal axis plays only a minor role in electrolyte regulation in this dilute medium.

There are a few observations on the response of these cells to collateral endocrine changes. They are slightly inactivated after radiothyroidectomy in the eel and trout, and thyroxine treatment led to slight stimulation of the eel ACTH cells, with nuclear hypertrophy, slight degranulation, and a few mitotic figures. In *Mugil,* too, the cells were stimulated by thyroxine and inhibited by thiourea. Removal of the enigmatic corpuscles of Stannius from eels led to changes in the ACTH cells that were variable and difficult to interpret. Also in the eel, the ACTH cells became more active with sexual maturation induced by injections of carp pituitary material, but interpretation of this effect is complicated since the carp material probably included ACTH, indicated by the fact that the interrenal tissue was more strongly stimulated than the degree of activation of the ACTH cells would explain. Boisseau described an annual cycle of activity in the ACTH cells of the male *Hippocampus,* related to the annual sexual cycle and characterised by hyperactivity of the ACTH cells associated with incubation of eggs in the marsupium. Boisseau also found that prolactin injections led to inactivation of the ACTH cells. This detailed work on *Hippocampus* is exceptionally interesting in indicating the way in which the ordinary complement of pituitary hormones may be adapted to control extremely specialised mechanisms: The maintenance of the marsupium (developed under testicular influence, in turn dependent on gonadotropins) is dependent upon the ACTH-interrenal axis and upon prolactin. Boisseau found evidence of complex interrelationships between prolactin, ACTH, and corticosteroids in *Hippocampus;* and the response of the η and ε cells to ACTH and prolactin administration may in themselves be specialised in connection with the highly unusual reproductive processes. Thus, prolactin treatment had no obvious effects on the ACTH cells in the eel, unlike *Hippocampus*.

The ACTH cells persisted, although in reduced numbers, in ectopic pituitary transplants in *Poecilia formosa* which continued to secrete small amounts of ACTH. However, these cells in the cel pituitary *in vitro* present an appearance of *increased* activity, showing degranulation, increased nuclear size, and rapid incorporation of radiouridine. In the trout the ACTH cells in culture display a marked migration of their granules toward the pole of the cells facing the neurohypophysis; this phenomenon is difficult to interpret, but recent unpublished work by Baker shows that, as in the eel, these trout cells displayed intense nuclear incorporation of radiouridine. Nevertheless, the granulation continues to accumulate at the pole of the cell, which suggests inhibition of release, in contrast to the degranulation of the eel cells. Thus there may well be marked species variations in the response of the ACTH cells to hypothalamic disconnection.

Growth Hormone Cells (Somatotrops, α Cells, GH Cells, or STH Cells)

The α cells are the most prominent acidophils in the pars distalis, especially in marine fishes with inactive prolactin cells, and occupy much of the proximal pars distalis. They have been described and recognised as a distinct cell type for many years and in many species. These are the cells that conform to the classic category of acidophil cells, staining intensely and selectively with orange G in the various trichrome and tetrachrome techniques. For many years the existence of two types of acidophils has been recognised in mammals and other tetrapods, one type staining selectively with orange G (α cells) and the other type (η cells) with azocarmine or erythrosin. However, the distinction is not always easy to obtain in all species and with all dyes and techniques; and although the distinction can be made in teleosts, here too it is delicate and not always easy. In the case of well-known pituitaries, such as *Anguilla* or *Poecilia,* this difficulty is not always important, since once established that one can in some circumstances make this tinctorial differentiation, one can always separate the α and η cells on the basis of their location and morphology; but with any new or little-known teleostean gland it is advisable to spend time and effort in establishing this point before making definitive identification of the α and η cells. It needs to be emphasised that the pituitaries of different fishes vary greatly in their staining reactions to any particular techniques, and one must experiment with dyes, timings, and other variables for each species.

In *Poecilia* the η-α tinctorial distinction can be made quite easily with the Azan or Aliz B method, the η cells staining red with azocarmine or erythrosin and the α cells staining orange or clear yellow (Aliz B). A similar distinction can be obtained in the eel, trout, salmon, *Xiphophorus*, *Fundulus kansae*, and *F. heteroclitus*. Emmart *et al.* emphasised the delicacy of the distinction, showing that the red-yellow differentiation obtained with PAS-erythrosin-orange G could be reversed by rinsing the slide initially in 1% boric acid. In *Poecilia* and *F. heteroclitus,* the combination Ox-AB-erythrosin-orange G usually gives a good α–η distinction, as well as staining the basophils in a sharply contrasting blue.

In *Poecilia* the α cells occur in the centro-dorsal region of the proximal pars distalis, intermingled with the much scarcer TSH cells. With the light microscope the cells are generally large and rounded in form, although frequently elongated and ellipsoid; typically, the cell is crowded with exceedingly fine orangeophilic granules, finer than the η granules and often packed so densely as to appear a homogeneous mass. The nucleus is ovoid, frequently eccentrically placed at one pole of the cell, and often seems squashed and distorted by the dense packing of the granules. In any one fish there may be great variations in the size and shape of the nucleus, and in the size of the nucleolus; the latter is generally prominent, however, and the cells usually show signs of considerable activity. The Golgi image, in the form of a ring or crescent, is usually very prominent, sharply outlined by the dense granulation.

A tendency to sexual dimorphism is evident in the α cells of *Poecilia:* In the adult female, the cells are numerous, often tightly packed together, densely granulated and very active in appearance, as already described, whereas the male fish usually presents few α cells, poorly granulated or even degranulated, with a central nucleus, nucleolus insignificant or invisible, and no sign of the Golgi image. The a cells mainly lie in the dorsally projecting fingers of the proximal pars distalis, hence in close proximity to the neurohypophysis. In the female, they frequently form a virtually continuous border between the neurohypophysis and proximal pars distalis, several cells in depth, forming a shell, dorsally concave, which is easily visible in the middle of a fresh gland, the α granules being translucent and appearing a dense white in contrast to the more transparent ventral gonadotrops. Similarly in mammals, acidophil granules appear opaque in fresh material. In contrast, in the male *Poecilia* the α cell border to the proximal pars distalis is thinner and discontinuous, the cells are poorly granulated and generally not visible in the fresh gland.

In the sexually mature female eel, as in *Poecilia,* the α cells display a sexual dimorphism. In immature individuals of both sexes, these cells form the greater part of the cords of the proximal pars distalis, since the TSH cells are located in the rostral region, and the gonadotrops are small and undifferentiated. At sexual maturity the cords become composed of a mixture of α cells with now more numerous and larger gonadotrops. In the mature male, the α cells are small, with indistinct boundaries, rather sparse granulation, and indistinct nucleolus and little or no visible cytoplasmic RNA, whereas in the female the α cells are much more numerous and often appear more active, with strongly marked cell boundaries, large nuclei and prominent nucleoli, denser granulation, prominent Golgi image and demonstrable cytoplasmic RNA. Thus in both *Poecilia* and *Anguilla,* the α cells are generally more numerous and more active in the mature female. In contrast, Schreibman found no indications of such a sexual dimorphism in *Xiphophorus.*

In both *Poecilia* and *Anguilla* the α granules are more refractile than the η granules. In contrast to *Poecilia,* the eel a granules are rather larger than those of the η cells, an observation confirmed by the electron microscope. Similar differences are found in comparisons of the two cell types in other teleosts, some fish resembling *Poecilia* in having smaller α granules, while others resemble the eel in having smaller η granules. This variability again emphasizes that granule size alone is not a criterion of cell type.

Like the η cells, the α cells are typical serous cells; that is, their granules do not contain glycoprotein and arc typically negative to PAS, AF, and AB. A faint staining with PAS can nevertheless be obtained, especially with the Ox-AB-PAS-OG procedure, in *Poecilia* and *Fundulus*, the eel and other species. This recalls the faint staining of acidophil granules with PAS in mammals, which may reflect the lipid content of these granules. This PAS reaction in the telcostean α cell is too weak to cause confusion with the strong reaction in the mucoid cells. The α cells in *Anguilla* contain SS/SH groups and are rich in protein, and in *Mugil* they incorporate radiocysteinc rapidly. Tritiated acetate and labeled amino acids were rapidly accumulated by the α cells of goldfish, evidence of intense protein synthesis. *Mugil* α granules are precipitated by weak (2.5%) trichloroacetic acid, in contrast to the η granules which precipitate only with a stronger (7.5%) solution; and Follenius found that the a cells were selectively labeled following intraperitoneal injections of D,L-noradrenaline-3H in *Gasterosteus*. The significance of this finding is uncertain at present.

α Cells have been recognised by nearly all workers on the teleost pituitary, and a resume of all their accounts would be unnecessary. Certain ultrastructural observations are important. In *Anguilla,* Knowles and Vollrath observed that the α cells appeared to be filled with spherical or slightly oval membrane-hound vesicles, about 400 mμ in diameter. No clear indications of endoplasmic reticulum or Golgi apparatus could be found, and there were few mitochondria. Follenius described ultrastructural variations which he thought might indicate a cycle of activity in the α cells of *Perca*. Cells rich in endoplasmic reticulum were poor in secretory granules, and vice versa; in cells that were primarily storing their secretory products, the endoplasmic reticulum was reduced to a few cysts, often perinuclear in position. Follenius described granules of secretion forming in the Golgi region (cf. mammals, Racadot *et al.*), and also extrusion of these granules from the cell surface in a process reminiscent of that described by Weiss for the teleostean α cell. Follenius emphasised that the α cells present greater variations in activity than the η cells, but he did not study the responses of the η cells to changes in salinity.

Evidence for the Secretion of Growth Hormone by the α Cells

A few years ago, Olivereau summarised the evidence that the teleostean α cell secretes growth hormone. *As* she said, there was then no direct proof of this ascription, although the similarity of the α cells to the somatotrops of mammals made it probable. In teleosts, the α cells are more numerous in salmon during the period of rapid growth in the smolt stage and were also very active in *Oncorhynchus* captured at sea during the adult period of rapid growth. The greater abundance and activity of these cells in the female eel may relate to the greater size attained by the female in freshwater, and the generally larger size of the female *may* similarly explain the sexual dimorphism of the α cells in *Poecilia*.

More recently, with the experimental evidence that has been derived for the functions of the other pars distalis cell types, the α cells have come to be accepted as somatotrops, as it were by a process of elimination. More specifically, when the cellular source of gonadotropin(s), TSH and ACTH were known, the two acidophils, α and η cells, remained as the producers of prolactin and growth hormone; we have seen that the η cells secrete fish prolactin, which leaves the α cells as the source of growth hormone. The functional distinction between the two acidophils was clinched by partial hypophysectomy in *Poecilia latipinna,* which showed that removal of most of the rostral pars distalis (η cells), while impairing prolactin secretion, had no deleterious effect on growth. This meant that growth hormone must be produced in the proximal pars distalis, and knowing the gonadotrops, TSH cells, and ACTH cells, this leaves the α cells as somatotrops. The scarcity and degranulation of α cells in the ectopic pituitary transplants that secreted little growth hormone in *P. formosa* is in agreement with this conclusion.

Unfortunately, little information is available about alterations in the α cells during the life history. The relationship of these cells to periods of rapid growth in salmon has already been mentioned. The α cells of the guppy, *Poecilia reticulata,* differentiate and attain their adult condition later than in the η cells; they are present and active in newly born and rapidly growing *P. latipinna fry*, and they have been described as staining more intensely in the elver than in the adult yellow eel. The α cells were enlarged and particularly active in melanoma-bearing hybrid *Xiphophorus*, a suggestion of a possible tumorogenic action of fish growth hormone.

Concerning the response of the α cells to collateral endocrine changes, Olivereau detected no action of thyroxine or radiothyroi,dectomy on these cells in the eel, and female *P. latipinna* display no marked α cell reactions to thyroxine or thiourea treatment. Radiothyroidectomy did not affect the α cells in trout. However, Sage, while agreeing that thyroxine alone had no action on the α cells of male guppies, *Poecilia reticulata,* found that thiourea caused increased storage of granules in the cells, indicating retention of growth hormone, and that thyroxine given in combination with thiourea reduced this storage of granules. Since thyroxine increased the growth of thiourea-treated guppies, Sage interpreted his findings as evidence that thyroxine in some way is necessary for the release of growth hormone from the α cells. However, on this hypothesis and with this type of evidence, the enhanced growth produced by thyroxine alone (and repeatedly observed in thyroxine-treated female *P. latipinna* also) cannot be attributed to increased release of growth hormone, since the α cells in the guppy, as in *P. latipinna*, showed no response to thyroxine by itself. Most probably thyroxine synergizes peripherally with normal levels of growth hormone to cause growth enhancement, in the same way that exogenous bovine TSH and hake GH synergize in hypophysectomised *Fundulus heteroclitus*.

Treatment with cortisol causes regression of the α cells in the eel, while surgical stress activates them, although not the stress of anesthesia and bleeding. Surgical removal of the eel interrenal did not appear to activate the α cells any more than the hamp operation, although the adrenocortical inhibitor SU 4885 slightly increased the α granulation. At sexual maturity the female eel α cells are reduced in number, which may correspond to a reduced growth such as commonly accompanies sexual maturation. Sokol described a marked cycle in the α cells of the female guppy in relation to gestation, the cells being mainly inactive during the greater part of the month-long gestation period, but showing transient activity at the time of parturition, while in the unrelated viviparous *Zoarces* Oztan described marked activation of these cells during gestation and regression following parturition.

Eels with small skin lesions exhibited hyponatremia accompanied by strong activation of the α cells, while, surprisingly, extreme activation of these cells was found in eels maintained in deionised water for up to 46 days. Prolonged exposure to deionised water leads to marked ionoregulatory disturbances in the eel, apart from hyponatremia, including a fall in plasma calcium with a rise in plasma potassium levels that seems to result from movement of potassium from muscles to plasma. Olivereau suggested that the increased secretion of growth hormone in deionised water might represent a response designed to oppose this internal shift of potassium, since the rise in plasma potassium occurs more rapidly in hypophysectomised eels than in intact animals, but for the moment the real significance of this reaction of the eel α cells is uncertain, and its explanation awaits a full-scale investigation of the possible role of growth hormone in teleostean electrolyte metabolism.

The α cells were reduced in number and relatively inactive in pituitary transplants in *Poecilia f ormosa*, and in correlation these cells frequently regressed in the trout pituitary cultured *in vitro*.

Gonadotrops (β and γ Cells)

When adequately identified, the gonadotrops of teleosts have usually been found located in the proximal pars distalis, and most often in the ventral part of this region. However, gonadotrops also spread into the rostral pars distalis at sexual maturity in the eel, salmon, and trout. They are typical mucoid cells containing granules that react strongly in the PAS procedure and are assumed to contain glycoproteins. For most practical purposes, this property amounts to a redefinition of the old category "basophil cells," just as the absence of typical PAS +ve granules defines the serous cells which are equivalent to the old category of "acidophil cells." With this definition in mind it is often convenient to continue using the terms "acidophil" and "basophil," as long as one remembers that the terms have no reference to affinities for acidic or basic dyes; these older terms are so deeply entrenched in general usage that their replacement by "serous" and "mucoid" cells, although frequently advocated, is unlikely to materialize. In nearly all cases reported, gonadotrop granules in teleosts stain with PAS, AF, and AB, and with Aniline blue in the common trichrome and tetrachrome techniques.

The appearance of the gonadotrops varies greatly during the cyclic development of the gonads, and the cells are best characterised by an account of their annual or shorter cyclic changes. However, one very important point to be dealt with first concerns the question of whether teleosts possess only one type of gonadotrops or two. Higher vertebrates, with the established possession of two gonadotropic hormones (FSH and LH), have generally been found to have two kinds of gonadotrops in the pituitary, the β cell secreting FSH and the γ cell secreting LH. Since there is in fact no very definite physiological or biochemical evidence for the presence of two distinct gonadotropins in teleosts, even in those fishes which have been found to possess two distinct kinds of gonadotrops, one may not apply the terms "beta" and "gamma" to these cells.

The eel and the Pacific salmon are two of the teleost species in which gonadotrops can be separated into two types, with characteristics summarised in figure elsewhere in this chapter. It will be seen that in these cases the distinction depends not only on tinctorial features but also on position, cell size, and granule size and morphological features such as vacuolation. The two types in the eel appear to be equivalent in both males and females. Measurements showed that the eel Type 2 cells tend to be smaller than the Type 1 cells, but with larger nuclei. At the electron microscope level, two kinds of gonadotrops have been separated in the eel, with granules of about 130 and 190 mμ diam, probably corresponding to the light microscope categories of Olivereau. Ox-AB-PASOG distinguishes two gonadotrops in the goldfish, as in the eel and salmon. Two types are also discernible in

Table 3.1 : Gonadotrops of the Mature Pacific Salmon, *Oncorhynchus* sp.

Feature	Type 1	Type 2
Size of cell	Large	Small
Location	Edge of cords	Interior of cords
Granules	Fine	Large
AB-PAS	AB - ve, PAS + ve (magenta)	AB + ve, PAS + ve (violet)
Aliz B	Pale blue	Dark blue
Postspawning evolution	Vacuolate and totally degranulate, finally containing OG + ve granules	Degranulate partially, do not vacuolate, finally still retain some aniline blue + ve granules

Mugil auratus : In this fish, one kind of gonadotrop is situated medially in the gland and is coloured a clear blue-green after Aliz B, and also stains with PbH. After spawning this cell becomes vacuolated. The second gonadotrop is placed more laterally in the gland, stains purple with Aliz B, is negative to PbH, and rarely becomes vacuolated. Different techniques applied to another mullet, *M. cephalus* also separated two gonadotrops: with the aldehyde thionine-PAS-naphthol yellow method (AT-PAS), one type lying at the periphery of the cell cords of the proximal pars distalis stains brick red, while the other type lies in the center of the cord and stains dark violet. The peripheral cells are also AB + ve, the central cells PAS +ve after OxAB-PAS-OG. In studying the precipitation of pituitary secretory granules by trichloroacetic acid (TCA), Leray found that while the acidophil granules were precipitated by lower concentrations of TCA, the central gonadotrops were not precipitated until the TCA concentration was raised to 10%, and the granules of the peripheral gonadotrops were only precipitated at higher concentrations, up to 15%; thus, the two gonadotrops in *Mugil cephalus* may be distinguished by their situation, staining reactions, and by the precipitation of their granules by TCA. The AT-PAS method also separates two gonadotrops in *Mugil capito*. Electron microscope studies indicate two types in the viviparous *Zoarces*. One type has granules 60-160 m_{μ} diam, which tend to accumulate at one pole of the cell, and the endoplasmic reticulum consists of vacuolar cisternae; this cell becomes highly activated after parturition, in relation to ovarian follicular development. The second gonadotrop has granules 80-240 mμ diam, an endoplasmic reticulum with cisternae as dilated sacs or tubules, and is numerous and active during pregnancy.

At the ultrastructural level, Knowles and Vollrath have given further details about the gonadotrops of the eel. The type with larger (190 mμ) granules may be an LH cell, and it persists with little change in seawater eels in which gonadal development was assumed to have started. However, the other type of gonadotrops, possibly an FSH cell, with granules of 130 mμ diam, apparently changes considerably as the eel becomes sexually mature; and Knowles and Vollrath reported that they could identify typical FSH cells and, in addition, cells probably derived from these, with larger granules (up to 180 mμ), and with big empty vesicles and an active type of endoplasmic reticulum. In the trout, Follenius described a cell type, probably a gonadotrop, which contained numerous granules of various sizes and showed extreme development of the endoplasmic reticulum in the form of smooth-walled vesicles. At certain stages of the activity cycle very large vacuoles appear in these cells.

From what has been written so far, the main tinctorial properties of the gonadotrops will be obvious. It is particularly interesting that PbH is so useful in distinguishing two types of gonadotrops, particularly when used in combination with PAS. This technique has also been valuable in study of the pars intermedia and ACTH cells, and it also has been used to separate two gonadotrops in mammals. More definitely histochemical investigations have been performed. Olivereau and Fontaine summarised their histochemical data on these cells in the mature eel: The PAS reaction indicates the presence of neutral glycoproteins, while the staining of the granules with Ox-AB at low pH (0.2) indicates their content of acid mucopolysaccharides, which is rather feeble however. The acidic nature of the gonadotrop granules is also indicated by their reactions with Gabe's AF after permanganic oxidation, and by their slight affinity for orange G in PAS-OG. In addition to these specific glycoproteinaceous secretory granules, the eel gonadotrops when mature also contain very large granules with variable tinctorial affinities (OG + ve, erythrosin + ve, PAS + ve, BA + ve, AF + ve, Aniline blue + ve), which may be lysosomes. Similar putative lysosomes occur in mature gonadotrops in *Poecilia.*

Olivereau found that eel gonadotrops are rich in SS/SH groups, a property they share with gonadotrops in other teleosts. In both the goldfish and *Poecilia sphenops,* the gonadotrops, and probably also the TSH cells (PAS + ve and AB + ve cells), incorporated I'S more rapidly than other cell types, probably forming sulfate esters of acid mucopolysaccharides; but in the goldfish the gonadotrops only feebly incorporated labeled amino acids. According to Leray the gonadotrops in *Mugil cephalus* display only minor content of SS/SH groups, in contrast to other fishes. The same author reported that carp gonadotrops are rich in sialic acid and that the pituitary content of sialic acid increases with sexual maturation and falls after spawning. Sialic acid is a constituent of mammalian gonadotropins and TSH, and can be demonstrated histochemically in both the TSH and FSH cells of the cat.

Gonadotrops have been identified and described in many teleosts, usually in relation to the sexual cycle. It is usually accepted that any pars distalis basophils that are quiescent or absent before sexual maturity, and which show pronounced secretory changes in correlation with the gonadal cycle, are gonadotrops. A complete summary of the many accounts of these cells in these terms would be impossible, and the reader is referred to some of the more detailed papers and to the review f Stahl. In addition to these correlative studies, there are reports of characteristic changes (resembling those seen in mammals) in telcost gonadotrops following surgical gonadectomy or administration of gonadal steroids.

The Gonadotrops of Poecilia latipinna. Like many cyprinodonts, the molly is viviparous, with approximately monthly cycles of gestation, parturition, and oocyte growth, and the young fish quickly reach sexual maturity, in a matter of a few months in males reared in good conditions. These facts make it an interesting species for studies on the gonadotrops. These cells were initially identified by comparing the pituitary in immature and mature fish. In the immature state, the proximal pars distalis contains only two chromophilic cell types, α cells and TSH cells, the latter identified by their responses to thiourea. On the ventral surface of this region, one finds many small chromophobic cells, with inactive nuclei, presumably undifferentiated cells. In the sexually mature fish, differentiated basophils are always found in this region, even during periods of sexual quiescence, when these cells assume an inactive appearance but are nevertheless recognizable as differentiated basophils. Thus, these ventral basophils would appear to be gonadotrops.

Confirmation of this identification comes from observations of changes in the ventral basophils during the monthly female cycle. The fully developed oocytes are fertilised in the ovarian follicles by spermatozoa which may he stored embedded in the ovarian epithelium for long periods. The ovarian follicle acts as a brood chamber for the developing embryo. The fertilised egg is not ovulated but is retained in the follicle, and the embryo then develops on top of the yolky egg. Many embryos usually develop synchronously, forming a brood. When the fry are fully developed the follicles rupture, and the brood is expelled via the oviduct over a short period, a few hours at most. Under laboratory conditions, with 9 hr illumination per day and at 25°C, female fish undergo regular cycles of brood production, giving birth about every month, although the cycle is readily interrupted by unfavorable external conditions. During the greater part of pregnancy the ovary contains only previtellogenie. (first phase) oocytes, along with the developing embryos; but at birth, or a few days earlier, some of these juvenile oocytes embark on rapid vitellogenic (second phase) growth and mature as a new crop of eggs ready for fertilisation within 5-8 days of birth of the previous brood. Hypophysectomy experiments have shown that pituitary hormones are not essential for gestation or parturition, but are essential, as in oviparous forms, for the second phase oocyte growth. Since this phase is usually completed during the cycle within a few days starting at or just before parturition, one would predict that elevated pituitary gonadotropic activity is confined to this limited *period of the cycle* and is curtailed or absent during the much longer period occupied by pregnancy. Study of the gonadotrops tends to confirm this prediction, these ventral basophils displaying marked cytological and morphological changes correlated with the ovarian cycle. These changes are based on hitherto unpublished observations on scores of female *P. latipinna* together with systematic study and measurement on 30 fish sacrificed at known stages in the monthly cycle. The range of morphological alterations displayed by the gonadotrops in the cycle is impressive and is illustrated in Figs. 5 and 6. As in many teleosts maximal activity in *Poecilia* gonadotrops is characterised by partial or total loss of the coarse glycoprotein granulation and development of vacuoles, in the presence of cytological evidence of marked synthetic and excretory activity (large nucleus and nucleolus, prominent Golgi image, and abundant cytoplasmic RNA). In addition to glycoprotcin secretory granules, the active cells contain small numbers of larger amphiphilic granules or droplets, which have a varied staining affinity (positive to PAS, orange G, and erythrosin). They are negative to Alcian blue, however, so that in Ox-AB-PAS preparations these large magenta granules (R granules) contrast with the smaller dark blue secretory granules (B granules). The R granules are absent or very rare in quiescent cells, but they become numerous in the active state and are often to be seen lining the vacuoles that are characteristic of late vitellogenesis.

Orange G positive granules, no doubt equivalent to the R granules of *Poecilia,* have been described in gonadotrops of *Xiphophorus,* being particularly numerous together with vacuoles in the gonadotrops of gravid fish, presumably corresponding to the early pregnancy condition of *Poecilia*. The putative gonadotrops of *Oncorhynchus* contain large and small granules, the former being PAS + ve but AB - ve, while the small granules are positive to both PAS and AB; also in *Oncorhynchus,* the type 1 gonadotrops of Olivereau lose their glycoprotein granulation after spawning and develop vacuoles containing large OG + ve granules, a feature very like the appearance of R granules in *Poecilia* gonadotrops in late vitellogenesis. A similar phenomenon is seen in the gonadotrops of *Salmo salar*. The gonadotrops of *Tor (Barbus) for* contain, in addition to glycoprotein granules, carminophilic granules that are particularly abundant in the vacuolated gonadotrops at the peak of their secretory activity. The development of vacuoles [which Olivercau and Fontaine consider may indicate impending cellular exhaustion] together with these acidophilic or amphiphilic granules and partial or total loss of glycoprotein granulation seems typical of the teleostean gonadotrop when highly active. As Olivereau has suggested, the amphiphilic granules may be lysosomes; similar bodies occur in gonadotrops of amphibians, and lysosomes, with staining properties similar to those of these bodies in fish and amphibian gonadotrops, are known to be present in the basophils of the human pituitary. Presumably lysosomes would serve to rid the cell of excess secretory products, which may explain the marked development of R granules during early pregnancy in *Poecilia,* when oocyte growth (= gonadotropin secretion) is arrested, and also the similar granules in the postspawning gonadotrop of the salmon.

The question of whether *Poecilia* possesses two distinct types of gonadotrops is unresolved. The behaviour and staining properties so far described apply to all the cells in the gonadotropic zone, and

PbH-PAS has so far been uninformative applied to *Poecilia,* all the gonadotrops staining with PAS and none selectively with the PbH. Sometimes it is possible to distinguish two apparently different cells with Ox-AB-PAS-OG at times of maximal gonadotropic activity, one type containing more R granules than the other, but in the absence of functional evidence for a clear separation there are no reasons for thinking that these are other than two stages in the activity cycle of one cell type.

The guppy, *Poecilia reticulata,* exhibits an ovarian cycle essentially like that of *P. latipinna*, and the gonadotrops also display a similar cycle, maximal activity of the gonadotrops being confined to a period starting in late pregnancy and finishing a few days after birth, the period when a new crop of oocytes is undergoing rapid vitellogenic development.

Little information is available about reactions of the gonadotrops to collateral changes in endocrine state. The eel gonadotrops were activated by treatment with an impure TSH, but this is probably to be attributed to the known contaminant, LH, since the gonadotrops of the eel are greatly stimulated by injections of human chorionic gonadotropin and carp pituitary homogenates or purified carp gonadotropin. It is unexpected that these exogenous gonadotropins, at the same time as they stimulate the eel gonad, should also cause activation of the eel pituitary gonadotrops; indeed, one might have expected that the gonadotrops would regress because of pituitary-gonad negative feedback. Probably a hypothalamic mechanism is involved, but the nature of this is not known. The quiescent gonadotrops of the immature eel are not affected by treatment with thyroxine or by radiothyroidectomy, injections of SU 4885, surgical interrenalectomy, sojourn in deionised water or seawater. Older work suggested that stress produced alterations in the gonadotrops of *Astyanax* and the goldfish resembling the changes following injections of ACTH or corticosteroids; such treatments generally lead also to inhibition of oocyte growth, suggesting failure of gonadotropin secretion. Atz presented in detail data showing that both ACTH and corticosterone administration in *Astyanax* led to transient inhibition of the gonadotrops followed by restoration, both at the beginning and end of the course of treatment; similar but less striking changes followed injections of saline twice daily, but not similar injections given once daily. These observations are probably best interpreted as a stress response, probably mediated by the ACTH-interrenal axis (for evidence bearing on this interpretation).

The gonadotrops regressed in ectopic pituitary transplants in *Poecilia Formosa*, in correlation with evidence of failure of gonadotropin secretion.

Thyrotrops (δ Cells or TSH Cells)

The thyrotrops form the second division of the pars distalis basophils (mucoid cells) or the third division in those forms in which two gonadotrops have been demonstrated. In mammals, which display a corresponding trio of pars distalis basophils, a good deal of effort has gone into the tinctorial differentiation of the three types. Separation of the two gonadotrops is discussed by Herlant, Purves, and Carlon. Separation of the TSH cells from the two gonadotrops is generally possible with a variety of staining techniques. In teleosts, it is generally found that the TSH cells closely resemble the gonadotrops in their staining properties, and most attempts to separate teleostean gonadotrops and thyrotrops purely by staining reactions have been inconclusive. However, there are exceptions. Barrington and Matty showed that in the minnow, *Phoxinus,* the gonadotrops are AF — ve but the thyrotrops are AF + ve; Matty and Matty confirmed this distinction in *Rutilus,* and described in other species both AF + ve and AF — ve basophils, without functional identifications. Similarly, in *Caecobarbus geertsi,* the putative thyrotrops are AF + ve, but the putative gonadotrops are AF — ve. More subtle colour distinctions occur in other teleosts, for example, in *Astyanax mexicanus* where both thyrotrops and gonadotrops stain with AF, but the gonadotrops stain more intensely by PAS. In the eel, the thyrotrops are more strongly AF + ve and AB + ve than the immature gonadotrops, and take on a slightly different colour after PAS-OG. Tinctorial distinction between thyrotrops and gonadotrops is also possible in *Poecilia,* in which the TSH cells stain a paler blue with Ox-AB at pH 0.2, and the gonadotrops, when active, contain many R granules which are present in small numbers, if at all, in the TSH cells. With Aliz B, the gonadotrops stain a deep blue, but the thyrotrops take on a slatey gray-blue.

Fortunately, the visual distinction between TSH cells and gonadotrops depends not only on such tenuous tinctorial reactions, but also on marked topographical separation of the two cell-types. Reference has alrea ly been made to the fact that the TSH cells lie in t(c rostral pars distalis, as in the eel, or in the proximal pars distalis, as in *Poecilia.* Even in the latter type, when both TSH cells and gonadotrops occur in the proximal pars distalis, there is usually a spatial separation between them, the TSH cells lying dorsally (e.g., *Poecilia* and other cyprinodonts) or antero-dorsally (cichlids and *Mugil),* while the gonadotrops tend to lie in the ventral region, sometimes forming a virtual gonadotropic zone as in *Poecilia.* However, this special separation may not always be clear (e.g., *Astyanax).*

In the immature male eel, the TSH cells are small angular elements, lying in the rostral pars distalis between the follicles of $_j$ cells, and sometimes infiltrating into the follicle walls, with very fine secretory granules staining with Ox-AB, Aniline blue, thionine, PAS, and AF. In immature female eels, the TSH cells are similar, but here they tend to form large cell cords between the follicles in the middle of the rostral region.

In *Poecilia* the thyrotrops lie in the finger-like inner prolongations of the proximal pars distalis into the neurohypophysis, mixed with the more numerous a cells. They are generally rounded or angular cells, more or less square in section, with a large rounded nucleus and a rather coarse granulation, positive to PAS, OxAB, AF, and Aniline blue.

In all teleosts in which the thyrotrops have been identified, they display the typical staining reactions of basophils. Surprisingly little information seems to be available specifically about the histochemistry of the cells. In immature *Mugil,* the thyrotrops incorporate labeled cysteine, and they also incorporate $^{35}SO_4Na_2$ in *Poecilia sphenops,* although more weakly than the gonadotrops. Correspondingly the TSH cells of the guppy contain SS and SH groups , although Olivereau could find no evidence of these groups in the thyrotrops of the eel. Certain cells in the rostral pars distalis of the goldfish, identified by Olivereau as TSH cells, rapidly incorporated glucose-^{3}H, probably an indication of their synthesis of mucopolysaccharides and glycoproteins. Otherwise, one must infer the histochemistry of the TSH cells from consideration of their general staining reactions, as done by Olivereau and Fontaine for the eel gonadotrops.

The TSH cells of the juvenile river eel are revealed by the electron microscope as containing small vesicles or secretory granules (ca. 140 mμ diam), and a diffuse endoplasmic reticulum with wide cisternae; in silver eels caught on their migration to the sea, the endoplasmic reticulum was more prominent and the secretory granules more concentrated, suggesting greater activity. The presumptive TSH cells of *Zoarces viviparus,* in contrast, contained large granules (diam about 400 mμ), with foamlike cisternae in the endoplasmic reticulum; in some cases, smaller granules of varied size (120-160 m,μ) could be found in these cisternae. The trout TSH cells displayed a well-developed endoplasmic reticulum, with numerous smooth-walled cisternae. The secretory granules varied greatly in size, from 100 to 800 mμ.

Nearly every student of the teleost pituitary has described basophils in the pars distalis, usually in the proximal region, but in some species in the rostral region also; however, most workers have not distinguished TSH cells from gonadotrops either tinctorially or functionally.

Evidence for the Secretion of TSH by the 8 Cells

The most detailed investigation of pituitary-thyroid relationships in teleosts is that of Olivereau on the eel. Massive doses of radioiodine resulted in partial or total destruction of the thyroid in male silver eels. The animals were sacrificed at different times after the radioiodine treatment, up to 7 months, and even at 1 month after radiothyroidectomy, the 8 cells were degranulated and hypertrophied, and displayed mitotic activity. The cells enlarged, losing their angular outline and the cytoplasm became foamy and vacuolated, and the nucleus and nucleolus hypertrophied. Although the η and ε cells were slightly affected in this experiment, the reactions of the δ cells were so strong and characteristic that there can be little doubt of their thyrotropic function; this marked degranulation of these cells in the radiothyroidectomised male eel is paralleled by a fall of nearly 80% in the TSH content of the pituitary of radiothyroidectomised female eels. The complement of radiothyroidectomy was the treatment of immature male eels with thyroxine, which produced marked histological signs of involution in the thyroid gland, accompanied by inactivation changes in the δ cells, which shrank and lost their granulation, finally becoming small inactive chromophobic cells with small nuclei and nucleoli. As in the case of the radiothyroidectomy experiment, collateral effects on the η cells and ε cells were observed, but the pronounced changes in the δ cells marked them as the thyrotrops.

In the trout, *Salmo gairdneri,* iodine deficiency stimulated the thyroid gland and led to an increase in the number of δ cells and in the size of their nucleoli; subtotal radiothyroidcctomy resulted in strongly hypertrophied δ cells, with enlarged nuclei and nucleoli and loss of glycoprotein granulation, again identifying these elements as thyrotrops. Subtotal radiothyroidectomy of the goldfish produced results in agreement with those in eel and trout, the δ cells becoming strongly activated. Further work on the goldfish demonstrated that partial or total radiothyroidectomy led to reduced thyroidal I'll uptake and transformed the granulated δ cells into hypertrophied and hyperplastic chromophobes with large nuclei and nucleoli.

Using *in vitro* techniques, Baker has extended and confirmed the identity of the TSH cells in eel and trout. When isolated from the hypothalamus in culture, trout δ cells degranulate fairly rapidly and gradually assume an appearance of increased activity. The addition of thyroxine to the culture medium specifically prevents this rapid loss of δ granules and leads more slowly to regression of the cells, which become smaller and exhibit nuclear and nucleolar atrophy. The eel delta cell *in vitro* degranulates more slowly and maintains an appearance of activity; addition of thyroxine to the medium from the start of culture results in nuclear atrophy. If trout glands are transferred from a thyroxine medium (granulated δ cells) to a control medium, the δ cells rapidly degranulate, showing that inhibition of their discharge by thyroxine is reversible; however, if the reverse transfer is made after the δ cells have been allowed to degranulate on the control medium, thyroxine is unable to regranulate the cells. In both eel and trout, δ cells maintained on thyroxine medium, with histological signs of regression, display impaired incorporation of uridine-^{3}H into the nucleus compared with cells on control medium. Exposure to thyroxine had no effects on any of the other pituitary cell types, confirming that the δ cells in both species are thyrotrops, and that thyroxine-TSH feedback appears to operate mainly at the level of the pituitary cell.

The δ cells of *Poecilia are* generally rather few in number and lie associated with the α cells. Their thyrotropic nature is indicated by their early differentiation, since they are the only pars distalis basophils in newly born fish, the gonadotrops only becoming differentiated much later. Taking immature fish (6 months old or less), with undifferentiated gonadotrops, Ball has studied the response of the δ cells to treatment with the antithyroid agent thiourea, administered in solution in the tank water. Results are summarised in table elsewhere in this chapter and clearly demonstrate the strong response of the δ cell to thioureainduced blockage of thyroid hormone synthesis. None of the other cell types responded to the treatment. Unexpectedly, treatment of adult fish with thyroxine produced no obvious regression of the δ cells, although reducing thyroidal 24-hr ^{131}I uptake to 0.5%. This lack of effect of thyroxine on TSH cell morphology in *Poecilia latipinna* agrees with the failure of thyroxine treatment to affect the TSH cells of the guppy, but contrasts with the distinct effects in the eel and *Mugil*, and with the rather slight changes in the thyroid and thyrotrops of *Astyanax*. Clearly the time course of the TSH cell response to different doses of thyroxine should be explored in *Poecilia*.

Leray and Blanc showed that treatment of young *Mugil auratus* with the thyroid inhibitor propylthiouracil (PTU) reduced levels of plasma protein-bound iodine (PBI)-that is, impaired the production of thyroid hormone-and produced signs of histological stimulation in the thyroid itself (i.e., raised levels of TSH). Correspondingly, the δ cells of the pars distalis became strongly activated, marking them as thyrotrops since no other pituitary cells were affected. Olivereau confirmed these observations in this same species, thiourea activating and thyroxine inhibiting the δ cells.

Other evidence bearing on the function of the δ cells comes from their marked activation in response to thiourea or thiouracil treatment in *Phoxinus*, *Astyanax*, and the guppy. Thyroxine administered with thiourea to the guppy reversed the activation of the δ cells induced by thiourea alone. Confirmation comes from Schreibman's work *on Xiphophorus,* in which the δ cells, centrally located in the proximal pars distalis and differentiated very early in life, were strongly hypertrophied and hypcrplastic in fish with thyroid tumors. The cavefish, *Caecoharhus,* has a quiescent thyroid, and also displays very inactive looking δ cells.

Reference has already been made to the early differentiation of the TSH cells in *P. latipinna* and *Xiphophorus,* long before the gonadotrops are differentiated, and these cells also differentiate early in the guppy and the cichlid *Herichthys*. The TSH cells sometimes exhibit functional changes during the life history. They are present only in small numbers in trout and the freshwater (parr) stages of the salmon, *Salmo salar,* in correlation with the very low content of TSH in these glands; however, in the salmon smolt, with a highly active thyroid, and about to migrate to the sea, TSH cells are numerous and highly active. In *Xiphophorus maculatus,* Schreibman described senility changes in aging females, in which the TSH cells became smaller, with smaller nuclei, and exhibit degranulation and vacuolisation; in the related guppy, thyroid activity decreases with aging, which supports interpretation of the senescence changes in the TSH cells *of Xiphophorus* as indicating reduced secretion of TSH. In young *Xiphophorus* the TSH cells look much more active. Silver eels exhibit vacuolisation and ultrastructural features of their TSH cells which perhaps indicate a greater activity than in the yellow stage.

Some collateral endocrine changes have been found to affect the TSH cells. They were slightly activated after castration in *Xiphophorus*, and another slight suggestion of gonadal influences on the TSII cells perhaps comes from Olivereau's finding that the cells are slightly less active in immature female eels than in immature males (although we do not know that the eel gonads at this stage secrete sex hormones). In the case of eels brought to sexual maturity by artificial treatment, the reactions of the TSH cells are difficult to interpret. When maturation was induced by injections of TSH contaminated with LH, the regression of the eel TSH cells is probably a response to thyroidal activation induced by the exogenous TSH; however, in male eels brought to maturity by human chorionic gonadotrops, the TSH cells were activated in parallel to activation of the thyroid, possibly the result of some unknown interplay between the exogenous gonadotropin and the hypothalamus. The inactivation of these cells in female eels matured by injecting carp pituitary extracts is in part at least attributable to carp TSH in the extracts. It is perhaps worth recalling that the question of whether or not sexual maturation in teleosts is *necessarily* accompanied by thyroidal hyperactivity is still open, with conflicting reports in the literature.

Eel TSH cells were not modified 8 or 9 days after surgical interrenalectomy but were quite strongly activated after 5 days treatment with the adrenocortical inhibitor SU 4885, and SU 4885 also activates the TSH cells after 5 days' treatment in *Poecilia,* although not at 2 days.

An unexpected reaction of the TSH cells of the eel occurs in response to injections of ovine prolactin. Olivereau found that chronic injections of prolactin resulted in marked histological activation of the thyroid in intact eels but not of hypophysectomised animals. After only 2 or 3 daily injections, the TSH cells in the intact animals were slightly hypertrophied, with occasional mitoses, and with further injections the TSH cells in the intact animals became extremely activated and displayed many mitoses. As Olivereau pointed out, the effects of the exogenous prolactin could result from its acting as a goitrogen, preventing the formation of thyroid hormone and, hence, by removing feedback inhibition, resulting in elevated TSH output; or, the ovine prolactin could somehow be causing elevated TSH output even in the presence of a normally functional thyroid, as if it were acting like a TSH-releasing factor. A similar action of ovine prolactin on the TSH-thyroid axis in the amphibian *Rana catesbiana* appears - to represent a goitrogenic effect, although this is not apparently true for thyroidal stimulation by prolactin in a urodele. Subsequent work on the eel showed that prolactin does not act like a goitrogen in this case, since in addition to the histological signs of thyroidal activation, prolactin also increased ^{131}I uptake by the gland; thus, the activation of the eel TSH cells by ovine prolactin presumably depends on some hypothalamic or intrapituitary mechanism not yet understood.

Atz observed striking changes in the thyrotrops of *Astyanax* following injections of cortisone or ACTH daily for 10 days. The TSH cells underwent a cycle of degranulation and vacuolation followed by regranulation, both at the beginning and the end of the injection period, and also displayed mitotic activity. The fact that similar effects were produced by daily saline injections suggests that Atz was observing a response to stress normally mediated by the ACTH-adrenal axis.

The TSH cells are better granulated in eels in seawater than in freshwater animals, probably indicating a lower rate of secretion in seawater related to the greater availability of iodine in this medium. However, Knowles and Vollrath reported that TSH cells in some eels caught at sea were more numerous than in river eels although in some seawater eels they could find no typical TSH cells. Silver eels during the downstream migration displayed very active TSH cells, judging by their extremely prominent endoplasmic reticulum. Thus, even in the one species, the relationship between TSH cell activity and external salinity is unclear. The eel TSH cells possibly tend to be less active in deionised water than in freshwater, although this was not a very marked reaction.

Reference was made earlier to the fact that in both eel and trout pituitaries, the TSH cells retain their activity *in vitro,* the trout cells actually degranulating and becoming apparently more active than *in vivo.* Comparable data comes from ectopic pituitary homotransplants in *Poecilia Formosa* which secreted TSH at normal levels, or in some individuals perhaps higher than normal levels, associated with the retention of very active TSH cells in the transplants, often degranulated but with very active nuclei. Comparable retention of TSH function and of active TSH cells has been observed in ectopic pituitary autotransplants in *P. latipinna.* The implications of these findings for the analysis of hypothalamic control of TSH function in teleosts are obvious.

Pars Intermedia

The pars intermedia partially or completely surrounds the distal and largest region of the neurohypophysial core and is usually extensively invaded by neurohypophysial tracts. Its relative size varies in different species. It may be the smallest region of the adenohypophysis, as in *Poecilia* and other cyprinodonts, or, together with the enclosed neurohypophysis, it may form as much as two-thirds the volume of the whole pituitary, as in salmonids. Cells from other regions of the adenohypophysis may invade the pars intermedia in small numbers, and pars intermedia cells are sometimes found in the pars distalis.

Two cell types may usually be distinguished by their staining reactions and often also by differences in shape and position. Staining characteristics vary with species; pars intermedia cells have been described as basophilic, staining for example with Aniline blue, or acidophilic (with azocarmine or orange G), or amphiphilic, taking -olh the blue and orange components in trichrome and tetrachrome techniques. Some species have been described as having only one chromophilic cell type in this region, together with chromophobes, or all the cells in the pars intermedia may appear chromophobic. In such cases, it is impossible to be certain that two cell types exist, although probably, as in the case of *Carassius*, more detailed investigations would reveal this to be so. Within any one species the staining properties may vary, depending on the physiological conditions of the cells and also on the techniques applied. Thus, in *Perca fluviatilis,* the PAS + ve cells bordering the neurohypophysis have been described as basophilic or acidophilic. A more precise differentiation between two pars intermedia cell types is achieved by staining with PAS followed by lead hematoxylin (PbH). The resulting differentiation was first described in Mugilids by Stahl: One cell type, oval in shape and bordering the neurohypophysis, was PAS + ve, and a second type, lying further from the neurohypophysis but connected to it by cellular prolongations, was PAS - ve but PbH + ve. Several teleostean pituitaries have since been examined by this technique, and generally the two types of cells have been differentiated. The PbH + ve cells frequently have a "club" shape as in *Mugil,* and they are often chromophobic by classic techniques.

Two cell types are not invariably present in the pars intermedia; for example, the PAS + ve cell appears to be missing from salmonids, in which,all the cells look alike and stain with PbH. However, most species studied display a PAS + ve cell type, often together with a chromophobe that no doubt represents the PbH + ve cell. Although PAS and PbH appear to be mutually exclusive in staining the two cell types, there seems to be no correlation between the reactions of the cells to trichrome stains on the one hand and to PAS on the other;

thus, the PAS + ve cells may be basophilic as in *Hippocampus, Phoxinus,* and *Anguilla,* or acidophilic as in *Ciehlasoma, Blennius,* and *Zoarces,* or amphiphilic as in *Poecilia latipinna.* It is our impression that the reaction of these cells in mixed staining techniques is more than usually sensitive to slight variations in timing, strengths of dye solutions, etc. Similarly the PbH + ve cells have been variously described as chromophobic (most often), acidophilic, or weakly basophilic.

Instrusive islets of the amphiphilic PAS + ve cells occur in the pars distalis of *Poecilia latipinna* and *P. formosa* particularly in the rostral region. The number of such islets increases with age. Similar islets of PAS + ve cells, presumably pars intermedia cells, are found in the rostral pars distalis of other cyprinodonts. The PbH + ve cells have not been found outside the pars intermedia in *Poecilia.*

Pars intermedia histochemistry has been neglected. The PAS + ve cells of *Poecilia sphenops* incorporate $35SO_4Na_2$ rapidly, and in *Mugil* their granules are precipitated by a much lower concentration of trichloroacetic acid (0.5%) than the granules of any other cell type in the gland.

AcidophiPc droplets, apparently colloidal, often occur between the pars intermedia cells. Similar material in the pars intermedia of other vertebrates has often been interpreted as an accumulation of hormonal material or as the products of cellular degeneration.

Few ultrastructural studies have been made on this region. The presence of two distinct cell types is confirmed in most studies, and in *Anguilla* a third cell type has been reported, but it is not clear what this corresponds to in light microscope studies. With the electron microscope, the two main cell types are most readily distinguished by differences in size of their osmiophilic secretory granules, although mitochondrial size and details of the endoplasmic reticulum are other criteria. Granule size and shape differ in different species and do not correlate with granule staining reactions. Thus, in *Zoarces* the PAS + ve cells have smaller granules (160-200 mμ) than the PbH cells (up to 400 mμ) , and the same is true of the eel (180 mμ vs. 320 mμ). In contrast, in *Hippocampus* the PAS + ve cells have the larger granules (250-300 mμ vs. 200 mμ). However, as in pars distalis cells, the granule size may vary under different conditions even within one cell type; for example, in the elver of *Anguilla* the average granule size is smaller in both cell types than in the older freshwater eel. The significance of these differences is unknown.

Follenius, on the basis of his ultrastructural observations on the guppy, *Perca* and *Salmo,* considers that the two cell types of the pars intermedia arc not distinct, but probably represent different stages in the functional cycle of a single cell type, an interpretation at odds with the conclusions of most workers at the light microscope level; the independence of the two cell types is indicated by their different reactions in the eel kept in deionised water.

Evidence for the Functions of the Pars Intermedia Cells. To understand the difficulties attending any allocation of function to the pars intermedia cells, it is necessary to appreciate that there is at present no general agreement about the total physiological role of this part of the gland. By homology with the tetrapod pars intermedia, it is to be expected that this region is the site of synthesis of melanophore-stimulating hormone (MSH, intermedin), and there is now good experimental evidence to support this idea. Extracts of teleostean pituitaries affect the distribution of pigment in both the melanophores and erythrophores of fishes, and experiments with separate extracts of anterior and posterior halves of the gland show MSH activity to be most abundant in the pars intermedia region. Long-term *in vitro* culture of anterior and posterior halves of the trout pituitary showed that 11-12 times more MSH is produced by the pars intermedia than by the rest of the gland; the small amount of MSH secreted *in vitro* by the anterior region may be attributed to the inclusion of some pars intermedia cells in these fractions.

The existence of two distinct cell types suggests the production of two distinct hormones by this region. On other grounds, it has been pro*posed that two antagonistic melanophorotropic hormones are secreted* by the fish pituitary, one (MSH) causing melanophore dispersion, the other (melanophore-concentrating hormone, MCH) inducing melanin concentration in the middle of the pigment cell. Stahl suggested that the two pars intermedia cell types might produce these two factors. This two-hormone hypothesis is not universally accepted, and some workers consider the experimental data, which originally led to the two-hormone concept in amphibians and fishes, are explicable in terms of only one hormone.

This lack of certainty about whether the pars intermedia produces one or two melanotropic factors is reflected by uncertainty about the functions of the two cell types of this region: At present, if only one hormone is secreted by the pars intermedia we cannot say which cell produces it; and if the other cell type does indeed secrete a separate hormone, we cannot he sure of its functions, except to say that there are some indications that such a second pars intermedia factor might have function(s) quite different from pigmentation control. The conditions under which the pars intermedia cells show cytological evidence of functional alterations are, therefore, of quite unusual interest.

Attempts have been made to locate the cellular origin of MSH by observing cytological changes in the pars intermedia in response to adaptation to a black or white background. In *Cichlasoma* and *Blennius* adapted to a black background for several weeks, the PAS + ve cells became larger and better granulated, with larger nuclei and nucleoli than in fish maintained on a white background, in which these cells regressed. While these observations point to the secretion of MSH by the PAS + ve cells, other work produced conflicting data. Thus, Knowles and Vollrath kept eels for several weeks in complete darkness, and then transferred them to an illuminated black or white background for 4 hr before sacrifice. There were many PAS + ve cells in the pars intermedia of animals transferred to the black background, and in these eels the melanophore's were in a dispersed state, indicating high levels of MSH secretion. In contrast, the eels transferred to a white background displayed concentration of melanophores (i.e., low levels of MSH secreted), and in the pars intermedia few PAS + ve cells could be found. Since in such a short-term experiment the scarcity of PAS + ve cells in the white-background fish must have been owing to discharge of secretory granules, we would interpret these observations as indicating that MSH is *not* secreted by these cells. Unfortunately, in this work there is no detailed information about the state of the pars intermedia in the eels in total darkness, but it seems that they displayed PAS + ve cells in the pars intermedia, which lends support to our interpretation. In other teleosts *(Ameiurus, Carassius,* and *Spinachia),* adaptation to black or white backgrounds did not lead to cytological alterations in the pars intermedia. Chavin, however, reported an increase in MSH content and hypertrophy of the amphiphils in the pars intermedia of black-adapted *Carassius,* which would indicate that these cells, PAS - ve and PbH + ve, secrete MSH. However, this interpretation is not entirely secure since Chavin stated that the gonadal state of his fish was variable, and there does exist some evidence of a correlation between the number and size of these PbH + ve cells in *Carassius* and the degree of gonadal maturation. Nevertheless, the different MSH contents of the pituitary of blackadapted and white-adapted fish is interesting and highly suggestive, and similar experiments should be repeated to avoid the complication of gonadal variations.

Melanin dispersion in fishes may occur independently of background adaptation. It should perhaps be emphasised that teleostean melanophores are under dual control, neural and humoral, and in some species (e.g., *Anguilla)* humoral control predominates, so that even innervated melanophores will respond to MSH, while in other species (e.g., *Fundulus* and *Poecilia)* the neural component predominates and the melanophores must be denervated to demonstrate experimentally their response to MSH. In both *Anguilla* and *Poecilia,* the adrenocortical inhibitor SU 4885 causes melanin dispersion, and in *Anguilla* (but not in *Poecilia)* this is accompanied by degranulation and hypertrophy of the PbH + ve cells, the PAS + ve cells remaining unchanged. However, in both species the ACTH cells were activated by the drug, and since mammalian ACTH possesses intrinsic MSH-like properties, the melanophore response in *Anguilla* could be an effect of endogenous ACTH. Furthermore, the darkening in *Poecilia* in response to SU 4885 involved innervated melanophores, which in this species are not directly responsive to exogenous MSH or ACTH, and it is probable that some narcotic property of the drug was responsible for this effect in *Poecilia,* if not, indeed, in *Anguilla.*

In *Salmo salar,* the cells of the pars intermedia were inactive in the parr during the winter months, became more active in the spring, and were extremely active during the smolt stage (i.e., the stage of migration to the sea), undergoing intense mitotic activity so that the whole region enlarged. Olivereau correlated this with the rapid deposition of melanin pigment in the smolt pectoral fins. In the adult salmon migrating upstream from sea to spawning grounds, the pars intermedia appeared to be degenerate, but was restored in spawning and postspawning fish, although not as active as in the smolt. The blind cavefish, *Caecobarbus geertsi, is* characterised by a total lack of melanin pigmentation, and Olivereau and Herlant found in correlation that the cells of the pars intermedia were inactive and "embryonic" in character,

Increased activity of the pars intermedia and increased melanin dispersion occur in *Anguilla* during a prolonged period in deionised water, the whole pars intermedia becoming hypertrophied. The PAS + ve cells undergo a rapid and intense degranulation and hypertrophy, while the PbH + ve cells only degranulate slowly. Since melanin dispersion also develops slowly, Olivereau tentatively suggested that the PbH cells might be the source of MSH. This conclusion would disagree with the results already cited from background adaptation of *Cichlasoma* and *Blennius.* Some role of the pars intermedia in electrolyte regulation may be inferred from these results of Olivereau, and it is worth recalling that osmotic stress is often associated with changes in the pars intermedia of tetrapods. More recently, Olivereau has reported that gradual transfer of *Mugil auratus* from seawater to freshwater results in activation of the PAS + ve cell in the pars intermedia, the PbH + ve cell remaining in its normal state of high activity; there are no data about pigmentation in these conditions, but the results perhaps support those on the eel in suggesting some nonmelanophorotropic function for the PAS + ve cells, again in opposition to the results from the background adaptation experiments of Baker and Chavin. Another confusing observation is that the PAS + ve cells in the pars intermedia of the elver (young eel) are activated during the upstream migration from the sea; since at the same time as they move into freshwater these animals also become more heavily pigmented, it is impossible to know whether this indicates an osmoregulatory or a pigmentary function for the PAS + ve cells.

A variety of observations seems to relate the pars intermedia to reproductive changes. For example, in *Zoarces viviparus* the PAS + ve cells increase in number and activity during oogenesis and during the breeding season, and in *Carassius* the PbH + ve cells become more numerous during gonadal maturation while the PAS + ve cells increase after the reproductive period. In the herring, *Clupea,* intense secretory activity has been seen in the pars intermedia during the spawning period. It is not yet possible to offer functional interpretation of such observations, but the variety of conditions affecting the activity of the pars intermedia cells underlines the possibility that the hormone (s) produced by this region might be concerned in functions other than control of pigmentation.

Both trout and eel pituitaries will synthesize and release MSH *in vitro* after separation from the hypothalamus. In the case of the eel, a cytological study of the cultured glands has failed to determine the source of MSH, since both the PAS + ve and the PbH + ve cells underwent gradual degranulation but retained an active appearance. In the trout gland *in vitro,* the rapid decline in the stored MSH content of the pituitary is associated with degranulation of the single PbH + ve cell type, in the presence of active-looking nuclei and nucleoli. In contrast, both cell types in the pars intermedia of ectopic pituitary transplants in *Poecilia formosa* presented a histological picture of involution.

The Neurohypophysis in Teleosts

The neurohypophysis is composed of axonal nerve fibers, mostly nonmyelinated, originating from neuronal cell bodies in the hypothalamus. These nerve fibers extend down the pituitary stalk (neurohypophysial tract) into the pituitary gland to form the main bulk of the neurohypophysis (neurohypophysial core) which forms the center of the whole gland. Many fibers terminate in the neurohypophysial core in close relationship to blood vessels, and many other fibers extend beyond the core and invade all regions of the adenohypophysis, an arrangement peculiar to teleosts. Scattered ependymal cells and glial cells, collectively termed "pituicytes," occur throughout the neurohypophysis but are especially abundant in the posterior part of the core close to the pars intermedia. Tubelike extensions of the third ventricle bordered by pituicytes may penetrate into the posterior neurohypophysial core, forming a hollow center in the fingerlike projections of the neurohypophysis into the pars intermedia.

A special feature of the fish pituitary is the extensive penetration of the pars intermedia by the neurohypophysis, the two regions mixing in complex interdigitation. Teleosts have specialised further in having the anterior part of the neurohypophysis penetrate the pars distalis with fingerlike processes, although in a less intricate manner than in the pars intermedia. This ramification of the neurohypophysis into the adeno hypophysis is not present in salmonids at hatching, and thereafter it develops slowly and is only very slight even at 4 months in the trout. Similarly in the elver, there is only slight indication of the extensive penetration of the pars intermedia by neurohypophysial processes, so characteristic of the adult eel.

Two types of neurohypophysial fibers may be distinguished with the light microscope, depending on whether or not the material contained in the fiber will stain with the classic neurosecretory stains, AF, chrome alum hematoxylin (CAH), aldehyde thionin (ATh), and AB. The two fiber types appear to correspond to the Type A ("stainable") and Type B ("nonstainable") fibers distinguished at the ultrastructural level by Knowles. Neurosecretory material in both "stainable" and "nonstainable" fibers may accumulate in the neurohypophysis in large masses termed "Herring bodies."

The neurohypophysis and the neurosecretory cells of the hypothalamus (mainly but perhaps not all, collected into groups termed "hypothalamic nuclei") thus form a functional unit concerned in the synthesis transport and release of neurosecretory materials. They also display, anatomically at least, the fundamental feature of neurosecretory elements, in that they form "a final common pathway linking the nervous and endocrine systems" (to which the reader is referred for an authoritative discussion of neurosecretion in general).

Various octapeptides with characteristic biological properties have been isolated from the pituitary in all vertebrate groups and in many cases have been shown to be associated with the hypothalamus and neurohypophysis. It is these octapeptides which are believed to be synthesised in the hypothalamic nuclei and passed down to the neurohypophysial core, probably contained in or associated with ultrastructural neurosecretory granules, which are generally considered equivalent to the stainable material seen with the light microscope. In teleosts, two octapeptides have been found in the pituitary, arginine vasotocin (AVT) and isotocin (IT), and these are thought to be the neurohypophysial principles of these fisl es.

"Stainable Fibers, Probably Knowles' type A

The stainable fibers probably mostly originate from the nucleus preopticus (NPO) of the hypothalamus, and they contain neurosecretory material that stains with AF, CAH, ATh, and AB. The majority of these fibers probably terminate in the distal part of the neurohypophysial core surrounded by the pars intermedia, ending in association with pituicytes or blood vessels or with the pars intermedia cells. In addition, fibers with these staining reactions have been observed closely associated with cells of the pars distalis. In the eel, Olivereau distinguished two kinds of neurosecretory fibers with the light microscope; one type containing classic neurosecretory material and mainly passing toward and penetrating the pars intermedia, the other type with material PAS + ve but AF - ve, which penetrates the pars distalis. She tentatively equated these with Type A and Type B fibers originating from the NPO and NLT. Similar distribution of the two kinds of fibers can be seen in *Poecilia*.

After hypophysectomy, classic neurosecretory material accumulates along the course of the severed nerve axons in the infundibular area. Transection of the NPO-pituitary tracts in *Lepidogobius* also led to accumulation of AF + ve material proximal to the cut, and eventually to loss of this material from the neurohypophysis. Thus the neurosecretory material is certainly formed in the hypothalamus and transported to the pituitary along the axons, as in other vertebrates. Exposure of hypophysectomised *Porichthys* to continuous light induced movement of AF + ve material from the NPO to the cut ends of the axons in the infundibulum. The ends of the severed axons in hypophysectomised goldfish and *Porichthys* regenerate so that the severed neurohypophysial stalk eventually forms a kind of isolated neurohypophysial core with axons terminating on blood vessels as in the normal condition. In hypophysectomised *Clevelandia,* the neurosecretory material that accumulated proximal to the cut ends of the axons eventually disappeared, probably into blood vessels that associated with the fibers.

The correspondence of this light microscope category with the Type A fibers defined by ultrastructure is almost certain, though perhaps not definitively established. Knowles defined Type A fibers as containing osmiophilic granules more than 100 mμ in diameter, in contrast to Type B fibers with smaller granules; and in *Perca* and *Salmo* Type A fibers contain granules the same size as those in the cell bodies of the NPO. Several kinds of Type A fibers have been described differing in the size of their granules, their destination within the pituitary, and in their response to experimentation.

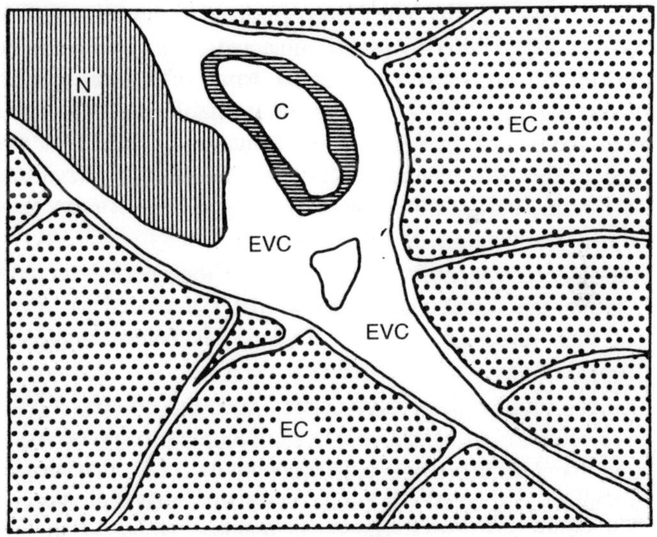

Fig. 3.2 : Diagram to illustrate the relationship between a neurosecretory fiber terminal (N), intrinsic endocrine cells (EC), and a blood capillary (C) in the proximal pars distalis of the eel; extravascular space (EVC).

Within the pituitary of *Anguilla* and *Conger,* the Type A fibers mostly terminate in or close to the pars intermedia, or on extravascular channels between the endocrine cells, or make direct contact with the pituicytes, or occasionally with the pars intermedia cells. A Type A3 fiber has been distinguished in the eel pars distalis and appears to be distinct from the A1 and A2 fibers found in the neurohypophysial core close to the pars intermedia. The Type A3 fibers in the pars distalis of the eel, like the more numerous Type B fibers in this region, terminate on extravascular channels, very close to the intrinsic endocrine cells.

Light microscope accounts of changes in the neurosecretory material in the neurohypophysis nearly all relate to the (presumptive) Type A fibers in the distal neurohypophysial core, close to the pars intermedia. The amount of stainable material in this region commonly alters in response to osmotic stimuli. In many teleosts, transfer to a hypertonic medium results in loss of this material, both from the neurohypophysis and also from the NPO.

Eels adapted to seawater have less stainable material in the neurohypophysial core than freshwater eels; on the other hand, the total amount of stainable material in the hypothalamo-neurohypophysial tracts did not appear to alter in response to changing salinity in the eel or the goby, *Clevelandia ios*, nor did the material in the neurohypophysial core change appreciably in *Mugil* subjected to progressive reduction in salinity. *Gasterosteus* transferred from one-third seawater to freshwater exhibited a rapid loss of stainable material, while in hypotonic media, or in response to dehydration, stainable material increased in the pars magnocellularis of the NPO of *Misgurnus,* but decreased in the parvocellular elements, suggesting different functions for the two components of the NPO.

In the neurohypophysis of eels kept for long periods in deionised water, Olivereau found that stainable material tended to increase at first but then decrease, the neurohypophysis being compressed by hypertrophy of the pars intermedia.

Changes such as these have usually been interpreted as indicating release or retention of neurohypophysial factors in response to osmotic demands, there being some experimental evidence that neurohypophysial factors, especially arginine vasotocin, may be concerned in ionic regulation and water balance in teleosts. However, in some of the experiments in which fish were transferred between different salinities, the osmotic conditions imposed must have been very severe, and it is possible that the neurohypophysial changes could have been nonspecific responses to stress. Olivereau and Olivereau recently reported that surgical stress led to rarefaction of stainable material in the eel neurohypophysis, and stress has also been reported to reduce the material in the rat neurohypophysis. Interpretation is all the more difficult, since ample evidence shows that many stress conditions in mammals, apart from osmotic changes, lead to discharge of the neurohypophysial hormones. Furthermore, it must be remembered that the amount of stainable material in the neurohypophysis results from a balance between synthesis and release. In order properly to interpret alterations in the amount of visible material, one requires information about the cytological features of the neuron cell body that would indicate low or high synthetic activity, such as nuclear and nucleolar size and Golgi status. Failing this kind of cytological information, it is desirable to have studies on the biological activity present in the neurohypophysis parallel to the changes in the stainable material. Such studies are almost nonexistent in fishes, but one is outstanding. Lederis showed that a high vasopressor activity is associated with those fractions of *Gadus* pituitary that contain particles resembling neurosecretory granules. In *Salmo gairdneri,* Lederis further found that on transfer to seawater there was a transitory decrease in the number of granules in some of the neurohypophysial fibers, which was correlated with a 50% reduction in the pituitary content of AVT. This suggests perhaps that this particular neurohypophysial principle is associated with the granules, although as Lederis emphasised the parallelism between AVT content and disappearance of these granules was very imperfect, only relatively few fibers displaying loss of granules, and this author emphasised that lack of stainability or of electron density are not necessarily related to absence of neurohypophysial principles.

Apart from alterations in neurosecretory material in relation to osmotic changes, the number of synaptic contacts between Type Al fibers and pituicytes (in neurohypophysial projections into the pars intermedia) was observed to increase when *Anguilla* was transferred to seawater.

There are other observations which relate fluctuations in the amount of stainable material to the reproductive cycle. Sawyer and Pickford reported a depletion of pituitary content of isotocin, but not AVT, during the reproductive season of female *Fundulus heteroclitus,* although not in the male. Neurohypophysial factors play a role in spawning behaviour in this fish, and in studying the pituitary Sokol found a decrease in the stainable material associated with the spawning period in both sexes. Similar observations were made in both sexes of *Tor (Barbus)* for. A reduction in the amount of stainable material in the neurohypophysis occurs during oviposition in *Oryzias,* which may be correlated with the demonstration that injections of neurohypophysial preparations will stimulate oviposition in this fish. Olivereau found a rarefaction of stainable material in the neurohypophysis of female eels brought to sexual maturity by treatment with a partially purified carp gonadotropin extract, and a similar change occurred in male eels brought to maturity by injections of human chorionic gonadotropin. The significance of this change is obscure, since the mechanism whereby exogenous gonadotropins activate the eel gonadotrops is unknown. Oztan observed depletion of AF + ve material and of the ultrastructural granules from the NPO cells of *Zoarces viviparus* during the process of gonad maturation and the accumulation of these materials in the cells during early pregnancy.

Several workers have been interested in the functional relationship between the neurohypophysis and the pars intermedia, these two regions being so intimately associated in fishes. In the eel, Knowles and Vollrath observed that terminations of neurosecretory fibers are associated with extravascular channels bordering the pars intermedia, these channels probably draining into blood capillaries. The pars intermedia cells appear to release their products into these extravascular channels, and the authors suggested that the Type A fibers might also discharge into the channels and influence the activity of the pars intermedia cells by this route. The same Type A fibers were also observed to make synaptic contact with neighboring pituicytes, and these synapses increased in eels briefly exposed to a white background. The pituicytes surround the tuhelike extensions of the third ventricle, and in these white-exposed eels there was an increase in the PAS + ve material in these tubes. Because of these findings, it has been suggested that this PAS + ve material may be secreted by the pituicytes to act as a feedback link between the neurohypophysis and the NPO by way of the third ventricle. This is an interesting idea, but it requires more evidence before it can be considered as more than a working hypothesis; some support comes from observations in higher vertebrates, in which an association between neurons and ependymal cells, and secretion by the ependymal cells into the third ventricle, have been observed. At any rate, it is now becoming clear that pituicytes (ependymal and glial elements) may play a more important role in neurosecretory control than was hitherto supposed.

Salmo and *Perca* exhibit an anatomical relationship between Type A fibers and pars intermedia cells similar to that in the eel, but other teleosts *(Poecilia reticulata, Xiphophorus, Phoxinus,* and *Tilapia)* show nerve fibers ending directly on the pars intermedia cells, with no basement membrane or extravascular channel in between.

Apart from the eel, changes in background colour or in illumination have been found to alter the neurohypophysial structure in other teleosts. In *Zoarces,* continuous illumination in April resulted in depletion of AF + ve material and of ultrastructural granules from the cells of the NPO, together with an increase in nuclear size, compared with animals kept in the dark, although a similar experiment on September fish gave different results, with no alterations in the nuclei and accumulation of material in the cells. In *Porichthys* also, constant illumination led to a depletion of AF + ve material from the NPO. Whether these changes related to alterations in pars intermedia activity and pigmentary conditions is not known.

An isolated report suggests some relationship between the neurohypophysis and the thyroid, although this is difficult to interpret in physiological terms: Injections of TSH, or thyroxine, encouraged the movement of neurosecretory material in *Salmo* from the NPO to the neurohypophysial core.

New fields for investigation are indicated by the observations that electrical stimulation of the olfactory tract of goldfish caused depletion of stainable material from the NPO and its axons; and by the report of wide diurnal variations in the amount of AF + ve material in the goldfish NPO, variations that do not seem to be regular (rhythmic), although generally the cells stain more deeply during the day than at night. In the light of this last finding, some of the earlier work may need reconsideration, since investigators have not usually paid attention to the possibility of diurnal changes in the neurohypophysis.

There have been few observations on the Type A neurosecretory material among the cells of the pars distalis, where the function of the fibers is presumably the control of adenohypophysial activity. Knowles and Vollrath described Type A3 fibers in the rostral and proximal pars distalis of eel and found that the number of these fibers, and the amount of stainable material, was greater in the silver eel than in younger animals. They suggested that this might be correlated in some way with the greater activity of the TSH cells and gonadotrops in the silver stage. In mature seawater eels, these authors observed a decrease in the AF + ve material and in Type A3 granules (140 mμ diam) in the proximal pars distalis, but fibers with smaller granules (100 mμ) were now found in this region. The significance of these changes is not known, and Knowles and Vollrath did not identify the fibers with 100 mμ granules as being distinct from the Type A3 fibers of the river eel.

"Nonstainable" Fibers, Probably Knowles' Type B

The nonstainable fibers probably originate from the nucleus lateralis tuberis (NLT) of the hypothalamus and do not stain with the classic neurosecretion stains, although they have been described as AF + ve by some workers. They can be stained with acid dyes such as azocarmine, eosin, light green, phloxin, and also with PAS Aniline blue or silver impregnation; similar staining reactions are seen in the cells of the NLT. Their pathway within the pituitary has proved difficult to follow with the light microscope, but they have been traced between cells of the eel proximal pars distalis and the trout pars intermedia, and pass toward the cells of the proximal pars distalis in *Salvelinus*.

At the electron microscope level, these nonstainable fibers appear to correspond to the Type B fibers in the eel neurohypophysis, which contain granules of less than 100 mμ diam, the granules having a central electron dense material which does not fill the bounding membrane and is less osmiophilic than the Type A granules. Similar granules are found in the NLT of eel. In *Salmo,* however, granule size in the Type B fibers is less than in the NLT, so that Follenius was not certain that the NLT is the origin of the Type B fibers in this species. Moreover, in some fishes (e.g., *Phoxinus)* the NLT is said to be absent, although Type B fibers are present. It is not certain, therefore, that all Type B fibers arise from the NLT in all teleosts.

Nonstainable and Type B fibers appear to be more numerous in the anterior part of the neurohypophysial core and in the pars distalis than in the more posterior regions and the pars intermedia, although a few Type B fibers are found in the eel pars intermedia, just as some type A fibers are found in the pars distalis. Type B fibers make both neurovascular and neuroglandular contacts in both the pars intermedia and the pars distalis. In the eel, the contacts in the pars distalis are all neurovascular; that is, the neurons discharge into an extravascular space, close to the endocrine cells but separated by a distance of about 300 m,μ or less. However, in *Hippocampus,* both Type A and Type B fibers were found to make direct contact with several kinds of endocrine cells in both regions of the adenohypophysis. The contacts were synaptoid for both fiber types in the pars intermedia, and Type B fibers were shown to make synaptoid contact with three types of pars distalis cells; but satisfactory evidence of synaptoid contacts between Type A fibers and pars distalis cells could not be obtaine. Direct contact between both types of fibers and pars distalis cells have also been seen in *Tilapia,* as well as terminations on the basement membrane adjacent to the endocrine cells. These findings suggest that activity of the adenohypophysial cells could be regulated by a direct dual neurosecretory innervation, with various kinds of contacts between the nerve fibers and the endocrine cells.

Knowles and Vollrath found that there appeared to be an increase in the activity of Type B fibers innervating the proximal pars distalis of seawater eels in which the gonadotrops were further developed and more active than in the river eel; this is particularly interesting in that there has been evidence for some years that the NLT rather than the NPO is concerned in regulation of gonadotropic activity in teleosts.

Blood Supply to the Teleostean Pituitary

Key references to the blood supply of the teleostean pituitary are Bretschneider and de Wit, Green, Da Lage, Barrington, Lenys, Jasinski, Follenius, Bhargava, and Wingstrand.

The most striking characteristic of the blood supply to the gland is that the blood reaching the adenohypophysis has passed through capillaries in the neurohypophysis; and further, there is no independent venous drainage of the neurohypophysis apart perhaps from a posterior connection to the hypophysial vein in some species. In these respects, the vascularisation of the fish gland differs greatly from the tetrapod condition.

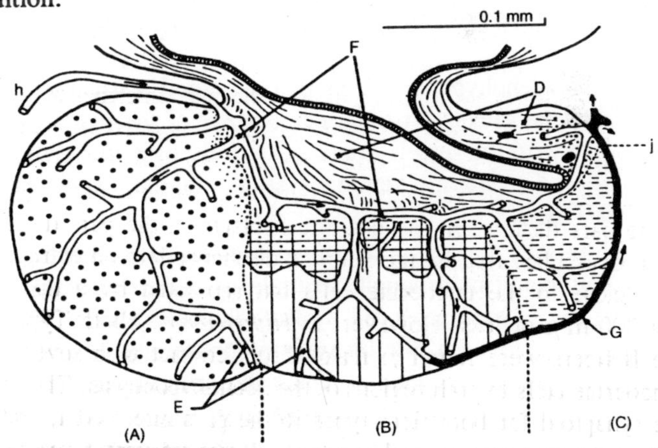

Figure 3.3 : Diagram to illustrate the main features of the blood supply to the cyprinodont pituitary. Anterior to the left. (A) rostral pars distalis, (B) proximal pars distalis, and (C) pars intermedia. Neurohypophysial core (D). Blood enters the gland from the hypophysial artery (h), and passes to the primary longitudinal plexus (F) in the neurohypophysial core. From here it is distributed to the adenohypophysis in the secondary centrifugal plexus (E), and is then collected into a superficial venous network (black) (G) and passes to the hypophysial vein (j). Modified from Follenius.

Most students of the subject have described a collection of capillaries in the neurohypophysis, forming a vascular plexus in the neurohypophysial core, close to the adenohypophysial boundary, or actually at the neuro-adeno interface. This plexus in the eurohypophysis is the primary longitudinal plexus or system of Follenius. In the adult *Phoxinus,* and no doubt in other species, it includes a large central longitudinal sinus. From this plexus a series of capillaries passes into the adenohypophysis and forms an elaborate network of capillaries and sinuses between the endocrine cells; Follenius has termed this the "secondary centrifugal system." According to some workers, the capillaries of this system arc confined to prolongations of the neurohypophysis penetrating between the endocrine cells. In contrast, in the eel, according to Knowles and Vollrath, these capillaries lie within a series of intervascular spaces or channels into which the neurosecretory fibers discharge and against which the endocrine cells lie, in both the pars intermedia and pars distalis. Whatever the details, the vessels of the secondary centrifugal system obviously come into intimate association with the adenohypophysial cells and are the only source of blood for these cells.

In salmon and eel and in *Hippocampus* the size and number of capillaries in the adenohypophysis varies with age and physiological state, suggesting the possibility of a hemodynamic control of secretory activity.

From the adenohypophysial capillaries, blood is usually collected into a superficial plexus on the outer surface of the gland, and this superficial network of vessels drains posteriorly by a few veins which empty into the systemic venous system via the hypocranial or hypophysial veins.

It will be seen that this modern work indicates a centrifugal flow of blood from the primary longitudinal plexus in the neurohypophysis, contrary to the scheme proposed by Bretschnieder and de Wit. Thus far, there seems to he a fair measure of uniformity among teleosts studied; however, the manner in which the primary capillary plexus receives its blood supply varies a good deal, and is very interesting because of the question of whether or not the teleostean adenohypophysis receives blood via a hypothalamo-hypophysial portal system such as is found in tetrapods, in which a primary capillary network in the median eminence region (just anterior to the pituitary stalk) is intimately associated with neurosecretory nerve endings, and blood passes from this network along the hypophysial portal veins to the capillaries in the pars distalis. This system in tetrapods is now believed to mediate hypothalamic control of the pars distalis. There exists an anatomical basis for similar hypothalamic control in teleosts in the form of the terminations of neurosecretory fibers on or close to the adenohypophysial cells and on vessels of the primary longitudinal plexus. There has, however, been considerable interest in the question of whether, in addition, elements of a hypothalamohypophysial portal system occur in these fishes; that is, in essence, whether any part of the adenohypophysis receives blood that has first been exposed to neurosecretory terminals in the hypothalamus.

Since the adenohypophysis in teleosts obtains its blood from the primary longitudinal plexus in the neurohypophysis, the essential question concerns the source of the blood in this plexus. In different fishes, the details vary. In most cases, a paired or single hypophysial artery (derived from the anterior carotid system) passes directly to the neurohypophysis without contributions from hypothalamic blood vessels. In *Hippocampus* and *Cichlasoma*, other branches from the internal carotid may first pass through the rostral pars distalis before entering the primary longitudinal plexus. In most cases blood from the carotid system via the hypophysial artery may be the only source of the blood in the primary capillary plexus, apart from a very few capillaries from hypothalamic vessels. In other species, the contribution from hypothalamic vessels may be more important, and it is these exceptional cases that have aroused interest in the past. In *Phoxinus,* a median ventral infundibular artery (= hypophysial artery?) supplies the ependymal floor of the infundibular recess, and just above the pituitary stalk it forms connections with a pair of ring vessels derived from the ventral hypothalamic artery. Bhargava disagrees with Barrington's description of these connections as a vascular bed, and instead he implies that they are few in number and essentially form looplike connections between the ring vessels and the ventral infundibular artery, the latter then passing into the neurohypophysial core to supply the central sinus which in the adult represents the primary longitudinal plexus. Barrington has described CAH + ve neurosecretory cells lying in the ventral hypothalamus in close association with the capillary connections ("vascular bed") and drew attention to a resemblance to the tetrapod median eminence, but Bhargava finds that these cells, which occupy the region where the NLT is found in other species, are not in fact particularly close to the capillary loops. Pending further investigations, there seems no good reason to think that *Phoxinus* exhibits any kind of rudimentary portal system. Another slightly unusual arrangement exists in the bitterling, *Rhodeus,* in which the hypophysial artery breaks up into a superficial capillary network (rete mirabile) on the anterior face of the pituitary stalk (hypothalamic floor), from which arise the capillaries which form the primary longitudinal plexus in the neurohypophysial core. This rete mirabile is totally superficial and not associated with any special nerve endings such as characterize the tetrapod median eminence. In *Corydora* and *Cichlasoma* meningeal vessels drain into the primary longitudinal plexus, and these receive contributions from capillaries in the hypothalamus which in *Cichlasoma* are closely associated with scattered hypothalamic cells containing discrete AF + ve droplets. In *Channa punctatus,* all the blood in the pituitary is said to derive from vessels in the hypothalamus, but no details are available about possible neurosecretory associations with these hypothalamic vessels.

While these scattered observations might be interpreted as indicating some kind of portal connection between hypothalamus and adenohypophysis, the important point to emphasize is that in no teleost has any close association between hypothalamic neurosecretory nerve endings and blood capillaries been observed outside the pituitary itself; thus, whether or not these various hypothalamic capillary contributions to pituitary vascularisation represent, as it were, a foreshadowing (or reminiscence) of the portal condition, there is no reason to regard them as functional teleostean portal systems. One must agree with Follenius that insofar as there may be a vascular link between hypothalamic neurosecretion and adenohypophysial cells in teleosts, that link is more likely to be found in the secondary centrifugal plexus rather than in the extrapituitary vessels supplying the primary longitudinal plexus; this matter is considered in the next section.

Hypothalamic Control of the Adenohypophysis

The limited information available about the behaviour of the teleostean adenohypophysis when deprived of direct hypothalamic connections was reviewed; from the work on pituitary transplants and *in vitro* culture, some degree of hypothalamic control of the gland was indicated, although the control seems less important than in higher vertebrates. It will be seen that anatomically there appear to be at least two routes by which hypothalamic control could be imposed on the adenohypophysis.

Neurosecretory Terminations within the Adenohypophysis

From ultrastructural and histochemical information, it appears that neurosecretory fibers of *Type A* or Type B come into more or less inti-° mate contact with all the adenohypophysial cells, Type A apparently predominating in the pars intermedia, and Type B in the pars distalis, but with overlapping distributions in each case. Type A fibers probably originate in the NPO, and most Type B fibers probably originate in the NLT. The nature of the contact between the nerve fiber and endocrine cells ranges from immediate physical contact, synaptic in some cases, as in the pars distalis of *Hippocampus, Tilapia, Phoxinus,* and the guppy, and the pars intermedia of *Conger, Gadus, Phoxinus,* and the guppy; through cases where the nerve fiber ends on a single or double basement membrane which separates the neuron from the endocrine cells *(Tinca, Perca, Salmo,* and *Tilapia) ;* to the condition in which the nerve fiber ends in an extravascular space which separates the fiber terminal from the endocrine cells. It should be noted, however, that Knowles and Vollrath have questioned the nature of the boundary between endocrine cells and nerve fibers that Follenius and Porte termed a "basement membrane," suggesting that it may in fact be a double membrane enclosing an extravascular space, as in *Anguilla.*

Neurosecretory Terminations on Capillaries in the Neurohypophysial Core

Apart from the neurosecretory terminations within the adenohypophysis, many workers have concluded that neurosecretory fibers make synaptic contact with the walls of the capillaries of the primary longitudinal plexus and its branches into the secondary centrifugal plexus. According to Follenius it is Type B fibers that terminate in the anterior part of the primary capillary plexus, close to the pars distalis, while Type A fibers from the NPO mostly terminate on capillaries in the more posterior part of the neurohypophysial core, close to the pars intermedia. Nucleus preopticus fibers terminate on capillaries in the neurohypophysial core in *Lepidogobius*, and Honma and Tamura found that both NPO and NLT fibers terminated on neurohypophysial blood vessels in *Salvelinus*. Nucleus preopticus fibers terminate on neurohypophysial capillaries and also among proximal pars distalis cells in *Porichthys*.

A Possible Third Route via the Third Ventricle

Cellular processes from cells of both the NPO and the NLT have been observed to penetrate the ependyma and project into the third ventricle, with evidence of secretion into the ventricle. Similar appearances have been described in other vertebrate groups. Stahl and Leray suggested that cells of the NLT might release their secretion into the third ventricle, to be transported thence by capillaries to the pituitary, or by being absorbed by the specialised ependymal cells lining the infundibular recess which appear to send fibers containing PAS + ve or AF + ve material toward the pituitary. In view of the evidence that NPO and NLT fibers penetrate the pituitary directly, it is difficult to assess the meaning of these appearances in terms of hypothalamic control. We should note, also, that some workers consider the specialised ependymal cells as secretory (into the third ventricle) rather than absorptive.

Thus, anatomically, the neurosecretory fibers of the neurohypophysis display features that strongly suggest they are concerned with regulation of adenohypophysial function. In addition, the neurohypophysis contains the biologically active octapeptides, AVT and IT. These or similar principles have well-documented systemic hormonal functions in tetrapods, and there is some evidence, perhaps not conclusive, that AVT and IT have peripheral actions in teleosts. How these neurohypophysial principles relate to the fiber types of the neurohypophysis is not certain, although the work of Lederis suggests that in *Gadus* and *Salmo* AVT is associated with Type A granules; this agrees with the suggestion of Knowles that Type A fibers may be concerned with elaboration of peptide molecules, whereas Type B fibers may secrete some nonpeptide material, possibly aromatic amines in line with indications from other vertebrates. The problem is to reconcile the histological and ultrastructural evidence that Type A fibers are involved in the regulation of adenohypophysial activity with the indications that the peptides which these fibers probably secrete may have peripheral activities. At the present time too little evidence is available to permit further analysis of this problem. More work is needed, particularly to demonstrate definitively the functional significance of the neurohypophysial invasion of the pars distalis, so characteristic of teleosts. Direct innervation of the pars intermedia is found throughout the vertebrates, but the anterior neurohypophysis-pars distalis association is a teleostean speciality, and it may be that the extensive neurosecretory innervation of the pars distalis cells in this group evolved as an adjunct or supplement to the capillaries of the secondary centrifugal plexus. It is worth recalling the comparison, frequently made, between the median eminence of other gnathostomes and the anterior neurohypophysial core of teleosts, a comparison all the more apt in the light of recent information, both being regions where hypothalamic neurosecretory fibers terminate on capillaries which convey blood to the pars distalis; it may be that the anterior capillaries of the secondary centrifugal plexus are functionally equivalent to the hypophysial portal vessels, conveying products of neurosecretory cells to the pars distalis, the median eminence and portal plexus being as it were enclosed within the pars distalis.

Comparisons between the tetrapod and fish pituitary regions are complicated by the presence in nearly all fishes of the saccus vasculosus, a thin-walled folded sac growing out from the third ventricle (infundibular recess) posterior to the pituitary. This structure, despite its close proximity to the pituitary (particularly marked in *Acipenser* and elasmobranchs) seems to have no functional connection with the gland. In the vertebrate embryo there arises from the postero-ventral floor of the hypothalamus a posteriorly directed downgrowth, the saccus infundibuli. In fishes, this gives rise to the saccus vasculosus, and the fish neurohypophysis arises more anteriorly from an area of the hypothalamic floor that in tetrapods and lungfishes gives rise to the median eminence. In tetrapods, the saccus infundibuli gives rise to the pars nervosa, the main part of the neurohypophysis. Thus embryologically, the neurohypophysial core of fishes is not comparable to the tetrapod pars nervosa, which is more closely comparable to the fish saccus vasculosus; but in functional terms the pars nervosa may be likened to the posterior part of the fish neurohypophysis, particularly considering the extensive innervation of the pars intermedia by neurohypophysial fibers in many tetrapods. On both embryological and functional grounds, the tetrapod median eminence may be comparable to the anterior region of the teleostean neurohypophysial core, and to the shallow anterior part of the neurohypophysis dorsal to the pars distalis (the infundibular floor) in the primitive bony fishes.

THE PITUITARY GLAND IN PRIMITIVE BONY FISHES

Bony fishes belong to the class Actinopterygii, and the vast majority belong to the most recently evolved group of actinopterygians, the superorder Teleostei. In addition, there are a few surviving members of the more ancient superorders Chondrostei and Holostei. The pituitary gland in these primitive forms is of obvious interest in possibly indicating the evolutionary developments which lead to the extreme specialisations exhibited by the teleostean gland. Unfortunately, published information is almost entirely limited to anatomical and histological descriptions, with little experimental data to help identify the various cell types.

Superorder Chondrostei

Order Palaeoniscoidei (Polypterus and Calamoichthys)

General accounts of the pituitary in these fishes have been given recently by Dodd and Kerr and Wingstrand, and Kerr has made a detailed histological study. The most remarkable feature is the presence in the adult of a persistent remnant of the ventral part of Rathke's pouch, the hypophysial duct, which opens ventrally into the roof of the mouth and dorsally into an expanded space in the ventral part of the pars distalis. From this space, short diverticuli penetrate the pars distalis, forming vesicles in sections. Kerr equates this ventral part of the gland with the rostral pars distalis of teleosts, although the cells lining the vesicles and duct are peculiar, elongated chromophobes with some basophilic secretion at their distal apices, unlike the acidophilic $_q$ cells around the follicles of primitive teleosts with which the duct cells are comparable on Kerr's interpretation. Associated with the vesicles and their elongated lining cells are typical basophils (Type 3 basophils), which are PAS + ve, AB + ve (weakly), but AF + ve. Type 3 basophils also occur in the dorsal region of the gland, which is probably equivalent to the proximal pars distalis of teleosts.

In this dorsal region, Kerr described a very small scattered acidophil, and two further basophils, most clearly separated by the fact that Type 1 is —ve and Type 2 +vc to AF without oxidation. There are unfortunately no grounds for allocating functions to these various cells. In the pars intermedia, Kerr described a single PAS + ve basophil, arranged in cords around the complex processes of the neurohypophysis. The PbH PAS technique confirms that the pars intermedia of *Calamoichthys* contains a single cell type, PAS + ve but PbH — ve.

The pars intermedia is the only part of the adenohypophysis to be invaded by neurohypophysial processes, although Kerr did see a single instance in which a short process of nervous tissue, laden with neurosecretory granules, projected a little way into the proximal pars distalis, perhaps a "foreshadowing" of the telcostean condition. In the pars intermedia region the inpushings of the neurohypophysis are hollow, the lumen communicating with the third ventricle and lined by ependymal cells (pituicytes), recalling the arrangement in the eel. Cords of ependymal cells extend beyond the lumen in the middle of

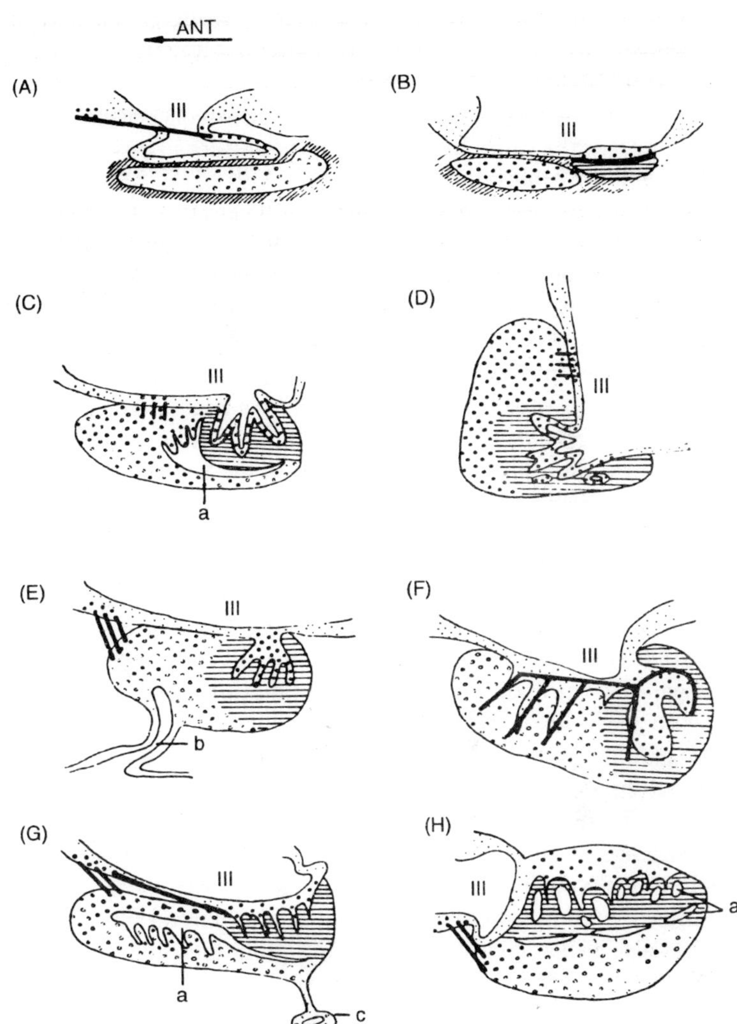

Figure 3.3 : Diagrammatic midsagittal sections to illustrate the main features of pituitary structure in the different fish groups. In all cases, anterior is to the left. (A) myxinoid, (B) lamprey, (C) Acipenser, *(D) Amid,* (E) Polypterus, (F) *teleost, (G) elasmobranch, and (H) dipnoan. Small dots, nervous tissue; large solid dots, stainable neurosecretory material; large open dots, pars distalis; horizontal lines, pars intermedia; thick black lines, blood vessels which appear to convey neurosecretory products to the adenohypophysis or (myxinoid) to the neurohypophysis; oblique hatching, connective tissue; a, hypophysial cavity; b, hypophysial duct; c, ventral lobe of elasmobranchs; and III, third ventricle.*

the distal nervous processes. The neurohypophysis contains typical stainable neurosecretory material, apparently derived from the NPO. A main tract of fibers from the NPO leads into the neurohypophysial core, and the stainable material can be seen to concentrate close to the pars intermedia cells. Lagios showed that fibers in this region also contact capillary walls.

A subsidiary tract from the NPO lies ventral to the main tract in the hypothalamic floor and shows prominent stainable material just anterior to the tip of the pituitary. In this region the neurosecretory fibers arc closely mingled with a plexus of blood capillaries, and branches from this plexus pass into the pars distalis across a connective tissue pad at the front end of the gland. As Kerr pointed out, there is little reason to doubt that here we have a typical median eminence and extrapituitary portal system. This is confirmed by ultrastructural studies, which showed that in the median eminence of *Calamoichthys* there are neurosecretory fibers with granules 100-150 mμ diam which terminate on extravascular spaces around capillaries in a manner indistinguishable from that in the amphibian median eminence. According to Lagios, the portal vessels contribute the sole vascular supply to the pituitary of *Calamoichthys,* but Kerr thought that part of the rich vascular plexus in the neurohypophysial tracts in the pars intermedia of *Polypterus* and *Calamoichthys* derived from vessels in the wall of the brain.

In addition to the portal vessels in the connective tissue pad, the connective tissue sheet between the pars distalis and the infundibular floor (= anterior part of neurohypophysis) is highly vascular, and its capillaries are in close proximity to the neurosecretory fibers in the infundibular floor. It is possible that some neurovascular exchange may occur here, as appears to be the case in holosteans, capillary branches also penetrating the pars distalis from the connective tissue plexus.

Order Acipenseroidei (Acipenser and Polydon)

The sturgeon pituitary has been studied many times, particularly by Russian workers, but there is still little detailed information about its histophysiology. The most notable anatomical feature is the persistence of a large central hypophysial cavity or cleft, apparently a remnant of the dorsal part of Rathke's pouch, although unlike the palaeoniscids the sturgeons do not retain a hypophysial duct to the roof of the mouth. The hypophysial cleft largely separates the posterior pars intermedia from the anterior pars distalis, and also largely splits the latter into dorsal and ventral portions. The cleft gives off numerous tubular extensions, particularly dorsally but also anteriorly and ventrally. In addition to these tubules there are in the anterior and ventral pars distalis closed vesicles, with no connection to the cleft. In the absence of detailed embryological information their morphological status is uncertain, but Kerr assumes they are derived from the cleft tubules. These vesicles are presumably homologous with the follicles of the rostral pars distalis of holosteans and primitive teleosts. *Acipenser fluvescens* shows acidophils, often elongated, oriented around these vesicles, together with equally abundant rounded basophils and chromophobes. However, in the rostral region of *A. stellatus,* where the vesicles predominate, Barannikova described strongly acidophilic cells around the vesicles, together with amphiphils and only a few basophils. Both authors agree that the vesicles usually contain basophilic colloid as in holosteans and teleosts.

The ventral pars distalis, beneath the hypophysial cleft, was considered by Kerr to be an extension of the rostral pars distalis, with the same cellular composition, but Barannikova distinguishes this zone from the rostral region, since its acidophil cells stain more weakly and it contains many more basophils and small chromophobes. In the dorsal pars distalis, above the cleft, Kerr described regular columns of cells, arranged around tubular extensions of the cleft, containing acidophils and basophils. Barannikova states that this region contains the faint acidophils, basophils, and chromophobes also found in the ventral zone, together with strongly staining acidophils. By bioassay of dissected glands of *A. stellatus,* Barannikova located gonadotropin secretion in the ventral and dorsal zones and showed that the activity was stronger in the ventral zone; the basophils of these two zones displayed secretory changes correlated with gonadal development, identifying these cells as gonadotrops. The gonadotrops display marked development of acidophilic granules during their activity cycle, reminiscent of the R granules of *Poecilia* and other teleosts.

Since the gonadotrops are associated in both dorsal and ventral zones with a weakly staining acidophil, apparently distinct from the rostral acidophils, we may provisionally conclude that these two zones in *A. stellatus* correspond to the teleostean proximal pars distalis. However, as in holosteans, there is no very sharp separation of the two regions of the pars distalis, with broad mixed zones exhibiting all the cell types.

The pars intermedia is very large, and consists almost entirely of large basophils together with small numbers of chromophobes and acidophils. The neurohypophysis is separated from the adenohypophysis by a connective tissue layer, except in places over the pars intermedia where long hollow neurohypophysial processes penetrate deeply among the endocrine cells, with ependymal cells lining the cavities of the processes in the usual arrangement. No processes penetrate the pars distalis. Neurosecretory axons from the NPO and possibly also from the NLT terminate on blood vessels in the neurohypophysial processes. The connective tissue sheet between the infundibular floor (anterior neurohypophysis) and pars distalis is vascularised, and as in holosteans neurosecretory fibers from the NPO may terminate in relation to these capillaries, forming a kind of median eminence.

Superorder Holostei *(Amia* and *Lepisosteus)*

An account of the anatomy and histology of the pituitary in both genera is available, and one of us has investigated the glands of *4 Amia* and *3 Lepisosteus* using the modern techniques.

Neither genus preserves a hypophysial cavity or duct in the adult. The gland lies attached along most of its length to the infundibular floor (anterior neurohypophysis) and in *Amia* is relatively shorter and deeper than in *Lepisosteus.* Behind the pituitary is a large saccus vasculosus. Both genera display to a marked degree an intermingling of all the cell types in the pars distalis, as in the sturgeons, so that it is difficult to divide the area clearly into rostral and proximal regions.

In both genera, the rostral tip of the gland consists of the characteristic closed vesicles or follicles, containing basophilic material (PAS + ve, AB + ve, and AF + ve), and it seems possible from embryology that these do indeed represent diverticuli from the hypophysial cleft. In *Amia,* the follicles are lined by elongated acidophils, which are erythrosinophilic in the Aliz B technique and resemble the η cells of the eel *(acidophil 1).* Also in the follicle walls, but not usually reaching the lumen is a second cell type, *acidophil 2,* again erythrosinophilic but also taking Alizarin blue, so that it usually stains a blue-red in contrast to the scarlet of acidophil 1; the acidophil 2 is strongly PbH + ve, and it may be the corticotrop. A few basophils and acidophils from the main (proximal) pars distalis may be mixed with the follicles, especially ventrally. Also ventrally, the follicles are joined by a very peculiar cell type. This is strongly PAS + ve, Af + ve, AB + ve, and PbH + ve, and this is termed *basophil 1;* however, after Aliz B, its coarse refractile granulation is brilliantly stained with erythrosin. In addition, there are usually present a few large Aniline blue + ve granules which, together with the large size of the granules and difference in cell shape, distinguish basophil 1 in this technique from acidophil 1. A mantle of basophil 1 cells extends laterally and ventrally round the gland, enclosing the rest of the pars distalis and the pars intermedia, containing an admixture of acidophils 1 and 3.

The proximal region of the pars distalis is composed of dorso-ventrally oriented cords of cells. They include an *acidophil 3,* sharply separated from acidophils 1 and 2 by staining with orange G after Aliz B, and possibly an a cell (like the teleostean a cell, it has a slight PAS affinity) ; this acidophil is most numerous dorsally and centrally, by the connective tissue interface with the infundibular floor (anterior neurohypophysis), but it occurs throughout the region. Two types of basophil are present in the proximal pars distalis cords. *Basophil 2* is smaller, with rounded contours, stains a clear blue in Aliz B, is slate blue in Ox-ABPAS-OG, is AF + ve and PbH + ve; this cell type occurs mostly in the dorsal and central parts of the region. *Basophil 3* is larger, with angular contours, and stains lavendar in Aliz B (it contains scattered red granules as well as dull blue granules), with a few large blue granules close to the nucleus in many cases. It stains magenta with Ox-AB-PAS-OG, and is more strongly AF + ve and PbH + ve than basophil 2. It occurs mainly in the ventral and lateral region. These basophils 2 and 3 probably correspond to the two basophils described by Kerr.

The pars intermedia is elaborately invaded by branching processes of the neurohypophysis, which are hollow (at least proximally), with a lining of ependymal cells. Two cell types are distinguishable in the pars intermedia cell cords: a predominant PbH + ve cell, with the club shape often found in its teleostean homolog, the prolongation contacting the neurohypophysial interface; and a rather scarce PAS + ve cell, which tends to be rounded. The PbH + ve cell is blue in Aliz B preparations but of widely varying shades. The PAS + ve cell is amphiphilic, its colouration varying from red to mauve after Aliz B.

The neurohypophysis is essentially the thin floor of the infundibulum, separated from the adenohypophysis by a connective tissue sheet, except where it deepens and penetrates the pars intermedia as the neurohypophysial core. The penetration is by a hollow downfolding of the infundibular floor, forming a large infundibular funnel which penetrates deeply to the middle of the neurohypophysis. No processes appear to penetrate the pars distalis. The neurohypophysial core is rich in neurosecretory material, AF + ve, AB + ve, PAS + ve, PbH + ve, Aniline blue + ve, and CAH + ve. Sathyanesan and Chavin have traced fibers with these staining properties from the NPO into the neurohypophysial core, and observed some of these fibers terminating on blood vessels in this region in *Lepisosteus;* in this genus, they also found fibers from the NLT, AF - ve, but CAH + ve, directed toward the pituitary.

A simple portal system appears to be present; whether or not it is extrapituitary depends on how one interprets the status of the infundibular floor above the pars distalis. In this region, the connective tissue sheet separating the nervous tissue from the pars distalis is extremely vascular, with a plexus of capillaries and sinuses. From this plexus branches pass up into the neural tissues and down to vascularize the pars distalis. Quantities of AF + ve neurosecretory material occur close to these capillaries in the neural tissue just above the posterior part of the pars distalis, suggesting that NPO fibers may terminate on capillaries of the mantle plexus and their product may be transported into the pars distalis. In addition, in the more anterior part of the infundibular floor, above the anterior proximal pars distalis and the rostral zone,

Herring bodies and neurosecretory grains of "nonstainable" type lie close to the mantle plexus capillaries, presumably deriving from the NLT, which displays similar material in its neurons.

Thus, in all the primitive bony fishes, a portal system exists in the form of capillaries linking the infundibular floor, rich in neurosecretory terminations (= median eminence) and the pars distalis capillary network.

It is easy to conceive how the teleostean condition could be derived, if this infundibular floor and its mantle plexus were to be folded into the pars distalis, resulting in an enclosed median eminence (anterior neuro hypophysial core) with its primary longitudinal plexus.

The pituitary of *Lepisosteus,* although of different shape, resembles the *Amia* gland in essentials. Main points of difference are: the pars intermedia is larger, with wider neural interdigitations and a small infundibular funnel; the rostral follicles are smaller, and intermingle more freely with proximal pars distalis cells. *Lepisosteus* displays acidophils 1, 2, and 3, and the mantle of basophils 1 seen in *Amia;* however, basophils 2 and 3 are less easily separable, apart from a much sparser granulation in basophil 2 and its more central location. The neurohypophysis and infundibular floor are essentially as in *Amia.*

In both genera, the mantle plexus in the infundibular floor appears to be supplied by arteries which enter the infundibular region anteriorly between the optic nerves and the pituitary, although additional sources may be present.

THE PITUITARY GLAND IN LUNGFISHES

As surviving, though distant, relatives of the animals ancestral to the tetrapods, the pituitary in lungfishes (Dipnoi) is of great interest, but there have been few studies on the gland using modern cytological techniques. The recent work of Wingstrand, Kerr and van Oordt, and van Oordt and Kerr all describes the pituitary in the African lungfish, *Protopterus sp.*

The organisation of the pituitary resembles that of amphibians rather than other fishes. The adenohypophysis includes a distinct pars intermedia, dorsally placed and intimately associated with the neurohypophysis, and a ventral pars distalis, separated from the intermedia by the hypophysial cleft. In the pars distalis, as in amphibians, the cell types are intermingled and not grouped into zones as in so many fishes.

Five cell types can be distinguished in the pars distalis. Although the lack of direct experimentation makes it impossible to associate each of these firmly with the secretion of a particular hormone, in their staining reactions and distribution these cells resemble the cells of the amphibian pars distalis for which there is adequate experimental backing for functional identification. On the basis of these similarities, and considering the time at which the dipnoan cell types appear during ontogeny, Kerr and van Oordt have made tentative functional identification of some of these cells.

Two acidophils can be identified. Type 1 acidophil is erythrosinophilic after Aliz B and is distributed throughout the gland. Type 2 is orangeophilic and is confined to the posterior region. Presumably these cells secrete prolactin and growth hormone. Three types of basophils occur. Type 1 and Type 2 both have the typical staining reactions of basophils (PAS + ve, AB + ve, AF + ve, Aniline blue + ve), but can be separated by details of shape, distribution, and granulation. Type 3 basophils are restricted to the anterior tip of the gland and are violet after Aliz B, PAS + ve but AF — ve and AB — ve. The Type 1 basophil appears very early in development and may be a thyrotrop. The Type 3 basophil appears later in ontogeny, and strongly resembles the amphibian LH cell in location and staining properties. The Type 2 basophils are found only in adult fish, in which they may be very abundant, and are possibly FSH cells. As far as one can compare the descriptions, Godet suggested a similar functional identification on the basis of the tinctorial properties of the cells. He observed that all three basophils regress during estivation, and correlated this with regression of the thyroid, cessation of spermatogenesis, and atrophy of sex accessories.

The pars intermedia and neurohypophysis are closely associated to form a dorsal neurointermediate lobe, separated from the pars distalis by the neurohypophysial cleft. *Protopterus* has a thick intermedia, composed of hollow tubules interdigitating with the neurohypophysial processes. The cavities of the intermedia tubules often communicate with the hypophysial cleft, from which they are probably ultimately derived, but in the adult many are closed. In *Neoceratodus* the hypophysial cleft also sends diverticuli into the pars distalis, as in *Acipenser*, and in *Lepidosiren* the pars intermedia is just a thin layer of cells. Two pars intermedia cells occur in *Protopterus:* one is weakly PAS + ve and AB + ve but strongly AF + ve, while the other is strongly PAS + ve but AB — ve. The latter cell type is more abundant in younger fish. There is some indirect evidence that MSH is secreted by this region as in other fishes. Total hypophysectomy leads to melanin concentration, but removal of the pars distalis alone causes melanin dispersion together with hypertrophy of the pars intermedia. This hypertrophy together with persistent skin darkening suggests an enhanced secretion of MSH, and it is possible that as in many other vertebrates the pars intermedia is innervated by inhibitory hypothalamic nerve fibers which are damaged when the pars distalis is removed.

Adaptation to aerial life in the cocoon during estivation is accompanied by melanin dispersion in *P. aethiopicus,* and by degranulation of the pars intermedia cells, although no histological details were given. Under these conditions, the cavities of the intermediate lobe tubules enlarge.

The dipnoan neurohypophysis is comparable embryologically to the pars nervosa of tetrapods rather than the neurohypophysis of other fishes. It consists basically of a system of hollow tubules of nerve fibers, shown in development to arise as outpushings of the infundibular cavity, but in the adult this relationship is lost and the tubules mostly become solid. Ependymal cells, originally surrounding the tubule cavities, extend in the nervous processes together with abundant AF + ve neurosecretory material. In the adult, the intertwining of the nervous processes with the cords of the pars intermedia are very complex. The neural processes contain more stainable material in aquatic *Protopterus* than in aerial (estivating) fish, and the amount of this AF + ve material decreases in fish transferred from water to aerial cocoons. When air is blown into the mouths of aquatic fish, there is a marked increase in the amount of neurosecretory material after 3 hr. It is not clear in all these experimental conditions whether an increase in the amount of material should be taken to mean increased synthesis or decreased release of the neurohypophysial principles.

Arginine vasotocin has been identified in the pituitary of *Protopterus,* and it has been shown to have diuretic and natriuretic effects when injected.

Anteriorly, the interdigitating neurohypophysis connects to the thin floor of the infundibulum which, as in the primitive bony fishes, is separated from the pars distalis by vascular connective tissue. This region passes at the anterior margin of the gland into a distinct median eminence, comparable to that of urodele amphibians. Its capillary plexus is supplied by fibers from the NPO, and portal vessels pass from this plexus to supply the pars distalis.

THE PITUITARY GLAND IN ELASMOBRANCHS

The adenohypophysis in this group is at first sight very unlike that in other fishes. It is divided into a very large pars intermedia, elaborately entwined with the neurohypophysis to form a neurointermediate lobe; and an elongated pars distalis, composed of a dorsal lobe (rostral lobe or anterior lobe) extending forward close beneath the infundibular floor, and a ventral lobe, attached to the dorsal lobe by a stalk and associated with the floor of the cranium. In sharks and certain rays, the entire pars distalis is hollow, the anterior part of the dorsal lobe containing vesicles or tubules communicating with the central hypophysial cavity, and the posterior part (tail of dorsal lobe) consisting simply of folds of tissue around the hypophysial cavity. In other skates and rays, the hypophysial cavity may be small or obliterated, and the dorsal lobe is a compact mass of cords and clusters of cells. The ventral lobe in all cases is hollow, described as containing vesicles. In the aberrant chimaeroids, the gland displays a hollow dorsal lobe, and a separated structure, composed of follicles of cells, which may represent a detached ventral lobe. All these spaces and vesicles may contain a colloid, which is PAS + ve, AF + ve, and AB + ve. The suggestion has been made that this colloid in the vesicles of the ventral lobe may represent a store of gonadotropins and thyrotropin, but Mellinger believes that secretion of the colloid throughout these hypophysial spaces is a nonspecific function of the cells lining the cavities, including the endocrine cells and certain nonendocrine cells. One is reminded of the follicles of the rostral pars distalis in some actinopterygians, which frequently contain a PAS + ve colloidal material.

Experiments involving surgical hypophysectomy (total or partial) and replacement therapy with mammalian hormones have suggested that the ventral lobe secretes gonadotropin (s) in dogfish and skate. Assays of different regions of the gland for TSH also locate thyrotropic function in the ventral lobe. ACTH activity has been located in the head of the dorsal lobe by bioassay.

The limited information available about cell types in the gland does not always correlate with this physiological data. In the head of the dorsal lobe are found strongly PAS + ve cells. Mellinger terms these the "T cells," and showed in various species that they are AF - ve and AB - ve and acidophilic. The T cells he equates with the α cells of Della Corte and Chicffi and Chieffi ; the Italian workers thought that these cells produced TSH, but Mellinger believes that they secrete growth hormone. The T cells occupy the region in which deRoos and deRoos located ACTH activity. In the tail of the dorsal lobe, Mellinger finds Q cells, PAS + ve to a variable extent, AF - ve and AB - ve and acidophilic, which he suggests secrete ACTH. Some of these possibly correspond to the Italian workers' a cells, which they suggested might secrete growth hormone, and some of their 8 cells, which they thought might produce prolactin. In the ventral lobe, there is more agreement about attribution of functions. Mellinger describes PAS + ve, AF + ve, and AB + ve cells in this lobe, which he divides on the size of their granules into V cells (gonadotrops) and X cells (thyrotrops) ; the Italian workers described here a single basophil which they thought secreted FSH.

In the chimaeroid, *Hydrolagus,* there are in the head of the dorsal lobe acidophils, AF + ve cells and chromophobes, while the tail of this region contains acidophils, bcdophils, and chromophobes. The putative ventral lobe displays small cells containing PAS + ve granules, an AF + ve cell type, also with PAS + ve granules, and chromophobes.

While the functional identity of the various elasmobranch cells is far from established, taking the physiological localisations of functions together with the histological data, it is clear that in the sharks and rays the dorsal and ventral lobes, taken together, do indeed have the functions of the pars distalis, as suggested in the nomenclature we have adopted. Alternative schemes, in which the dorsal lobe is termed the "anterior lobe" or the "rostral and proximal pars distalis", or simply "rostral lobe" all imply in one way or another potentially misleading comparisons with parts of the gland in tetrapods and teleosts.

The pars intermedia lies below the thin neurohypophysial layer and is penetrated by neurohypophysial fibers to a variable extent, forming a neurointermedia lobe. The intermedia cells may be arranged in distinct lobules separated by highly vascular connective tissue, or the cell cords may fuse to form a mass of cells with an irregular plexus of blood vessels. The penetration of nerve fibers between the intermedia endocrine cells may be slight as in *Squalus* and *Etmopterus* or very extensive as in *Scylliorhinus Torpedo,* and *Raia*. Most workers agree that only one cell type occurs in the pars intermedia, weakly PAS + ve and acidophilic. Knowles differentiated peripheral and central cells, on the basis of shape and ultrastructure, in *Scylliorhinus,* and suggested they may have different functions. In the peripheral cells, he distinguished a synthetic region at the apex of the cell and a hormone release region at the opposite pole close to a blood vessel. Knowles described Type A neurosecretory fibers from the NPO terminating on the synthetic pole of the peripheral cells, and Type B fibers, possibly from the NLT, or arising within the neurohypophysis, terminating on the release pole of the cell. Knowles interpreted these findings as indicating that hormone synthesis and hormone release are under independent neurosecretory control. Other workers have observed terminations of Type A fibers on the pars intermedia cells and on blood vessels in this region, and Meurling described AF — ve fibers, possibly equivalent to Knowles' Type B, ending on pars intermedia cells and blood vessels in *Etmopterus*. However, Mellinger described only Type A fibers in the neurointermediate lobe of *Scylliorhinus,* and further showed that destruction of the NPOhypophysial tract led to activation of the pars intermedia cells and release of excessive MSH, indicating that the lesion had removed inhibitory control of both synthesis and release. Chevins found in *Raia* that not only tract section and ectopic transplantation of the neurointermediate lobe but also destruction of the NPO itself led to excessive MSH secretion, which suggests either a single inhibitory control of both synthesis and release, contrary to Knowles' hypothesis, or that both Type A and Type B neurons originate in the NPO, which is contrary to the histological and ultrastructural evidence.

Large osmiophilic and acidophilic globules occur in the elasmobranch pars intermedia, as in other vertebrates, and have been interpreted as the products of cellular degeneration or as a hormone store.

As in other groups, the pars intermedia has been shown to secrete MSH, the evidence coming from surgical removal of the neurointermediate lobe, bioassay of different pituitary regions, and lesioning of the NPO-neurohypophysial tract, which caused melanin dispersion associated with hyperactivity of the pars intermedia cells. Apart from this last study, there have been virtually no observations on natural or experimentally induced changes in the pars intermedia cells other than reports of the appearance of giant cells in this region in female *Scylliorhinus* and *Torpedo* in relation to the sexual cycle. Knowles did not observe these cells in his work on *Scylliorhinus,* and their significance is unknown.

The anatomy of the neurohypophysis has been studied in many elasmobranchs, key references being Scharrer, Meurling, Mellinger, Mellinger *et al.* Follenius, Knowles, and Polenov and Belenky. As in other fishes, the neurohypophysis develops from the thin floor of the infundibulum, dorsal to the pars intermedia. Above the pars distalis, the infundibular floor remains thin, similar to that in primitive bony fishes. The neurohypophysis may be very small, or together with the intermedia it may form two large lateral lobes, as in *Scylliorhinus*. Neurosecretory fibers, AF +ve and AF –ve, myelinated and nonmyelinated, penetrate downward to form a relatively thin nervous layer above the intermedia, and also penetrate among the intermedia endocrine cells to some extent. In older specimens of *Squalus,* Meurling described diverticuli from the infundibular cavity lined with ependymal cells, which penetrate with nervous processes into the pars intermedia, an arrangement similar to that in the eel and primitive actinopterygians. The more dorsal region of the neurohypophysis is often poor in stainable neurosecretory material, the greatest accumulation of which is in the region immediately dorsal to the pars intermedia. Some species display a distinct membrane between the neural tissue and the intermedia cells, containing a network of blood capillaries; in these cases there is little penetration of nerve fibers between the endocrine cells. In other species, the membrane has largely disappeared, and the fibers may penetrate between the cell cords either in broad bundles (e.g., *Scylliorhinus)* or irregularly.

Knowles' demonstration of Type A and Type B fibers in the

dogfish neurointermediate lobe has been discussed above. Meurling also traced AF - ve fibers, presumably corresponding to Type B, to terminations on pars intermedia cells. The origin of Type B or AF – ve fibers is uncertain, although some at least probably come from the NLT. The AF +ve (= Type A?) fibers have been traced back to the NPO. The NPO Type A fibers have been traced into the neurointermediate lobe to terminations on pericapillary spaces, pituicytes, and gland cells; and other Type A fibers pass to the ventral hypothalamus (anterior infundibular floor) just above the dorsal lobe of the pars distalis, where they become associated with a capillary network to form a median eminence. Type B fibers have also been traced to the median eminence. The existence of these two fiber types has been established in the hypothalamo-neurohypophysial tract and neurointermediate lobe of *Raia*, in addition to the dogfish, *Scylliorhinus*.

There are no published accounts of induced changes in the amount of material in the neurohypophysis of these fishes. Peptides with the usual properties occur in the neurohypophysis of elasmobranchs, although apparently differing from the principles in other groups. Perks and Dodd showed that after section of the hypothalamo-neurohypophysial tract of *Scylliorhinus* the oxytocic activity of the neurointermediate lobe eventually disappeared, and Chevins found that typical AF + ve neurosecretory material eventually disappeared from the tract and neurointermediate lobe of *Raia* following ablation of the NPO.

The blood supply to the elasmobranch pituitary has attracted great interest and is reviewed by Meurling. The vascular supply in chimaeroids is described by Sathyanesan, Jasinski and Gorbman, and Meurling. Apart from the dorsal pars distalis in the chimaeroid, *Hydrolagus*, it appears that each lobe of the pituitary receives arterial blood directly from either the vertebral or internal carotid arteries. In addition, there is evidence for a hypophysial portal system, although its details differ in different accounts. All authors agree that there is a region in the anterior infundibular floor (anterior hypothalamus) where neurosecretory axons from the NPO (and some from the NLT) are grouped around a capillary plexus in a way that suggests a neurohemal organ. Ultrastructural studies confirmed the presence in this region of neurosecretory terminations on the capillary walls, strongly resembling those in the mammalian median eminence. From this primary plexus in the median eminence, most of the blood passes backward in capillaries, some of which supply the tail of the dorsal lobe, while a few (Mellinger) or many pass back to supply the neurointermediate lobe. The head of the dorsal lobe is said by some authors to receive portal blood from the anterior part of the median eminence, but other workers have not described this. Most workers have said that the ventral lobe receives no blood from the portal vessels, but in *Raia* sp. Chevins finds that blood enters the ventral lobe from sinuses in the dorsal lobe, so that at least some of this blood must be portal in origin.

In the chimaeroid, *Hydrolagus,* the portal system is similar. The dorsal lobe, however, does not receive any direct arterial blood, and numerous portal vessels from the elongated median eminence supply all parts of the dorsal lobe. No portal vessels enter the neurointermediate lobe in this fish, in contrast to sharks and rays, and the putative ventral lobe receives no portal blood at all. As these authors point out, this may well be the most primitive of all vertebrate hypophysial portal systems.

All authors agree on the absence of direct neurosecretory innervation of the pars distalis, in contrast to teleosts, so that the portal system would seem to be the only route for hypothalamic control. The existence of a portal supply to the neurointermediate lobe is puzzling in view of the rich neurosecretory innervation of this region, and the functional significance of this part of the portal system awaits future work. The portal supply to the pars distalis is clearly similar to that in primitive actinopterygians and dipnoans, and as in those groups is to be correlated with the absence of direct innervation of the pars distalis.

THE PITUITARY GLAND IN CYCLOSTOMES

In this group, we find that the most primitive of all pituitary glands has a much simpler structure than that of gnathostomes. Both neural and glandular components can be recognised, but direct or vascular communication between the two is very limited. Many of the cells in the adenohypophysis appear chromophobic, especially in young animals, which hampers investigation of the histophysiology. Another great difficulty in studies on these animals is that their primitive evolutionary position makes it hazardous to assume that all the hormones secreted by the gnathostome pituitary are also produced by the cyclostome gland. In fact, evidence for the secretion of honnones other than gonadotropin(s), MSH, and AVT in lampreys is very insecurely based.

The pituitary in myxinoids (hagfishes) appears to be more primitive than that of lampreys (Petromyzontidae). The adenohypophysis of *Myxine* has been described by Olsson, Matty, Adarń, and Olsson *et al.* It consists of follicles and clusters of cells embedded in connective tissue below the hypothalamus, and there is no clear cytological differentiation between pars distalis and pars intermedia. Some workers report that the connective tissue septum between neurohypophysis and adenohypophysis may be missing posteriorly, so that the two components here are in contact, but this has not been observed by other workers. In the anterior region of the adenohypophysis, most of the cells are chromophobic, but a few basophils and acidophils have been recognised. Two types of basophils have been differentiated, both PAS + ve: one is weakly AB + ve and is grouped in follicles sometimes associated with accumulation of intercellular PAS + ve colloid, and with cytoplasm rich in SS/ SH groups. The second basophil stains with AF, forms signet ring cells after gonadectomy, and may be the gonadotrop. Two erythrosinophil cell types are present, one with fine granulation which responds to adrenocortical inhibitors, cortisol and thiourea, which may be the ACTH cell; and a second with coarser granulation is activated by reserpine treatment, and may secrete prolactin. The predominant cell type in the caudal region is PAS + ve and is presumed to be the source of MSH, although there is no evidence for the secretion of this hormone in myxinoids.

The gland in lampreys is organised in a more familiar pattern. Key references are Roth, van de Kamer and Schreurs, Oztan and Gorbman , Evennett, Larsen, Riihle and Sterba, and Bage. The pars distalis is embedded in connective tissue, which separates it from pars intermedia and neurohypophysis; this region is divisible into rostral and proximal parts by analogy with the teleost gland. The posterior pars intermedia is separated from the neurohypophysis only by a vascular plexus.

In the rostral pars distalis are found chromophobes, and basophils which are PAS + ve, AF + ve, and contain SS/SH groups. In various lampreys, these basophils increase in number and staining affinities at metamorphosis, when the entire pars distalis may increase in size. Following metamorphosis these cells exhibit changes which are not easy to interpret, but which have been taken to indicate a gonadotropic function; but these changes have not always been observed. The electron microscope shows that there are two kinds of basophils in this region, differentiated by granules size.

The proximal pars distalis exhibits acidophils and basophils, but the majority of cells are chromophobic and appear most active at metamorphosis in *Lampetra planeri* or during the anadromous migration in *L. fluviatilis*.

Changes in the acidophils have been described, but are impossible to interpret. The basophils are PAS + ve and AF + ve, and they are not numerous. They have been described as most active during metamorphosis or in the spawning migration and as reduced in activity after spawning. Roth and Evennett observed that these PAS + ve cells in the proximal pars distalis became increasingly chromophilic during the gonadal maturation, a change that could be prevented by gonadectomy. Since partial hypophysectomy indicated that gonadotropic function located in the proximal pars distalis, Evennett concluded that these PAS + ve cells are gonadotrops. On the other hand, Larsen, working on the same species as Evennett *(L. fluviatilis),* found that both the rostral and the proximal regions secreted gonadotropins; thus, the basophils in both regions could be gonadotrops. The changes described in the rostral basophils would fit this idea, although their activation during metamorphosis could indicate a thyrotropic function, the thyroid developing at this time from the larval endostyle. At the ultrastructural level, Bage differentiated three types of chromophils as well as chromophobes in the proximal region.

The pars intermedia has been shown to secrete MSH. There is no significant penetration of this region by the neurohypophysis. Many of the intermedia cells are chromophobic, but some authors have recognised a single chromophil, PAS + ve and azocarmine +ve (i.e., amphiphilic). Others have described two chromophils: one AF + ve and PAS — ve and elongated toward the neurohypophysis although situated ventrally, the other AF + ve and PAS + ve and situated close to the neurohypophysis. Little certain information is available about functional changes in these cells, although they have been examined in lamprey larvae made pale by constant illumination, but with no clear results. Various other changes have been described in the pars intermedia in relation to metamorphosis, migration, and spawning, but they are difficult to interpret.

The neurohypophysis of myxinoids is peculiar, consisting of a dorsoventrally flattened hollow sac above the pars intermedia region, which communicates with the third ventricle only by a narrow aperture. Nearly all the neurosecretory fibers from the NPO terminate in the highly vascular dorsal wall of this sac, there being few nerves or blood vessels in the ventral wall. Many of the axons contain granules varying in size from 100 to 200 mμ, together with much smaller vesicles. In lampreys, the neurohypophysis is merely a slight thickening of the floor of the infundibulum above the pars intermedia. It is composed of fibers from the NPO, many of which terminate around ncurohypophysial blood vessels or between the ependymal cells. There is no evidence of neurohypophysial penetration between the adenohypophysial cells.

Exposure of larval lampreys to continuous light decreased the amount of neurosecretory material in the NPO and proximal axons, while continuous darkness had the reverse effect. Neurosecretory material became scarce during metamorphosis of *L. planeri,* and AF + ve material completely disappeared during gonad maturation. Sterba and Bruckner made ultrastructural studies on the neurohypophysis of *L. planeri* after hypothalamic lesions; they found that the axons degenerated, liberating the elementary granules which were phagocytosed by ependymal cells and then released into the third ventricle. The authors obscurely interpreted these phenomena as evidence for a normal feedback route from neurohypophysis to cerebrospinal fluid, as suggested by Knowles and Vollrath for the eel.

The blood supply to the cyclostome pituitary, which differs in the two groups, has been reviewed by Gorbman. In myxinoids, the adenoand neurohypophysis are independently vascularised by branches of the internal carotid. In addition, in *Myxine glutinosa* the dorsal region of the neurohypophysis receives blood by a portal vessel from what appears to be a median eminence in the floor of the hypothalamus just anterior to the pituitary. A similar portal system has been observed in the hagfish, *Polistotrema,* supplying the neurohypophysis from a neurohemal area just behind the optic chiasma. No portal vessels or blood from the neurohypophysis seems to supply the adenohypophysis, the two components being largely separated by connective tissue. Blood leaves the adenohypophysis by a posterior vein which then passes through the neurohypophysis before leaving the gland. The neurohypophysis is also drained by a more anterior hypophysial vein.

In the lampreys, the pars distalis has an arterial supply and venous drainage separate from the neurohypophysis and pars intermedia. Several small capillaries from the internal carotid enter the pars distalis, and several small venules drain this region. The curious portal supply to the neurohypophysis seen in the hagfishes is absent from lampreys, this region being solely vascularised by a branch of the internal carotid which forms a vascular plexus between the neurohypophysis and the pars intermedia. Blood is drained from this region by a ventral hypophysial vein.

Obviously, these ancient animals have highly peculiar vascular supplies to the pituitary. In hagfishes there seems to be neither neural nor vascular links between the neurohypophysis and adenohypophysis, while in lampreys the pars distalis is similarly isolated from the neural component but there appears to be the possibility of neurohypophysial-pars intermedia communication *via* the nerve terminations on the vascular plexus between the two regions. The functional significance of the extraordinary hypothalamo-neurohypophysial portal link in myxinoids is totally obscure.

4

Hormones of Neurohypophysis

Studies of the neurohypophysis provide a good example of the interaction between mammalian and comparative physiology. They show clearly how each division can be of benefit to the other. It was Scharrer's pioneer work on the teleost neurohypophysis that introduced the concept of hypothalamic neurosecretion into mammalian research, and replaced the original idea of the posterior pituitary as a simple endocrine gland, by the concept of a complex neurohypophysial system, in which hormones, secreted by nerve cells in the hypothalamus, passed down the nerve axons, to be stored in the pars nervosa of the pituitary. In turn, studies of the mammalian neurohypophysial principles opened the way to new ideas concerning the nature of the hormones present in fish. Investigation of the pressor, oxytocic, milk-ejection, avian depressor, and antidiuretic activities of the mammalian pars nervosa culminated in the purification, analysis, and synthesis of oxytocin and arginine vasopressin. However, the demonstration of an analog, lysine vasopressin, in the pig, opened up the possibility of species variability in neurohypophysial peptides. It gave new significance to Heller's early demonstration that lower vertebrate pituitaries, including those of teleost fish, contained a neurohypophysial principle which was particularly potent in promoting water reabsorption in the frog. Soon, this "water balance factor" was shown to be arginine vasotocin, a molecule containing moieties from both oxytocin and arginine vasopressin, and it became clear that the lower vertebrates, including the fish, contained a fascinating family tree of interrelated principles. Perhaps the least satisfactory phase of neurohypophysial studies has been in the elucidation of the function of neurohypophysial principles in the metabolism of fish. Although concepts derived from mammalian studies have been useful, they may have clouded our view to some extent, since it is probable that the peptides have new and, as yet, undiscovered functions in aquatic vertebrates. Perhaps further work will allow the fish to introduce yet new ideas into mammalian physiology, this time new ideas concerning function. In this vein, it is particularly interesting that recent work has suggested that arginine vasotocin may be present in the mammal during its term of "aquatic" existence, i.e., during fetal development in the uterus.

THE CYCLOSTOMES

The Structure of the Neurohypophysis of the Cyclostomes

The hagfish (Myxiniformes) and the lampreys (Petromyzoniformes) hold a place of special importance in comparative studies, since they are the only living representatives of the first vertebrate class to be found in the fossil record. Already, they possess a relatively well developed neurohypophysial system, with only general tendencies which might appear to be primitive. This implies that comparative studies can examine only the final molding of the system, and its origins will remain obscure. However, the fact that it exists in both the two living groups of cyclostomes, which may have originated separately from ancestral Ostracoderms, suggests that it was present also in these common ancestors. If this diphyletic origin is correct, the hagfish and the lampreys may not be as closely related as is often assumed. This consideration, together with their long history of separate development, their clear specialisations, and their degenerate features, makes it difficult to be certain which characteristics of their neurohypophyses are truly primitive, and which are developments peculiar to each group. Also, it suggests that they should be given separate consideration.

Myxiniformes

Despite the difficulties pointed out above, a number of workers have concluded that the Myxiniformes possess the most primitive pituitary of any living vertebrate. However, in his early studies, de Beer had suggested that their pituitaries might be degenerate; despite this, he was impressed by the development of the infundibular process and felt that it was better formed than in the lampreys. Unlike the lamprey, the infundibulum of the hagfish consists of a hollow, flattened sac, 1 mm long, and attached to the brain by a narrow stalk. Early accounts suggested that this process was poorly differentiated, but

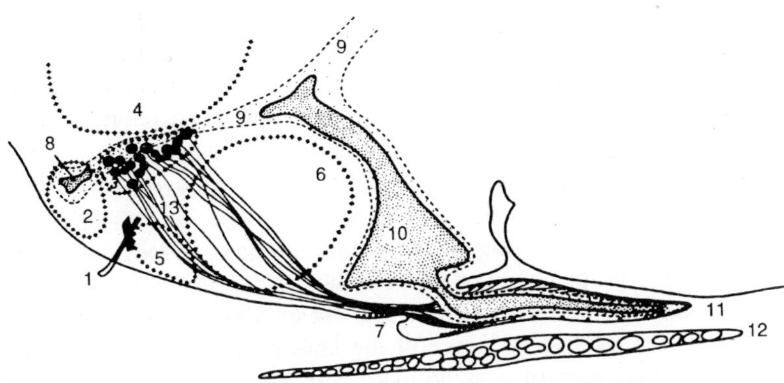

Figure 4.1 : The hypothalamus and pituitary of the hagfish, Myxine glutinosa, sagittal section. Note that the terminations of the preoptico-hypophysial tract are predominantly in the dorsal wall of the infundibular process, and the adenohypophysis is separated from the neurohypophysial system. (1) Optic tract (poorly developed) ; (2) preoptic nucleus, pars parvocellularis or anterior region; (3) preoptic nucleus, pars magnocellularis; (4) primordium hippocampi; (5) postoptic nucleus; (6) postoptic commissure; (7) possible median eminence, with a few neurosecretory endings (Herring bodies); (8) preoptic recess; (9) areas of remnants of buried ependymal cells; (10) third ventricle; (11) infundibular process; (12) adenohypophysis; and (13) preoptico-hypophysial tract.

recent work has shown that all the components of a neurohypophysial system are present in both *Myxine glutinosa* and in *Polistotrema stouti*. The following general account of the hagfish neurohypophysis is based on the work of these recent authors.

In the hagfish, the paired preoptic nuclei consist of ill-defined clusters of cell bodies, located dorsally to a poorly developed optic tract. The more ventral cells ("parvocellularis") lie close to the preoptic recess and do not appear to contain neurosecretion. In contrast, the dorsal cells ("magnocellularis") contain small quantities of fine, perinuclear neurosecretory granules which stain with Astra blue, but which are not detectable by the classic chrome-hematoxylin-phloxin stain of Gomori. Unlike neurosecretory cells in many other lower vertebrates, the preoptic cells of the hagfish do not send dendrites to the cerebral ventricle; however, it should be pointed out that the ventricles are greatly reduced in these species. Neurosecretion may leave the nucleus by two possible pathways. The first is probably unique to the Myxiniformes: It is possible that neurosecretory material may pass *directly* from the nucleus into a portal system which leads to the neural lobe of the pituitary. This is suggested by the presence of accumulations of secretion between and within the capillary walls of a vascular plexus which lies directly beneath the preoptic nucleus;

this plexus has been seen to drain to the neural lobe and pars nervosa. The second pathway is well established in many vertebrates; neurosecretion may leave the nucleus along the extremely thin processes which make up the preoptico-hypophysial tract of the hagfish. The axons disseminate widely at first. They are mostly empty of neurosecretion in this part of the tract, but fine droplets occur on rare occasions. The axons converge on the floor of the hypothalamus, and there is an accumulation of large droplets of neurosecretion, termed "Herring bodies," near to the origin of the infundibular process. This region is highly vascular, and both Olsson and Adam have suggested that it represents the median eminence of the higher vertebrates. However, Gorbman *et al.* Regard it as no more than the most anterior part of the neural lobe or pars nervosa. Some axons appear to form nerve endings in the narrow stalk of the infundibular process. However, most of the fibers of the preoptico-hypophysial tract continue to the end of the process, where they form an ill-defined pars nervosa within the terminal neural lobe. Olsson has likened this lobe to the equivalent structure in the embryo of the higher vertebrates. It is possible that some nerve fibers penetrate the inner ependymal lining of the neural lobe and deliver neurosecretory products into the infundibular cavity. However, the great majority of the neurosecretory fibers terminate in the *dorsal* wall of the lobe, mostly in the outer two-thirds of its thickness, and this might be regarded as the pars nervosa. Here, there is a relatively dense accumulation of neurosecretion. The neurosecretory droplets are closely associated with a rich vascular bed which penetrates into the *dorsal* surface of the neural lobe. This plexus drains into the general circulation, and it could serve as a route for the delivery of neurohypophysial peptides to the tissues of the body.

There is no good evidence for the passage of neurosecretory material through the ventral wall of the neural lobe to the poorly differentiated adenohypophysis. Herlant has suggested that the poor differentiation of the adenohypophysis may result from this lack of neural contact. The ventral wall of the neural lobe, which faces the islets of adenohypophysial tissue, is poorly supplied with either nerve terminations or neurosecretory granules. In many specimens of *Myxine* and *Polistotrema* the two regions of the pituitary are separated by a thick layer of connective tissue. This is penetrated only by a few, small blood vessels. However, Matty has described specimens of *Myxine* in which there is intimate contact between the caudal region of the neural lobe and the adenohypophysis. This apparent contradiction of a number of other workers could result from individual variation, but it could also be explained by Adam's observation that the connective tissue septum becomes thinned in specimens over 30 cm in length. If this degree of contact increased with age and length, it is possible that the neurohypophysis could come to influence the adenohypophysis in later life. However, Matty could find no evidence for the passage of neurosecretory droplets into the adenohypophysial tissues even though contact existed. Clearly, it is not possible to discount interactions between the two regions of the pituitary, but the present consensus is against their existence. It is worth noting that Gorbman *et al.* have pointed out that the neurohypophysial circulation of the hagfish is well adapted to the delivery of hormones to the general circulation.

Petromyzoniformes

The neurohypophysial system of the lamprey is markedly different from that of the hagfish. Unlike the hagfish, the lampreys do not develop a clear infundibular process. Nevertheless, studies of *Petromyzon marinus*, *Lampetra lamottei*, and *L. planeri* have shown that a complete neurohypophysial system is present in the slightly thickened floor of the hypothalamus. The preoptic nucleus is well marked, although less prominent and less vascular than in the teleost fish. It lies in a more rostral position than that of the teleosts, and it is found lining the preoptic recess, anterior to the optic chiasma. The preoptic cells are bipolar and of relatively uniform size; there is no division into parvocellular and magnocellular regions. Even in the youngest ammocoete larva so far examined, these cells contain neurosecretion, but it is different from that of the hagfish, since it will stain by both Gomori's chrome-hematoxylin-phloxin stain, and with aldehyde fuchsin. The preoptic cells of larval and adult *Petromyzon marinus may* lie close to the ventricle itself, or they may send short dendrites containing neuro-secretion between the ependymal cells, to the cerebrospinal fluid. This is not found in the hagfish. In addition to these connections with the ventricle above, a few preoptic nerve fibers appear to make direct or indirect connections with the adenohypophysis below. Roth has suggested that in adult *Petromyzon marinas* some nerve fibers which enter the region close to the preoptic recess may contact capillaries which supply the adenohypophysis. Oztan and Gorbman have reported that in the ammocoete larva of the same species, a small number of preoptic fibers terminate in the vascular connective tissue between the rostral adenohypophysis and the brain. It is even possible that these fibers, which bear beaded neurosecretory granules, may penetrate the adenohypophysis itself. In addition, a

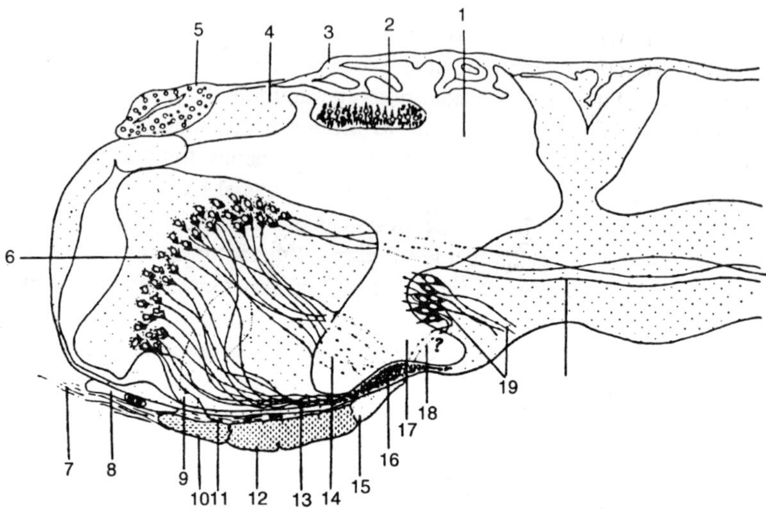

Figure 4.2 : The brain, hypothalamus, and pituitary of the ammocoete larva of the lamprey, Petromyzon marinus, *sagittal section. The general structure of the adult neurohypophysis is essentially similar. Note the lack of an infundibular process. (1) Third ventricle; (2) subcommissural organ; (3) choroid plexus; (4) habenula; (5) pineal body; (6) preoptic nucleus; (7) nasopharyngeal stalk; (8) blood vessel; (9) optic chiasma; (10) adenohypophysis (rostral zone) ; (11) preoptic neurosecretory axons ending close to adenohypophysis; (12) adenohypophysis (proximal zone); (13) preoptico-hypophysial tract (ventral division) ; (14) preoptico-hypophysial tract (lateral division); (15) pars intermedia; (16) pars nervosa; (17) infundibular cavity; (18) axons from posterior hypothalamic nucleus; (19) posterior hypothalamic nucleus (neurosecretory); and (20) preoptic axons connecting to the midbrain.*

few preoptic fibers appear to connect with the hindbrain. However, as Roth has pointed out in *Petromyzon marinus*, the majority of the axons of the preoptic cells are directed caudally, toward the pars nervosa. These axons carry beaded droplets of neurosecretion and form into a discrete preoptico-hypophysial tract over the rostral region of the adenohypophysis. At this point, the tract is relatively avascular, and it is separated from the adenohypophysis by an almost continuous sheet of connective tissue. The tract runs within the ventral hypothalamic wall. This wall is relatively thick and full of neurosecretion in *Lampetra planeri*, but in *Petromyzon marinus* it is thin, and contains few droplets or nerve endings. The axons of the preoptico-hypophysial tract terminate caudally in a slightly thickened area of the hypothalamic wall. They enter in a rostro-caudal direction, and also laterally, and they may be joined by a few axons from neurosecretory cells located in a posterior hypothalamic nucleus (nucleus lateralis tuberis?). In *Lampetra planeri* the thickened area is broadly continuous with the floor of the brain, but in *Petromyzon marinus* it has a distinctly globular structure with a central septum at its caudal extremity. Although it is not as clearly demarkated as in a mammal, this region may be regarded as the neural lobe or pars nervosa since it appears to store neurosecretion. A few of the neurosecretory axons which enter this region carry beadlike droplets of neurosecretion between the ependymal cells which line the ventricle. These ependymal cells, which contain aldehyde-fuschin-positive granules, send down processes into the underlying neural tissue and form a structural lattice for the pars nervosa. Occasionally, these ependymal cells appear to sink into the pars nervosa, where they have been described as pituicytes . Although a few neurosecretory fibers associate with the ependyma, it is clear that most terminate in swollen accumulations of neurosecretion, like Herring bodies, and these are grouped around blood vessels. Although the ventral surface of the pars nervosa interdigitates with the pars intermedia to give a serrated appearance, there is no evidence that any neurosecretory axons penetrate the intermedia tissue. The border between the two regions is marked by thin connective tissue, which is liberally supplied by a reticulum of capillaries. This reticulum is filled and drained by blood vessels which are independent of those concerned with the rostral regions of the adenohypophysis. Since there is no observable portal system in the lamprey pituitary, and these vessels both supply the pars nervosa and penetrate the pars intermedia, they are a potential vascular link between the two regions of the pituitary. It may be significant that van der Kamer and Schreurs have noted that these vessels enlarge during the metamorphosis of the ammocoete larva of *Lampetra planeri*. However, these same blood vessels also empty into the general circulation of the head, so that they form a potential route by which neurohypophysial peptides could reach all parts of the body.

The Nature of the Neurohypophysial Principles of the Cyclostomes

Myxiniformes

There is little knowledge of the principles present in the hagfish. This is because of the disappointingly low activities found in the glands, together with the difficulties of obtaining these deep-water animals. Such studies as exist are confined to one species, *Myxine glutinosa*. Herring's remarkable early work reported the presence of a vasopressor principle in this species. It was almost half a century before Adam observed a frog water-balance effect when extracts from *Myxine* were tested on *Bufoviridis.* In 1964, Follett and Heller were able to detect oxytocic and natriferic activity in an extract of 150 glands, but they failed to find a vasopressor effect. The total oxytocic activity obtained amounted to only 74 mU, and since no more than one quarter of this showed the lability to sodium thioglycollate treatment typical of neurohypophysial principles, they deduced that the hagfish pituitary contained no more than 0.13 mU/gland. Although it is possible that the reagent was partly expended on extraneous materials in so crude an extract, it is clear that the hagfish pituitary contains only traces of an oxytocic agent. The detection of frog water balance activity, together with oxytocic and natriferic effects, suggests the possibility that arginine vasotocin is present in the hagfish, as well as in the lampreys, but the present evidence is inadequate for anything more than speculation.

Petromyzoniformes

The neurohypophysial principles of the lampreys have received more attention than those of the hagfish. Herring showed that extracts from *Petromyzon fluviatilis* could cause a weak antidiuretic effect in the cat. These early studies were extended by Lanzing in 1954, when he detected frog water-balance and oxytocic activities in *Lampetra fluviatilis.* However, it was doubtful whether the agent which stimulated the guinea pig uterus was a true neurohypophysial peptide, since it was resistant to sodium hydroxide inactivation. In 1955, W. H. Sawyer was able to demonstrate that pituitaries from *Petromyzon marinus* contained oxytocic and vasopressor activities which were labile to sodium hydroxide treatment, and, in a series of studies carried out by W. H. Sawyer *et al.* milk-ejection, antidiuretic, frog bladder and hen oviduct stimulating actions were added to the properties of the extracts. Careful quantitative comparisons of these activities suggested that the extracts contained arginine vasotocin, a molecule which shared the ring structure of oxytocin with the terminal side chain of arginine vasopressin. This peptide might well serve as a potential "primitive" neurohypophysial principle. Purification of *Petromyzon marinus* extracts by gel filtration and the use of carboxymethyl cellulose columns yielded an active eluate with pharmacological properties similar to arginine vasotocin. A small "notch" in the eluted activity peak left some slight possibility that two similar peptides could coexist in the extracts. The purification procedure did not separate any oxytocinlike neutral peptide. However, it is possible that small quantities of a neutral peptide could have escaped detection during pharmacological studies, and these might have been lost during the prolonged storage of the extract prior to

Table 4.1: The Structure of the Neurohypophysial Principles of Fish and other Vertebrates

A. Basic Vasopressor Principles

1. Arginine vasotocin

Cys Tyr Ile Gln Asn Cys Pro Arg Gly(NH_2)
1 2 3 4 5 6 7 8 9

Exists in all fish; possibly in all vertebrates, at some stage of development

2. Arginine vasopressin

Cys——Phe——Cys——Arg——

Exists in most mammals, except the pig family

3. Lysine vasopressin

Cys——Phe——Cys——Lys——

Exists in mammals of the pig family

B. Neutral Oxytocinlike Principles

4. Oxytocin

Cys—— Ile ——Cys——Leu——

Exists in mammals, birds, reptiles, amphibia (?), Dipnoi (?), and holocephalians (?)

5. 8 Ile oxytocin ("Mesotocin")

Cys—— Ile-——Cys—— Ile ——

Exists in reptiles, amphibia, and Dipnoi

6. 4 Ser, 8 Ile oxytocin ("icthyotocin," "isotocin")

Cys——Ile · Ser——Cys——Ile——

Exists in teieosts, holosteans, and brachiopterygians

7. 4 Ser, 8 Glu(NH_2) oxytocin ("glumitocin")

Cys——Ile · Ser——Cys——Gln——

Exists in elasmobranchs, particularly skate

8. Unknown elasmobranch oxytocic principle (EOPI) (Ala, Val, Ser peptide(s)?). Exists in *Squalus acanthias*, and perhaps other sharks

purification. In essentially similar pharmacological studies on *Lampetra fluviatilis*, Follett and Heller showed that oxytocic, vasopressor, and natriferic activities were also in agreement with the presence of arginine vasotocin. Paper chromatography in butanol-acetic acid-water, 4:1:5, showed that the major activity peak ran with an R_f value of 0.25-0.35, which was similar to that of arginine vasotocin. Although, milk-ejection effects and antidiuretic activity were detected at R_f values between 0.5 and 0.8-the approximate position of oxytocinlike peptides-Follett and Heller did not consider that this was evidence for the presence of an oxytocinlike principle. However, their findings suggest that there should be some caution before the possibility of an oxytocinlike peptide is eliminated from cyclostome physiology. Most work has been done on adult animals during their spawning migration, and W. H. Sawyer has pointed out that oxytocinlike peptides might exist during other stages of the life cycle. The histological studies of van der Kamer and Schreurs have shown that neurosecretory material is remarkably dense in parts of the neurohypophysis during the ammocoete stage, but parallel pharmacological studies are lacking.

There have been no chemical studies of the structure of any cyclostome principal. Nevertheless, all present evidence indicates that the cyclostome pituitary contains an overwhelming preponderance of arginine vasotocin and that some cyclostomes may well be unique in possessing a single neurohypophysial principle in their pituitaries.

The Actions of Neurohypophysial Principles in Cyclostomes

Myxiniformes

Studies of the actions of neurohypophysial principles in the hagfish are few and brief. It is possible that the peptides influence salt and water metabolism. Chester Jones *et al.* Working with *Myxine glutinosa*, determined the effects of neurohypophysial principles on hagfish whose body fluids had been diluted or concentrated by immersion in diluted or concentrated seawater. Hagfish pituitary extracts, equivalent to 10 glands, or possibly 1.3 mU of oxytocic activity, were injected daily for 4 days into hagfish adapted to diluted (70%) seawater. The treatment resulted in a rise in the serum sodium level. A mammalian oxytocin-vasopressin preparation (Pituitrin) was injected at high dose levels (1000 mU), daily for 5 days, into hagfish adapted to concentrated (165%) seawater. This treatment resulted in a fall in serum sodium level. It is dangerous to generalize from such scanty data, but it seems possible that neurohypophysial peptides tended to return body sodium levels toward the natural value for normal seawater, presumably by affecting sodium fluxes. Although the hagfish is incapable of maintaining its internal environment in the face of external changes in osmolarity, this effect on sodium levels might delay, and therefore buffer, the changes which follow slight, transitory variations in the salinity of its relatively constant, deep-water environment. However, the original authors draw the careful conclusion that the neurohypophysial peptides can affect the relationship between the extracellular and intracellular distribution of water and electrolytes. There is a hint that the hagfish's own principle (s) are more potent in the hagfish than those of the mammals. A second possible effect of neurohypophysial principles on the metabolism of the hagfish is an influence on water balance. Adam observed that in *Myxine glutinosa*, the injection of high doses of a mammalian vasopressin preparation (Pituifral, 100-1000 mU) resulted in up to 7% rise in total body water. Clearly, there is need for more extensive investigations, with the use of the principle(s) native to the hagfish pituitary.

Since water, salts, and gases might be exchanged readily by the gills, it is interesting to note that the experiments of Somlyo and Somlyo have suggested that neurohypophysial peptides might influence the branchial circulation of the blood. They showed that isolated strips taken from the ventral aorta of *Eptatretus stoutii* contracted to extremely low doses of synthetic arginine vasotocin (10^{-10} *M*), and that higher doses of oxytocin (0.1-100.0 mU/ ml) and arginine vasopressin (1.0-100.0 mU/ml) would produce a similar response. The isolated dorsal aorta behaved in a similar way, but it was less sensitive, and it relaxed more readily. There are no studies of more general effects of neurohypophysial principles on the blood pressure or circulation of the hagfish.

At present, it is not possible to reach dependable conclusions on the function of the neurohypophysis of the hagfish. Further study is made more urgent by the demonstration of Gorbman *et al.* that the hagfish neurohypophysis is unlikely to be concerned in the local control of the adenohypophysis and is specially adapted to pour its secretions into the systemic circulation.

Petroomyzoniformes

Although experiments involving hypophysectomy have suggested that the pituitary of the freshwater lamprey was not essential to life, the extent to which the pars nervosa had been removed is uncertain, and, in any case, the intact preoptic nuclei might have continued to supply the necessary hormones.

Direct studies with neurohypophysial peptides have been confined almost entirely to the freshwater lamprey, *Lampetra fluviatilis*, and they have produced a number of negative results. The injection of pure arginine vasotocin in doses of 36 mμ moles/kg failed to cause any water-balance effect, and injections of 2.2 μg failed to change the volume of urine production over a 6-hr period. However, it is possible that the lamprey neurohypophysis may influence sodium metabolism, and this is reminiscent of the possible effects in the hagfish. The injection of a number of neurohypophysial peptides, including arginine vasotocin, into the peritoneum of *Lampetra fluviatilis*, resulted in an increase in the rate of sodium loss from the fish into its external medium. The site of action was in part the kidney, since injection of arginine vasotocin increased the sodium and potassium concentrations of the urine, and it is possible that a depression of tubular sodium reabsorption may be one factor involved. Although all the peptides tested provoked sodium loss, a comparison of equimolar doses suggested that arginine vasotocin, the natural principle of the lamprey, was only one-fifth as potent as mammalian oxytocin, on a weight basis. If comparisons were made on the basis of the oxytocic activity which was injected, arginine vasotocin was slightly less effective than oxytocin and considerably weaker than the potent 4 Ser, 8 Ile oxytocin-which is the natural oxytocinlike peptide of the teleost fish. This unexpected weakness of argininc vasotocin must lead to speculation on the possible existence of a neutral, oxytocinlike peptide in the lamprey pituitary. This second peptide might be present only during the marine period of existence, for Bentley and Follett have suggested that the actions of the principles on sodium metabolism would be of greatest use during the marine phase of the life cycle. Further, Dodd *et al.* have pointed out that the high doses required to produce a sodium response could reflect a reduced tissue sensitivity which had occurred when the lamprey first entered freshwater. Morris has suggested that there is a breakdown of osmotic controls at this time. This could also be connected with the low levels of neurohypophysial peptides found in pituitaries taken from lampreys adapted to freshwater (1.2 mU/ gland, *Lampetra fluviatilis*). In fact, the doses of neurohypophysial principles needed to evoke a sodium loss are greater than the total hormonal activity available in the lamprey's pituitary. However, it must be remembered that a low hormone content in the pituitary does not necessarily imply a low rate of loss into the circulation, since the stored peptide is only a reflection of the balance between hypothalamic supply and pituitary loss. This same consideration may account for the failure of experiments to demonstrate a significant difference between glands dissected from lampreys living in freshwater and those immersed for a short period in more concentrated saline (4 hr in 45% seawater).

A completely different action of neurohypophysial peptides has been suggested by the fact that arginine vasotocin, in doses similar to those which affect sodium metabolism, will cause a rapid rise in blood sugar and in muscle glycogen. An injection of arginine vasotocin amounting to 1.2 mμmoles, or 150 mU oxytocic activity, resulted in at least 40% rise in the blood glucose level, possibly by a fat mobilizing action. The dose used was high, and the same considerations which were dealt with under sodium metabolism apply here. Again, the physiological significance of the response is uncertain.

The lamprey neurohypophysis may be of local as well as of systemic importance, but the only evidence available, at present, is from histology. W. H. Sawyer *et al.* have pointed out that the close relationship between the neurohypophysis, the intervening blood capillaries, and the pars intermedia suggest that neurohypophysial peptides could pass to the adenohypophysis and modulate its function. It is possible that this action could be of special importance at metamorphosis and spawning. Van der Kamer and Schreurs have shown the presence of a considerable accumulation of neurosecretion in the ammocoete larva of *Lampetra planeri;* however, there is a dramatic loss of this material at metamorphosis. During metamorphosis, neurosecretion is found mainly in the nerve terminals which border capillaries that enter the adenohypophysis. It is also found in the fibers which reach the ependyma. Metamorphosis is accompanied by activity of the pars intermedia, and the lamprey takes on an adult colour pattern. After spawning, the neurohypophysis is almost empty of granules. These observations are highly suggestive of involvement of the neurohypophysis in metamorphosis and spawning, but they do not prove that there is a direct relationship. There is a small possibility that the changes in neurosecretion are linked to the emergence of the larva from the dark mud. Oztan and Corbman have seen that the distribution of neurosecretion in *Lampetra lamottei* and *Petromyzon marinus* is influenced by light. When these species are placed in continuous darkness there is a dense accumulation of neurosecretion in the preoptic nucleus and its axons. When the lampreys are returned to continuous light, the material is depleted. However, unlike the changes seen in metamorphosis, the nerve terminals of the pars nervosa are not affected. It is possible that these various observations indicate a neurohypophysial control of the pars intermedia, but more direct information is needed before any conclusions can be drawn.

CARTILAGINOUS FISH: THE ELASMOBRANCHS

The Structure of the Neurohypophysis of the Elasmobranchs

The pituitary of the elasmobranchs appears to be markedly different from that of the cyclostomes, and it is undoubtedly more advanced. However, there are a few surprising aspects which are reminiscent of both the hagfishes and the lampreys.

As early as 1685, Collins had illustrated the pituitary of the "skait"; he wrote, "The lower region of the brain of a skait seemeth to be composed of three ranks of processes, and an odd one." The "odd one" was an unusually large, spherical, centrally located process which lay beneath the caudal extremity of the saccus vasculosus; it was clearly the neurointermediate lobe of the pituitary. Despite this early start, evidence for the existence of a neurohypophysis in the elasmobranchs was slow to take shape. Many early workers failed to find any area which could be termed a pars nervosa. A few investigators accepted the thin infundibular lamina of the ventral surface of the brain as the only possible candidate for the neural lobe. However, in both Pokorny and de Beer observed areas of neuroglia-type cells within the pars intermedia of *Raia clavata* and other species. The reason for the early difficulties became clear when Scharrer used Gomori's hrome-hematoxylin-phloxin technique to demonstrate the presence of a complete neurohypophysial system in *Scyliorhinus stellaris* (*syn. Scyllium stellare*). He showed that the pars nervosa was a diffuse structure which penetrated throughout the tissue of the pars intermedia. Soon afterward, these observations were extended to *Raia clavata*, *Dasyatis marinas*, *Torpedo ocellata*, and *Scyliorhinus caniculus*. At about the same time, van der Kamer and Verhagen gave a detailed description of the histology of the pars nervosa of *Scyliorhinus caniculus*. They observed the presence of both pituicytes, and of secretory "parenchymatous pituicytes," which they considered to be possible sources of hormones. In 1960, Perks, Dodd, and Dodd, working on the same species, confirmed the general structure of the neurohypophysial system by the use of the chrome-hematoxylinphloxin stain, and showed that the neurosecretory droplets were probably rich in sulfur, since they stained by the performic acid-Alcian blue method of Adams and Sloper. The vacuolated parenchymatous pituicytes did

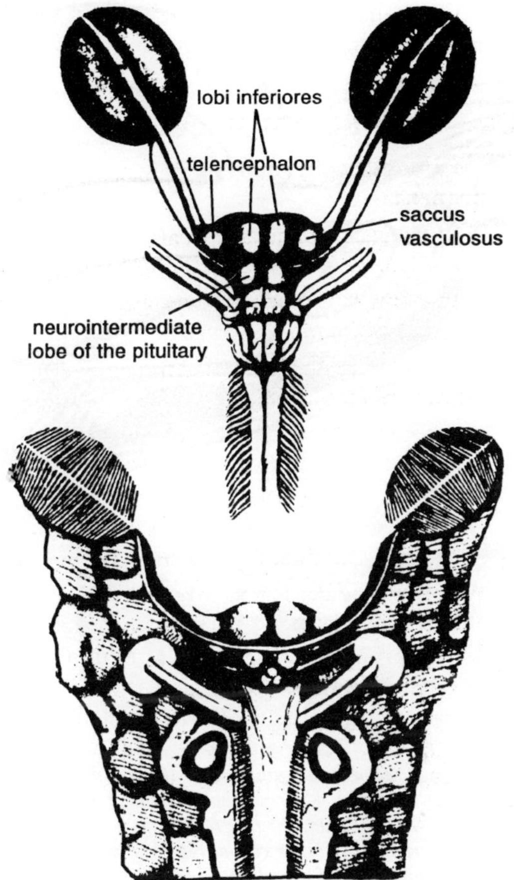

Figure 4.3 : "The Brain of a Skait". The lower diagram shows the ventral surface of teh brain of a skate. The neurointermediate lobe of the pituitary is shown as a large, centrally located process, which lies between teh dark lateral extensions of the saccus vasculosus.

not show any evidence for a high sulfur content, and therefore they were unlikely to be a source of neurohypophysial peptides. In, Braak was able to trace the neurosecretory tracts of *Spinax niger* by the elegant technique of staining and clearing whole mounts. In the following year, Knowles applied electron microscopy to the study of the diffuse neural lobe of *Scyliorhinus stellaris*. However, the most important advance of recent years has been the demonstration of a hypothalamo-hypophysial portal system in the elasmobranch pituitary. Meurling has made careful comparisons of both the portal system and

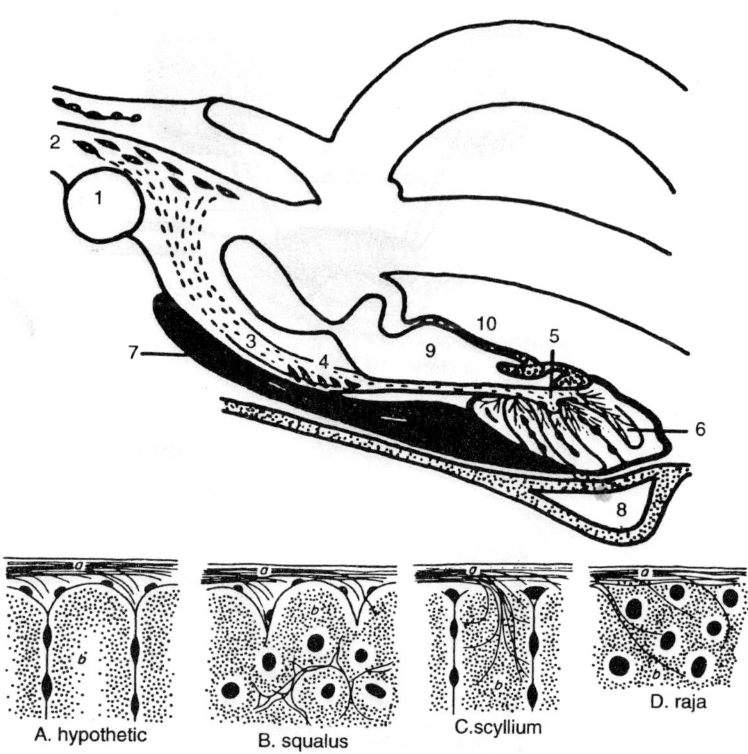

Figure 4.4: The hypothalamus and pituitary of the elasmobranchs; diagrammatic representations. Upper diagram: a parasagittal section of the infundibular region of Scyliorhinus caniculus. *(1) Optic chiasma; (2) preoptic nucleus; (3) preoptico-hypophysial tract; note that it becomes discrete in the caudal region of the hypothalamus; (4) median eminence, containing intrusions of portal blood vessels; (5) pars nervosa, note that many axons ramify onward into the pars intermedia; (6) pars intermedia; (7) adenohypophysis; (8) ventral lobe of the pituitary, embedded in cartilage; (9) infundibular cavity; and (10) saccus vasculosus.*

the neurohypophysis in different species, and the following general description is based largely on his excellent work.

The preoptic nucleus of the elasmobranchs is found dorsal to the optic chiasma and close to the ependymal lining of the ventricle. It is shorter than that of most teleosts, and its position varies from being relatively rostral in the sharks (especially in *Squalus acanthias*), to being remarkedly caudal in the skates. The cells are often found lying with their long axes parallel to the ventricular lining. They are relatively uniform in size, and there is no division into a pars magnocellularis and a pars parvocellularis, as seen in the telcosts. The preoptic cells are usually large and distinct, as in *Etmopterus spinax* and *Scyliorhinus caniculus*, but in some species such as *Squalus acanthias* they are less well marked. They contain innumerable small granules, which stain by Gomori's chrome-hematoxylinphloxin method, and by other so-called neurosecretory stains. The droplets appear to contain a high sulfur content. In *Scyliorhinus caniculus* there is a great variation in the amount of neurosecretion present in the preoptic nucleus, and indeed, throughout the entire neurohypophysis. If the neurointermediate lobe of the pituitary is removed, or the preoptico-hypophysial tract is cut, the preoptic nucleus becomes completely empty of neurosecretory granules. In normal specimens different cells within the same nucleus appear to be in different stages of secretion. Studies with the light microscope have suggested that neurosecretory granules appear in the walls of peripheral cytoplasmic vacuoles, and then spread throughout the cytoplasm, becoming particularly dense in the poles of the cells. Beads of neurosecretion can be found in dendrites which pass to the third ventricle, but most appear to leave the cells along the thick cellular processes, which narrow down to form the unmyelinated fibers of the preoptico-hypophysial tract. The axons of this tract run down behind the optic chiasma in a diffuse manner; at first they lack neurosecretory granules and give only a uniform staining, which makes them difficult to follow. However, in the caudal postoptic lamina the axons converge and form a discrete tract close to the midline. Here the tract shows densely packed neurosecretory granules and Herring bodies in many-but not all-of its fibers. In this region, portal capillaries penetrate the tract, and a median eminence is formed (*Scyliorhinus caniculus*, *Squalus acanthias*, *Raia batis*, *R. radiata*, Meurling; *Scyliorhinus caniculus*, Mellinger). In *Scyliorhinus caniculus*, neurosecretory material has been seen close to the blood vessels which invade the tract, and light and electron microscopy has shown that isolated neurosecretory axons, surrounded by a coat of glial cells, leave the main tract and form terminals on the penetrating capillaries: These terminals contain a central mass of synaptic vesicles and abundant neurosecretory granules, and it is considered probable that neurosecretory materials may pass from them into the capillaries. The capillaries drain into a complex of portal vessels which are the sole blood supply of the rostral lobe of the adenohypophysis; this arrangement strongly suggests a mechanism for control of the adenohypophysial cells. However, in *Scyliorhinus caniculus*, *Pristiurus melanostomus*, and various species of *Raia*, portal vessels also supply the neurointermediate lobe of the pituitary, a situation strangely similar to that of the hagfish.

The majority of the axons of the preoptico-hypophysial tract do not terminate in the median eminence but pass through the thin hypophysial stem toward the pituitary. In *Raia radiata*, this thin stem contains a wide band of nerve fibers, which may fall into as many as six different types, although some may be intermediate forms or variations along a single fiber. The axons are partly localised into particular groups. The lateral region contains mainly nonneurosecretory axons, which are of largely unknown significance. The medial region of the stalk contains predominantly Gomori-positive neurosecretory fibers. The electron microscope has shown that these contain large, electron-dense neurosecretory vesicles, 2000-3000 A in diameter, probably comparable to the Type A elementary vesicles described in the neurointermediate lobe of *Scyliorhinus caniculus*. In addition, there are axons which contain smaller, electron-dense vesicles, 1000-1300 A in diameter, which have a distinct external membrane, and may be comparable to the Type B elementary vesicles of *Scyliorhinus caniculus*. In some cases, these latter fibers form terminal swellings within the stalk. It has been suggested that they may affect the nearby pars medialis of the adenohypophysis, or perhaps liberate neurovascular transmitters into the nearby portal vessels, which could mediate an influence on the pars intermedia. As the neurosecretory fibers continue in a caudal direction, glial elements and secretory cells become less frequent, and it is probable that the secretory cells are lost before the axons of the preoptico-hypophysial tract enter the neurointermediate lobe of the pituitary. Here, axons bearing granules and Herring bodies, and some devoid of secretion pass below the single or multiple layers of ependymal cells which line the ventricle. The overlying ependymal cells send down fibers which divide the tract into bundles. In *Squalus acanthias* and *Scyliorhinus caniculus* some ependymal cells which contain vacuoles are found enclosed within the tract itself. In some species the neurosecretory axons terminate in a fairly well defined pars nervosa, but in others the terminations penetrate diffusely throughout the pars intermedia of the pituitary. The welldefined pars nervosa of *Squalus acanthias* may be a primitive feature. In this species the neurosecretory axons carry variable numbers of granules of different sizes, and they are sometimes in the form of beads. The axons are confined mainly within a lamina of variable thickness which covers the dorsal surface of the pars intermedia. This lamina sends down solid and tubelike processes into the intermedia tissue, and these projections, together with the dorsal neural covering, constitute the major portion of the pars nervosa. In the primitive sharks, *Hexanchus*,

Hexaanchus, and *Chlamydoselachus*, similar rodlike and tubelike processes are well developed. Within the pars nervosa of *Squalus acanthias* most axons terminate in an outer layer of ependymal fibers which follows the boundary with the pars intermedia. This layer contains a dense accumulation of neurosecretion. It contacts an external reticular membrane, which encompasses a network of capillaries. This neural lobe plexus forms the boundary between the pars nervosa and the pars intermedia. It drains into the sinuses of the intermedia tissue and could form a neurovascular link between the two areas of the pituitary. This is reminiscent of the situation in the lampreys. However, neurosecretory axons do enter the pars intermedia of *Squalus acanthias* to a small extent. Sometimes isolated axons penetrate the boundary membrane and take an irregular course between the intermedia cells. More often, and particularly in the neural processes, groups of neurosecretory axons pass through gaps in the reticular membrane and run between the intermedia cells or within the connective tissue which divides the cells into cords. Other species show a greater penetration by the pars nervosa. *Etmopterus spinax* is essentially similar *to Squalus acanthias*, except that the pars nervosa is covered by a connective tissue membrane which contains extensive fenestrations in the posterior region of the gland. Here, nerve axons penetrate the pars intermedia; some are neurosecretory and form occasional terminal swellings against the cells, but others carry no neurosecretion and terminate in a nerve plexus around the blood sinusoids of the pars intermedia. In *Scyliorhinus caniculus*, *S. stellaris*, and *Mustelus mustelus*, the neural lobe is still recognizable, but there is no membrane and no typical, continuous neural lobe blood plexus to limit its boundary. In *Scyliorhinus caniculus* nerve fibers penetrate between the intermedia cells, and neurosecretion of high sulfur content is found packing closely around these cells. Electron microscopy has shown that the neurosecretory nerves make close contacts, suggestive of synapses, with the intermedia cells. In *Scyliorhinus stellaris* (syn. *Scyllium stellare*) a high proportion of the neurosecretory axons enter the pars intermedia and secretion is stored everywhere between the cells. Electron microscopy has shown that the neurosecretory fibers can be divided into two types. The first, Type A, contains large electron dense vesicles and makes contact with the "synthetic" region of the intermedia cells. The second, Type B, contains small, irregular neurosecretory elements and terminates on the "storage and release" region of the cells. This suggests the possibility of a separate control of synthesis and release from the intermedia cells. In *Pristiurus melanostomas* there is an intimate fusion of neural and intermediate components, and in many places a neural lobe is no longer distinguishable; no neural lobe plexus has been described. In the skates and rays (*Raia batis*, *R. oxyrhynchus*, *R. fullonica*, *R. radiata*, *R. clavata*, *Dasyatis* sp., and *Torpedo* sp.) there is complete fusion of neural and intermediate elements. The only region which could possibly be termed a neural lobe is the small anterior area of the infundibular floor, where the preoptico-hypophysial tract enters the neurointermediate lobe. There is no neural lobe plexus, but blood vessels are found close to the ventricular surface, and the entire neurointermediate lobe contains a rich plexus of sinuses. A combination of light and electron microscopy has suggested that the bundles of nerve fibers which penetrate between the strands of intermedia cells in *Raia clavata* and *Trygon pastinaca* are made up of both neurosecretory axons and nerve fibers which do not carry such granules. The nerve axons are free of connective tissue or blood vessels, but they are accompanied by neurogliatype pituicytes. These pituicytes are closely associated with the nerve terminals. The neurosecretory axons terminate as large swellings, which contain irregular mitochondria, vesicles resembling synaptic vesicles, and fine "elementary vesicles" which represent the neurosecretory granules of light microscopy. These elementary vesicles are 1100-2800A in diameter, and it seems probable that they possess an outer lipoprotein membrane with electron dense material inside. Within the nerve terminals there are similar structures with contents of lower electron density, and others which appear empty, except for a few granules; it is probable that these represent various stages in the loss of neurosecretory materials. Since no granules can be found outside the terminals, Polenov and Belenki suggest that the neurosecretory neurons lose components of low molecular weight from their elementary vesicles. Many of the neurosecretory terminals make close contact with intermedia cells; none appears to associate directly with the blood sinusoids in *Raia clavata* and *Trygon pastinaca*.

There is no doubt that neurosecretory nerves make direct contact with the cells of the pars intermedia in many species of elasmobranchs. However, vascular connections could also be important. The neurosecretory axons deposit dense neurosecretion close to the neural lobe blood plexus of species such as *Squalus acanthias* and *Etmopterus spinax*. The neural lobe plexus drains into the sinusoids of the pars intermedia, and this could allow neurosecretory materials to pass from the pars nervosa into the pars intermedia. In those species which lack a neural lobe plexus, direct neural contacts may be more important, but it is interesting to note that these are also the species which possess a neurointermediate lobe component of their pituitary portal system, so that humoral materials could be supplied directly from the median eminance. In *Scyliorhinus caniculus*, the neurointermediate lobe receives its entire blood supply from veins which leave the brain, and there is no direct arterial supply. Neurosecretory materials may not only reach the sinuses of the pars intermedia, but they may pass on from the sinuses, through the interorbital vein, into the general circulation. However, in some species, such as *Raia clavata*, *loss* into the systemic circulation does not appear to be as easy, since there is no neural lobe plexus, and the neurosecretory axons do not appear to contact the blood sinusoids of the pars intermedia. Nevertheless, the nerve terminals may well be sufficiently close to the rich blood sinuses of the intermedia tissue for their active agents to reach the general circulation.

The Nature of the Neurohypophysial Principles of the Elasmobranchs

Early work on the biological activities of the elasmobranch neurohypophysis gave inconclusive results, mainly because of the low levels of activity which were present in the pituitary gland. Herring failed to find pressor effects or consistent changes of urine flow after the injection of skate pituitary extracts into cats. However, he obtained an oxytocic action on the isolated rat uterus after the addition of extracts from the pituitary of *Raia clavata*. His most remarkable observation was that of a rapid ejection of milk, after the intrajugular injection of an extract of skate pituitary into a lactating cat. At the time, he did not recognize this milk-ejection effect as a property of a neurohypophysial principle. In, Hogben and de Beer showed that skate pituitary extracts produced a weak oxytocic effect on the isolated guinea pig uterus. Although Hogben found that the extract caused a weak depressor effect when injected into a duck, he did not emphasize that this was evidence for an elasmobranch oxytocinlike principle. In 1936, Geiling and Le Messurier gave a brief report of "a suggestion" of pressor activity, and an "inconclusive" antidiuretic effect from extracts of the pituitaries of dogfish and skate (*Squalus acanthias*, *Raia sp.?*). However, in, Heller detected and assayed a low level of antidiuretic activity in glands from European dogfish and skate (*Scyliorhinus caniculus*, *Raia sp.?*). Nevertheless, by, Waring and Landgrebe bad failed to confirm the presence of either oxytocic or vasopressor activities in glands taken from living elasmobranchs, and they suggested that the earlier positive results were caused by the presence of histamine. However, this suggestion was not in full agreement with all the known facts. It was not supported by Herring's early observation of milk-ejection activity nor by his detection of oxytocic effects on the isolated rat uterus, which was known to be insensitive to histamine. It was made even less likely when Maetz *et al*. observed both oxytocic and natriferic activity in extracts from the pituitaries of *Scyliorhinus caniculus* and *Hexanchus griseus*.

The consensus of these different observations suggested to Perks, Dodd and Dodd that the elasmobranchs might contain an oxytocinlike principle. For this reason they examined elasmobranch neurointermediate lobe extracts by a combination of different biological assays, and determined the properties of the active agents by the use of inhibitors known to destroy mammalian neurohypophysial principles. They detected milk-ejection, oxytocic, and antidiuretic activities in extracts from *Squalus acanthias*, *Scyliorhinus caniculus*, *Raia clavata*, and *Raia batis*. The activities were relatively weak. The oxytocic activities ranged between 7.0 and 54.2 mU/mg acetone dried powder, and the highest value so far reported is only 58.5 $\pm$ 10.8 mU/mg, for *Dasyatis sabina*. Pressor effects could not be detected. The active agents were distributed preferentially in the neurointermediate lobe. They shared with oxytocin inactivation by sodium thioglycollate, sodium hydroxide, ultraviolet light, and chymotrypsin. In contrast to the mammalian vasopressins, the elasmobranch oxytocic agent was resistant to trypsin, and therefore it was unlikely to contain arginine or lysine. Despite the general parallel with the properties of oxytocin, there were quantitative differences in the ratios of different biological activities. In *Squalus acanthias*, *Scyliorhinus caniculus*, *Raia clavata*, *R. batis*, *Dasyatis sabina*, *Sphyrna nwkarran*, *Eulania milberti*, and *Carcharinus leucas* the ratios of milk-ejection: oxytocic:antidiuretic activities were relatively consistent, and averaged 3.4:1:0.03. This contrasted with ratios of approximately unity for mammalian (ox) pituitary powders, and of 1:1: 0.0067 for synthetic oxytocin. This discrepancy between the elasmobranch and mammalian extracts was confirmed in *Squalus acanthias* by W. H. Sawyer *et al*. These workers extended the activities which could be found in the extract to avian depressor activity, frog bladder activity, and stimulation of the hen oviduct. In addition, they detected a weak pressor effect; although no pressor activity had been found by Perks *et al*. in their earlier experiments, it is probable that this was masked by the strong vasodepressor substances which were present in their extracts. Sawyer and his co-workers found a twofold potentiation of the oxytocic effect by the presence of magnesium ions; again, this indicated the presence of a peptide other than oxytocin. It was clear that the elasmobranchs containecj a new oxytocinlike principle of high milk-ejection potency, free, or nearly free, of any vasopressor peptide: This was the first demonstration of a naturally occurring, neutral analog of oxytocin.

The pharmacological studies left the possibility that the observed activities could be the result of a mixture of oxytocinlike peptides. Heller and his co-workers attempted to resolve this possibility by paper chromatography. They subjected extracts from *Squalus acanthias*, *Scyliorhinus caniculus*, *Raia clavata*, and *Negaprion brevirostris* to paper chromatography in butanol: acetic acid: water, 4:1:5. The oxytocic activities were resolved into two peaks. The first was a fast-running component, designated "E2." It represented the major oxytocic component of the elasmobranch pituitary, and

its oxytocic effect was potentiated twofold by the addition of magnesium ions. The presence of a peak of similar mobility-although not necessarily of the same chemical constitution-has been confirmed by a number of other workers, and extended to include *Squalus acanthias* (Pacific variety = *Squalus suckleyi*), *Dasyatis sabina*, *Carcharinus leucas*, *S phyrna mokarran*, and *Eulania milberti*. Perks has shown that the behaviour of this component is parallel to that of oxytocin in a large number of chromatography systems, and it would appear to be an oxytocinlike, neutral principle, or principles. However, it is not the same as oxytocin, since it is potentiated by magnesium ions, and it possesses other biological differences from the mammalian principle. In *Squalus acanthias*, this major component can be purified by column chromatography, and it appears to have a relatively high potency (70 U/mg, W. H. Sawyer, 110 U/mg, Heinicke and Perks). Column partition chromatography of the purified product, on a system which would separate oxytocin, 8 Ile oxytocin, and 4 Ser, 8 Ile oxytocin suggests that it is a homogenous peptidebut this has not yet been confirmed by amino acid analysis.

The second oxytocic peak resolved by Heller and his group was a slow-running component, which moved with the R_f of a basic principle, similar to arginine vasotocin or the mammalian vasopressins. It was designated "E_1." It showed a distinctive spectrum of biological activities, which included a remarkable sixfold potentiation of its oxytocic potency by the presence of magnesium ions. However, in *Negaprion brevirostris* the potentiation was closer to the twofold value of the fast-running E_2 peak. This suggested that the properties of the slow-running component could vary in different species. Moreover, there was reason to suppose that it did not exist in all extracts. Acher and his co-workers did not observe a peptide which combined the properties of the E_1 principle during their purification and analysis of skate extracts. Other groups, using the same paper chromatography methods as those of Heller and his co-workers, failed to detect a principle with the properties of the E_1 peptide. Perks and Swiatkiewicz *et al.*Working with *Squalus acanthias* and other species, could only resolve a single fast-moving oxytocic peak with a uniform twofold magnesium potentiation throughout its length. At first, overloading artifacts were avoided by the use of relatively low doses of crude extracts, but later, extracts were subjected to gentle preliminary purification by gel filtration on Sephadex columns, and it was found possible to chromatograph quantities as high as 7000 mU of oxytocic activity. Even at such high levels, no E_1-type oxytocic principle could be separated on the chromatograms.

In 1967, W. H. Sawyer carried out essentially similar experiments on *Squalus acanthias* and found that the great majority of the oxytocic activity ran as a fast-moving, homogeneous E_2-type peak. However, he detected a slow-running moiety, present in trace amounts, which corresponded to arginine vasotocin. Swiatkiewitz *et al.* confirmed this by demonstrating a slow-running region of high frog bladder and antidiuretic activity in their chromatograms; however, in their experiments, the level of this arginine vasotocinlike principle was never sufficient to show up as a second oxytocic peak. The presence of arginine vasotocin was not unexpected since it had been found in trace amounts (1% of the total oxytocic activity) in *Squalus acanthias by W. H.* Sawyer and in *Raia clavata by* Acher *et al.* Heller and his co-workers did not measure pressor, antidiuretic, or frog bladder activities in their slow-running E_1 peak. Their estimations of avian depressor activity could have been affected by contaminating vasodepressor substances-and these have been shown to exist in elasmobranch extracts, at least in terms of the rat blood pressure. Further, the high magnesium potentiation, which was the most notable characteristic of the E_1 peak, has been shown to be a variable and unreliable criterion for identifying elasmobranch principles. Finally, the ratio of milk-ejection: rat uterus oxytocic activity (-$Mg^{2}+$) of the slow-running E_1 peak of *Raia clavata*, which was 1.77 (1.47-2.07), is unlikely to be significantly different from the values of approximately 1.8 and 2.2 given for synthetic arginine vasotocin by Berde *et al.* and by W. H. Sawyer, respectively. Therefore, it is possible that the slow-running principle of Heller and his co-workers represents arginine vasotocin, perhaps contaminated by small quantities of vasodepressor substances, and by variable amounts of the major oxytocic principle, which might trail behind if any degree of overloading had occurred. However, it is still possible that the slow-moving moiety is a new peptide, which might fluctuate with the season, reproductive cycles, or other unknown factors. Further studies are needed to resolve this situation, but we can be certain that most elasmobranchs contain at least one neutral oxytocinlike principle, together with small quantities of arginine vasotocin.

The first purification and amino acid analysis of a major elasmobranch principle was that carried out by Perks and Sawyer in 1965. Extracts from *Raia ocellata* were purified by gel filtration and ion exchange chromatography on CM-Sephadex resin. The purified product appeared to have a specific oxytocic activity of 17.6 U/mg, which was lower than that of the principle later purified from *Squalus acanthias* (70 U/mg; W. H. Sawyer, 110 U/mg, Heinicke and Perks,), and lower than that of pure oxytocin (430 U/ mg) . The purified peptide from *Raia ocellata* gave a ratio of rabbit milk-ejection : rat uterus : antidiuretic activities of 8.5: 1: 0.04. At the time, this unusually high milk-ejection potency suggested that it was a new principle, different from that found in other elasmobranch species. Amino acid analyses showed that it contained cystine, tyrosine, isoleucine, serine, asparagine, proline, glutamine, and glycine, in ratios compatible with a neurohypophysial principle. Soon afterward, Acher and his co-workers purified extracts from *Raia clavata*, *R. batis*, *R. naevus*, and *R. fullonica*. They found small amounts of arginine vasotocin, together with a major oxytocic component which contained the same amino acids as were present in the peptide from *Raia ocellata*. The principle was identified as 4 Ser, 8 Gln oxytocin. At the time, the high magnesium potentiation of this material (10-fold) appeared to differentiate it from the principle of *Raia ocellata* (2.6-fold), but this criterion has been shown to be variable and unreliable in elasmobranch work. The other biological data given for 4 Ser, 8 Gln oxytocin were insufficient to establish any possible identity between this peptide and the principle of *Raia ocellata*. However, Heinicke and Perks have shown that the oxytocic peptide of *Raia ocellata* has the same pharmacological spectrum as that of *Raia rhina*, and W. H. Sawyer *et al.*have demonstrated that there is a close parallel between the biological properties of the principle of *Raia rhina* and those of synthetic 4 Ser, 8 Gln oxytocin, when they are assayed *directly* against one another. Therefore, it is probable that 4 Ser, 8 Gln oxytocin is the main oxytocic principle of all species of skate so far examined.

In contrast to the skate, direct assays of extracts from *Squalus acanthias* against synthetic 4 Ser, 8 Gln oxytocin have shown marked differences between the two active agents. This is particularly true of their rabbit milk-ejection and rat antidiuretic potencies. Preliminary amino acid analyses do not suggest the presence of 4 Ser, 8 Gln oxytocin, but they raise the possibility of a serine-valine containing peptide or of a mixture of such peptides. It is too early to make any clear statement on the nature of the principle of *Squalus acanthias*, but it is clear that it is different from that of the skates. This raises the interesting possibility that the skates and sharks possess different neurohypophysial peptides.

It is concluded that the elasmobranchs possess traces of arginine vasotocin, together with at least one major neutral peptide. While the neutral peptide of the skates is probably 4 Ser, 8 Gln oxytocin, that of *Squalus acanthias*, and perhaps of the sharks in general, is another, unidentified principle or principles, probably containing valine and serine, and, according to recent results, alanine in addition.

The Actions of Neurohypophysial Principles in the Elasmobranchs

Little is known of the functions of neurohypophysial principles in the elasmobranchs. The few studies recorded have utilised mainly mammalian principles, which are different from those of the elasmobranchs themselves. Although there is no reason to assume that the functions of the principles bear any close relationship to those seen in higher vertebrates, most studies have been based on our knowledge of the mammalian pattern.

There is no evidence that neurohypophysial principles affect the water or salt balance of the elasmobranchs. The immersion of *Scyliorhinus caniculus* in hypertonic seawater, enriched to 10% NaCl, failed to change the pituitary content of neurohypophysial peptides, as judged by oxytocic assay. However, it is possible that hypothalamic production kept pace with systemic loss, and therefore this negative observation is inconclusive. The injection of 20 mU/kg (rat uterus oxytocic activity) of a crude extract of *Raia clavata* pituitaries into four nurse sharks, *Ginglymostoma cirratum*, caused no significant changes in their body weights, and it would appear that there is no water-balance effect analogous to that seen in the frog. An antidiuretic action on the kidney is unlikely, since marine elasmobranchs do not suffer the extremes of osmotic conditions which may affect a terrestial vertebrate, and their urine production appears to be constantly and notably low (12.9 ml/kg/day, *Squalus acanthias*, Clarke,15.8 ml/kg/ day, *Mustelus laevis*, Kempton,1.8 ml/kg/day, *Scyliorhinus caniculus*, Perks, Dodd *et al.*). Since water and osmotic balance are achieved by urea conservation, it is possible that investigations of the actions of neurohypophysial principles on urea metabolism could prove well worthwhile, but at the present time there are no data available.

The fact that neurohypophysial principles affect sodium metabolism in the lampreys and in the teleost fish might suggest that they have a similar action in the elasmobranchs. However, neither 8 lysine vasopressin, nor an arginine-lysine vasopressin mixture appears to affect the sodium-secreting rectal gland of *Squalus acanthias*. Perhaps the use of the oxytocinlike elasmobranch principles-or even oxytocin itself-would have been more rewarding. It would have been closer to the physiological situation in the elasmobranchs, and also more in keeping with our knowledge of the teleosts, where oxytocinlike neutral peptides have strong influences on salt metabolism, but 8 Lys vasopressin is ineffective.

Clearly, milk-ejection activity is meaningless in an elasmobranch. An oxytocic action would be more understandable, but no author has reported stimulation of the gravid or nongravid elasmobranch oviduct by either mammalian or elasmobranch extracts. Dreyer applied an ox pituitary extract (10 mU/ml, perfusing medium) to the isolated oviduct of *Raia erinaca*, *Squalus acanthias*, *Mustelus canis* and *Prionace glauca* and failed to obtain a response. He found that extracts from *Squalus acanthias* were equally ineffective. Similarly, Perks found that oxytocin in doses up to 40 mU/ml, or an extract equivalent to one-half of a neurointermediate lobe of *Squalus acanthias* per 62 ml of bathing medium, had no effect on the isolated oviduct of mature *Scyliorhinus caniculus* nor on uterine strips from near-term, pregnant *Squalus acanthias*. Effects on other smooth muscles are not as consistently negative. Dreyer failed to find effects of pituitary extracts on the stomach or intestinal muscle of the same species as those in which he tested the oviducts. However, M. E. Sawyer utilised high doses of whole ox pituitary extract (5000 mU, injected intravenously, intact preparation; 25 mU/ml supporting medium, isolated preparation) and observed rhythmic contractions of mesenteric muscle of dogfish and skate (*Squalus acanthias*, *Raia sp.?*). Movements of the pyloric stomach were also observed.

Despite the inconclusive effects of neurohypophysial principles on smooth muscle, it appears that the peptides may be capable of producing a vasopressor effect in the elasmobranchs. A marked and prolonged rise of blood pressure, with inhibition of heart contractions, was found after the injection of 5000 mU, whole mammalian pituitary extract (0.5 ml of Burroughs Wellcome obstetrical Pituitrin) into *Raia erinacea*, *R. diaphanes*, *R. scabrata*, or *R. stabuli f oris*). A similar rise of blood pressure occurred in the ventral aorta of *Squalus acanthias* after the injection of various mammalian preparations. In contrast, Waring *et al.* found that arginine-lysine vasopressin had little effect on the elasmobranch blood pressure, although there was a rise in the frequency and amplitude of the heartbeat. This apparent disagreement could be explained if the vasopressor effects within the elasmobranch circulation were vested in the oxytocin present in the mammalian preparations. If this were so, the oxytocinlike principles which predominate in the elasmobranch pituitary might be important pressor agents within their own species. Direct experiments are needed in this field.

Since attempts to determine the systemic effects of neurohypophysial principles have been so frequently doubtful or negative, even when extraordinarily high doses of principles have been used, it is reasonable to consider the possibility that they have only local effects within the pituitary. The histological evidence given previously suggests that it would be possible for neurosecretory materials to pass from the median eminence through the portal system to the adenohypophysis, where they might perform a regulatory function. Since each of the species examined by Meurling possessed either a component of the portal system which reached the neurointermediate lobe or a neural lobe sinus which could facilitate the passage of substances from the pars nervosa to the pars intermedia, it appears possible that the neurohypophysis could influence the pars intermedia by a vascular route. Besides this, nerve axons pass directly to the pars intermedia, where they form terminals on its cells. Mellinger has shown that damage to the preoptico-hypophysial tract of *Scyliorhinus caniculus* is always followed by melanophore dispersion and a consequent darkening of the skin. This is explained by the removal of an inhibitory action on the intermedia cells. With the removal of this inhibition, the pars intermedia cells liberate melanocyte stimulating hormone (MSH) into the circulation, and this hormone causes dispersion of pigment throughout the melanocytes of the skin. From Mellinger's experiments, it seems possible that the inhibitory influence is mediated by the neurohypophysis, and it is worth remembering that Knowles has distinguished individual types of neurosecretory fibers which pass either to the "synthetic" or to the "release" poles of the intermedii cells. Moreover, Knowles has noted that secretomotor junctions are formed between the neurosecretory fibers and areas of the pars intermedia cells which are distinguished by a sharply delimited endoplasmic reticulum. There was some degree of correlation between the state of the reticulum and signs of hormone liberation from the neurosecretory fibers. Knowles suggests that this would be compatable with the view that a neurohypophysial principle directly affects MSH synthesis in the cytoplasm of the intermedia cell, in the region of the endoplasmic reticulum. Despite these histological indications, it cannot be certain that the inhibitory effect results from neurosecretory fibers or their peptides, since nonneurosecretory axons are also present in the preoptico-hypophysial tract. Indeed, in recent work, Meurling *et al.* have succeeded in preferentially cutting the neurosecretory fibers of the infundibular stem of *Raia radiata*, a species in which the neurosecretory axons are localised in a medial position. They found that the release of MSH was little affected. However, section of the lateral, possibly adrenergic nonneurosecretory axons caused profound effects in which colour responses to a white background were abolished. It is clear that further work is needed to clarify the true significance of the close relationship between the neurosecretory fibers and the intermedia cells.

CARTILAGINOUS FISH: THE HOLOCEPHALIAN

The Structure of the Neurohypophysis of the Holocephalians

This ancient group of cartilaginous fish occupies a puzzling position in the evolutionary series. Its members are of particular interest because in some ways they may be intermediate between the elasmobranchs and the bony fish.

The holocephalians have a flask-shaped pituitary which recalls that of the elasmobranchs. However, it has an individuality of its own. Fujita, working in *Chimaera monstrosa*, recognised the presence of a unique and only partly formed neurohypophysial system in which the neurosecretory cells were located in the neurointermediate lobe, but this was not confirmed by Altner, who found a complete preoptico-hypophysial system in the same species. It is possible that the relative absence of neurosecretory material noted by Fujita may have been a reflection of some physiological or pathological state, since Sathyanesan and Jasinski and Gorbman have demonstrated a well-defined neurohypophysial system, rich in neurosecretion, in the related species, *Hydrolagus colliei*. The following general description of the neurohypophysis of *Hydrolagus colliei* is based on the work of these investigators.

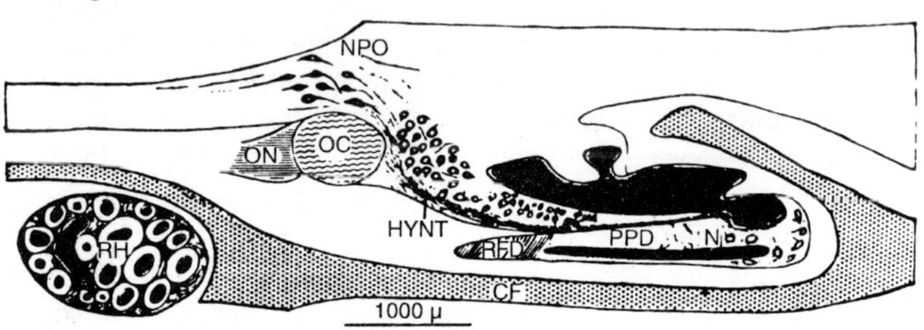

Figure 4.5 : The hypothalamus and pituitary of the Holocephalian, Hydrolagus colliei; *median sagittal section from a young specimen. Note the general similarities to the same region of the elasmobranchs. NPO = preoptic nucleus; HYNT = preoptico-hypophysial tract; ME = median eminence; NI = neurointermediate lobe; PPD = proximal pars distalis of adenohypophysis; RPD = rostral pars distalis of adenohypophysis; NLT = nucleus lateralis tuberis; OC = optic chiasma; ON = optic nerve; CF = cartilaginous floor of the cranium; and RH = Rachendach-hypophyse (ventral lobe of the pituitary).*

In *Hydrolagus colliei*, the preoptic nucleus lies anterodorsal to the optic chiasma, in a location similar to that of the elasmobranchs. It is diffuse in form but highly vascularised. The cells are irregular in shape, larger than those of elasmobranchs, and not divided by size into magnocellular or parvocellular varieties, as in the teleosts. The preoptic cells contain intensely staining, aldehyde-fuchsin-positive neurosecretory material. A few cells send beaded axons toward the forebrain, but the majority give rise to two groups of notably wide axons, which unite behind the optic chiasma to form a single preopticohypophysial tract. This tract is rich in neurosecretion, and accumulations similar to Herring bodies are found even in the preoptic region. The tract passes in a caudal direction, and runs below the elongated nucleus lateralis tuberis, which also contains aldehyde-fuchsin-positive material. In this region, the preoptico-hypophysial tract gives off ventral loops of nerve fibers, which terminate in swollen endings, full of neurosecretion, and closely applied to capillaries. These plentiful capillaries make shallow intrusions into the infundibular wall. They drain into a portal network, which is the sole blood supply to the rostral regions of the adenohypophysis ("pars distalis"). Both Sathyanesan and Jasinski and Gorbman regard this region of the infundibular wall as a functional median eminence, where neurosecretory materials might pass to the adenohypophysis and modulate its function. No neurosecretory axons appear to cross the vascular connective tissue between the infundibular floor and the rostral adenohypophysis.

Although a few neurosecretory axons terminate in the median eminence, most continue in a caudal direction and become more densely loaded with neurosecretion. Within the large neurointermediate lobe they terminate to form a pars nervosa, which contains a rich store of neurosecretion. It is remarkable for the presence of large ganglion cells, which have been seen in both *Hydrolagus colliei* and *Chimaera monstrosa*. The pars nervosa consists of tongues of neural tissue which penetrate intimately into the pars intermedia in a manner reminiscent of such elasmobranchs as *Scyliorhinus caniculus*. The lack of connective tissue between the pars nervosa and the pars intermedia allows fine axons bearing neurosecretion to penetrate the intermedia tissue. Here, they entwine the intermedia cells, or penetrate between them and terminate as swollen endings on the wide, irregular sinusoids of the highly vascular neurointermediate lobe. These sinusoids contain blood supplied from special hypophysial arteries, which are concerned with the neurointermediate lobe in particular. The sinusoids drain into the dorsal hypophysial veins, which lead into the general circulation. The anatomical and vascular connections described by both Sathyanesan and Jasinski and Gorbman suggest that the neurosecretory materials of *Hydrolagus colliei* could be concerned in both local control of the adenohypophysis and more general systemic functions.

In *Chimaera monstrosa*, Meurling has shown the presence of a pituitary portal system, which connects the median eminence to the adenohypophysis, and which might allow the activity of the adenohypophysial cells to be modulated. In this species, the vascular connections of the neurointermediate lobe are well adapted to passing neurosecretory materials through the intermedia tissue into the general circulation. The neurointermediate lobe shows a close intermingling of neural and intermediate elements and possesses not only wide blood sinuses but also a neural lobe plexus, which is absent from *Hydrolagus colliei*. Although this plexus might facilitate the passage of neurohypophysial agents out of the pars nervosa, Meurling was unable to find any accumulation of neurosecretion close to its blood vessels, as seen in the elasmobranchs *Squalus acanthias* and *Etmopterus spinax*. For this reason he considered that the neural lobe plexus did not serve for the release of neurohormones. However, he felt that this might occur more distally, so that neurosecretory materials could well have functions throughout the body.

The Nature of the Neurohypophysial Principles of the Holocephalians

In, W. H. Sawyer demonstrated that crude extracts from the pituitaries of *Hydrolagus colliei* showed a pharmacological spectrum which suggested the presence of oxytocin and arginine vasotocin. This was extended by the chromatographic separation of two principles, on both carboxymethyl cellulose and carboxymethyl-Sephadex columns. The basic principle, which accounted for only 4% of the total oxytocic activity of the crude extract, had pharmacological properties similar to arginine vasotocin. The second, neutral principle was subjected to partition chromatography on G-25 Sephadex resin; it moved as a homogeneous peak, which could be distinguished from both 4 Ser, 8 Ile oxytocin and 8 Ile oxytocin. Partition chromatography did not distinguish it from the neutral principle of *Squalus acanthias*, but it showed clear differences from this principle in its pharmacological spectrum. Partition chromatography did not distinguish the *Hydrolagus* neutral peptide from oxytocin, but in this case multiple biological assays against oxytocin itself showed a close parallel between the properties of the two active agents. It must be concluded that present evidence suggests that oxytocin is the natural, neutral principle of *Hydrolagus colliei*, but no final decision will be possible until this is confirmed by chemical analysis.

The presence of arginine vasotocin in *Hydrolagus colliei* is not a surprising finding, and the low proportion of this principle is reminiscent of the situation in the elasmobranchs. However, the possible presence of oxytocin in the holocephalians is a remarkable observation, since this principle is most closely associated with the mammals and its existence so early in the vertebrate tree is not in accordance with current theories of neurohypophysial evolution. It is another example of the individuality of the holocephalians.

The Actions of Neurohypophysial Principles in the Holocephalians

There is no direct evidence concerning the actions of neurohypophysial principles in the holocephalians. The most that can be suggested is that the apparent anatomical existence of a hypophysial portal system may indicate a local control of the adenohypophysis, while direct neural connections to the pars intermedia may suggest that there is a control of the intermedia cells. The direct drainage of the neurointermediate lobe sinuses into the general circulation is compatible with possible systemic functions of the holocephalian neurohypophysial principles.

BONY FISH: THE BRACHIOPTERYGIANS

The Structure of the Neurohypophysis of the Brachiopterygians

The correct classification of the brachiopterygians (or polypterine) bony fish is disputed, but it is likely that they are the closest living relatives to the early, extinct paleoniscoid stock, and a number of authors have recognised primitive characteristics in their pituitaries. However, most attention has been paid to the open oro-hypophysial duct which connects the adenohypophysis to the mouth, and the neurohypophysial system has not received careful attention until the recent work of Kerr. The following general account of the brachiopterygian neurohypophysis is based on the studies of Kerr, except where the work of other authors is indicated.

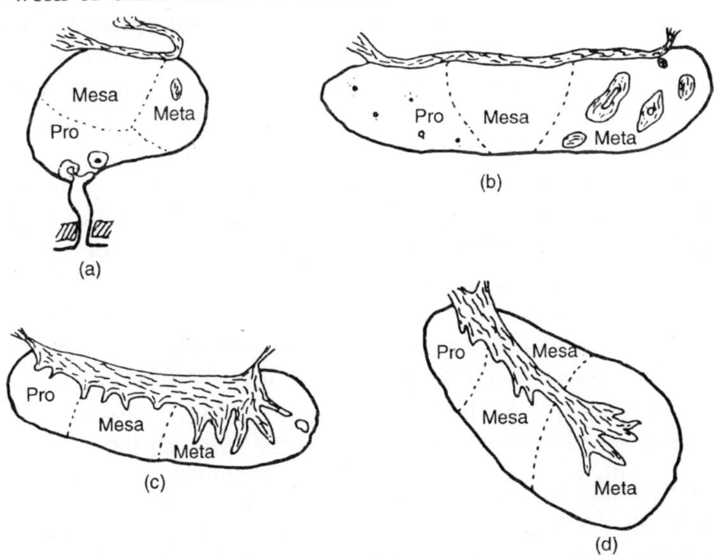

Figure 4.6 : The relationship between the pars nervosa and the adenohypophysis in various groups of bony fish: (a) Calamoichthys *(brachiopterygian),* (b) Lepidosteus (= Lepisosteus) *(holostean), (c) primitive teleost, and (d) advanced teleost. Neural elements are shown by short wavy lines. Pro indicates rostral pars distalis (proadenohypophysis) ; Mesa indicates proximal pars distalis (mesa-adenohypophysis); and Meta indicates pars intermedia (meta-adenohypophysis).*

The neurohypophysial systems of *Calanwichthys calabaricus*, *Polypterus senegalis*, and P. *bichir* are closely similar. The elongated preoptic nucleus is situated in a relatively rostral position and shows a weakly marked division into magnocellular and parvocellular regions (*Calamoichthys calabaricus*, *Polypterus bichir*, Charlton, species as above, Kerr,). The cells contain a considerable quantity of neurosecretory granules. There are no reports that they make connections with the ventricle, but they do form a preoptico-hypophysial tract. This tract is diffuse and ill defined at first, but it becomes discrete in the floor of the hypothalamus, just anterior to the adenohypophysis. One of the most notable features of the brachiopterygians (and some chondros-teans) is a tract which divides away from the main trunk in the infundibular floor; it forms a ventral branch, which is less well defined than the main tract. This branch becomes closely associated with a plexus of blood vessels, which may penetrate relatively deeply into the ventral surface of the infundibulum (species as above, Kerr). Neurosecretory granules have been seen within the palisade layer which borders these capillaries (*Polypterus sp.*, Wingstrand,). The capillaries appear to connect to the adenohypophysis and perhaps with the neurointermediate tissue; this strongly suggests that a median eminencepituitary portal system is present in the brachiopterygians. Further, this indicates that a pituitary portal system may have been a primitive feature of the bony fish.

The major trunk of the preoptico-hypophysial tract continues past the region of the median eminence and is deeply buried in the floor of the tubular infundibulum. It carries neurosecretory granules throughout its length, and these are particularly concentrated toward its ventral surface. Although a few fibers terminate in the saccus vasculosus, most penetrate into the caudal infundibular floor, which is thin in species of *Calamoichthys*, but may be thickened in species of *Polypterus*. The infundibular floor produces short, tubelike projections, which arise irregularly on both sides of the midline in a pattern which Kerr has considered to be primitive. The tubes, which penetrate into the underlying pars intermedia, soon consolidate into solid rods, which retain a central core of ependymal cells. They may contain large ganglion cells. The axons of the preoptico-hypophysial tract enter these projections, which are rich in neurosecretory material and constitute the pars nervosa. Herring bodies have been seen lining the outer surface of the neural processes, in close association with intermedia cells. The neural processes intermingle with the pars intermedia in a manner which may be more complex and intimate than that seen in the teleost pituitary. However, it is probable that there is little penetration into other regions of the adenohypophysis. Nevertheless, the pars intermedia, which is relatively small in some species, does not completely invest the neural processes in all specimens of *Polypterus*, and Kerr has noted one exceptional invasion of neural tissue into the anterior regions of the adenohypophysis of this species. This intrusion could foreshadow an evolutionary trend toward the extensive neural penetration seen in the

teleosts. The surface of each neural process is close to a rich capillary plexus. Some vessels pass between the neurointermediate tissue and other regions of the adenohypophysis, and they form a route by which neurosecretory materials might reach other areas of the pituitary. However, it is probable that neurohypophysial peptides could pass on into the general circulation.

The Nature of the Neurohypophysial Principles of the Brachiopterygians

Studies of crude and partially purified extracts from the pituitary of *Polypterus senegalis* have shown the presence of oxytocic, avian depressor, guinea pig milk-ejection, frog bladder, and natriferic activities. The high potency in promoting water reabsorption from the frog bladder, and the level of natriferic activity, indicate the presence of arginine vasotocin. In addition, a neutral, oxytocinlike peptide has been separated by both paper chromatography and purification on carboxymethyl cellulose columns. Early pharmacological data suggested that this neutral peptide was 8 Ile oxytocin, or a mixture of this peptide with 4 Ser, 8 Ile oxytocin. However, column partition chromatography, together with a direct comparison of the neutral peptide with synthetic 8 Ile oxytocin and synthetic 4 Ser, 8 Ile oxytocin, has strongly suggested that *Polypterus senegalis* contains 4 Ser, 8 Ile oxytocin. At present there is no chemical confirmation of the structure of the peptides of this species. However, the pharmacological data suggest the presence of arginine vasotocin and 4 Ser, 8 Ile oxytocin in the primitive brachioptergians, and this indicates that the pattern of the neurohypophysial peptides was set early in the evolution of the bony fish.

The Actions of Neurohypophysial Principles in the Brachiopterygians

There is no direct information on the actions of arginine vasotocin or of 4 Ser, 8 Ile oxytocin in this group. It is only possible to point out that the anatomical relationships of the pituitary are compatible with their possible importance in the control of the adenohypophysis. They could also reach the systemic circulation, and affect tissues throughout the body.

BONY FISH: THE CHONDROSTEI

The Structure of the Neurohypophysis of the Chondrostei

The Chondrostei, which include the sturgeons and the paddlefish, represent a distinct branch of the evolutionary tree. However, they may be closer to the main trunk of vertebrate evolution than the teleost fish. Their pituitaries are more primitive than those of the teleosts, and they are more closely allied to those of *Polypterus* and the brachiopterygians. In recent years, their neurohypophysial system has been described by Polenov and Barannikova, and the reports of these authors are the basis of the following general description.

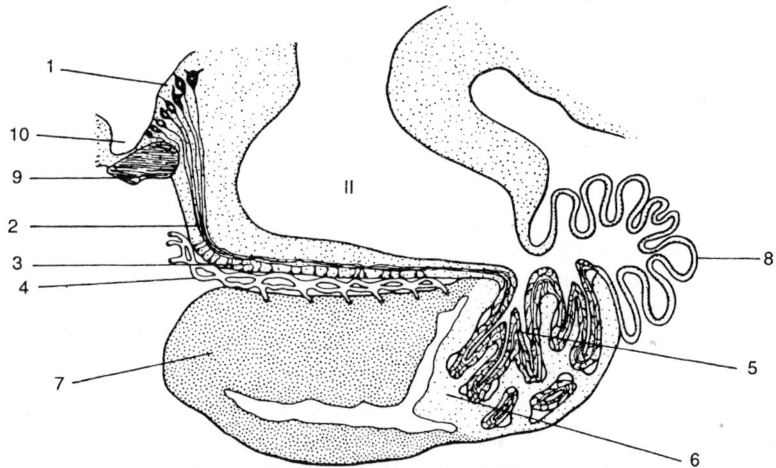

Figure 4.7 : The hypothalamus and pituitary of the sturgeons, Acipenser guldenstadt *Brandt and* A. stellatus *Pallas. Note the presence of a median eminence, and that the processes of the pars nervosa invade only the pars intermedia, which is divided from the rest of the adenohypophysis by a cleft. (1) Preoptic nucleus; (2) preopticohypophysial tract; (3) median eminence ("proximal contact region") ; (4) vascular network; (5) pars nervosa ("distal contact region") ; (6) pars intermedia; (7) pars distalis region of the adenohypophysis; (8) saccus vasculosus; (9) optic chiasma; (10) preoptic recess; and (11) third ventricle.*

In 1932, Charlton noted the presence of a relatively caudal and extensive preoptic nucleus in the paddlefish, *Polydon*. In the sturgeon, *Acipenser fulvenscens*, the nucleus lies in a similar position to that of the teleost fish. In *Acipenser giildenstadt* Brandt, *A. stellatus* Pallas, and *Huso huso L.* the nucleus is divided into a ventral region of small bipolar or unipolar cells, and a dorsal region of large multipolar cells. In *Acipenser fulvescens*, a number of the larger type of cells intrude among the smaller variety, so that the distinction into two zones is not as marked as in the teleost fish. The preoptic cells have single, eccentric nuclei, which are round or oval in shape. Their cytoplasm contains fine neurofibrils, Nissl substance, and clear vacuoles. More important, they contain neurosecretory granules, which are variable in size and less numerous than in the teleosts. According to Polenov and Barannikova, neurosecretion is formed in the perinuclear region and later spreads throughout the cytoplasm; although granules were found around the vacuoles, these authors do not consider that they were formed by the vacuoles themselves. The preoptic cells extend thick dendrites between the ependymal cells which line the ventricle of the brain. They give rise also to the preoptico-hypophysial tract. This tract carries neurosecretory granules mainly in its peripheral fibers and in some specimens granules may be found throughout its length. At first, in the region of the optic chiasma, the tract consists of symmetrically arranged bundles, densely packed with secretion, and close to the surface of the infundibulum. Over the adenohypophysis, these groups of fibers merge into a single, thin, broad bundle. In *Acipenser fulvescens*, this tract gives off a ventral branch, reminiscent of that seen in the brachiopterygians. This branch is a compact bundle of axons, which comes into close contact with the vascular connective tissue which lies between the infundibular wall and the adenohypophysis. Herring bodies are found in the infundibular floor. In contrast, Polenov does not describe a discrete ventral branch in *Acipenser giildenstadt* Brandt or *A. stellatus* Pallas. However, he gives a detailed description of a broad "neurosecretory contact area," where the loosely arranged, nonmedullated neurosecretory fibers of the preoptico-hypophysial tract give rise to an outer layer of fine radial axons. These axons form club-shaped, ribbonlike, or tapering terminal bulbs against the vascular connective tissue which separates the brain from the pituitary. There are no large Herring bodies in the area of contact, but the total amount of neurosecretory material is considerable because of the large size of the area involved. The connective tissue membrane becomes thinner wherever contacts are most intimate. It contains broad, sinusoidal capillaries, which penetrate the adenohypophysis in its caudal region. It is reasonable to suggest that this "contact area" is the homolog of a median eminence.

The majority of the axons of the preoptico-hypophysial tract continue caudally in the infundibular wall and spread out into a broad, thin bundle. At the end of the infundibular process, this neural lamella gives rise to large, hollow diverticuli and occasional solid processes, which form on either side of the midline. These protrusions branch and penetrate into the underlying intermedia tissue, and they constitute the pars nervosa. The pars nervosa is similar in form to that of the primitive brachiopterygian bony fish. Its penetration of the adenohypophysis is strictly restricted to the pars intermedia by the persistent adenohypophysial cavity, into which the neurointermediate complex protrudes. This primitive situation contrasts with the extensive general penetration found in the teleost fish. The pars nervosa of the Chondrostei contains large Herring bodies in which a central neural structure can be distinguished. These neurosecretory bodies are found in the tissues of the tubelike neural processes, within their ependymal lining, and even inside their central ventricular cavities. They also occur beneath the endothelial linings of the sinusoidal capillaries which cover the surface of the neural processes. These capillaries drain into veins which are separate from those of the rest of the pituitary, and empty through the cerebri media veins into the general circulation. It appears that neurohypophysial material could reach the rostral adenohypophysis through the primitive portal system, and it could also reach the general circulation through the sinuses of the neurointermediate lobe.

The, Nature of the Neurohypophysial Principles of the Chondrostei

The neurohypophysial principles of a number of Chondrosteans have received preliminary study by pharmacological methods. The levels of oxytocic and vasopressor activity detected in the pituitaries of *Polydon spathula* suggested that they contained only arginine vasotocin, and this was supported by the failure of paper chromatography to resolve a neutral principle. However, any neutral peptide accounting for less than 6% of the total oxytocic activity would have escaped detection. This is a serious possibility, since extracts from species of *Acipenser* were subjected to paper chromatography and contained a neutral principle which never amounted to more than 3.9% of the total oxytocic activity of the major hormonal constituent, which was probably arginine vasotocin. It can be concluded that the chondrosteans so far examined appear to contain mainly arginine vasotocin, with small, perhaps variable, amounts of an unidentified neutral principle. This is the reverse situation to that seen in the elasmobranchs.

The Actions of Neurohypophysial Principles in the Chondrostei

There are no direct experiments on the effects of neurohypophysial principles in the Chondrostei. Again the presence of a median eminence, and the close association of the pars nervosa with the pars intermedia, suggest that the neurohypophysis could be important in the local control of the adenohypophysis. However, anatomical considerations indicate that neurohypophysial principles could also escape into the general circulation. The only suggestion of their possible influence outside the pituitary comes from the histological studies of Barannikova and Polenov. These workers noted that there were seasonal changes in the distribution of neurosecretion within the neurohypophysis. Fish in the Volga showed neurosecretion localised mainly in the preoptic nucleus in the spring and mostly in the pars nervosa during autumn. Specimens taken in the North Caspian Sea, at the point of migration and spawning, possessed a preoptic nucleus rich in vacuoles and neurosecretion. Such observations are difficult to interpret. Barannikova and Polenov suggested that the neurohypophysis was concerned in "certain vegetative functions," and they appeared to be impressed by changes in the activity of the thyroid. However, comparison with other observations in the telepsts suggests that the principles in the Chondrostei might be important in spawning or egg laying.

BONY FISH: THE HOLOSTEI

The Structure of the Neurohypophysis of the Holostei

It is probable that the Holostei, the bowfins and the garpike, are more closely related to the teleost fish than are the other "preteleost" groupsthe brachiopterygians and the Chondrostei. Therefore, it is not surprising that their general pituitary structure has been said to be similar to that of the teleost bony fish. However, this is only partly true of the neurohypophysis, for the recent work of Sathyanesan and Chavin has shown that it has clear similarities to that of the other preteleost groups. The following general account is based on the results of these authors, except where otherwise indicated.

The neurohypophysial system is well developed in the various species of bowfin, *Amia*, and of garpike, *Lepidosteus*. The preoptic

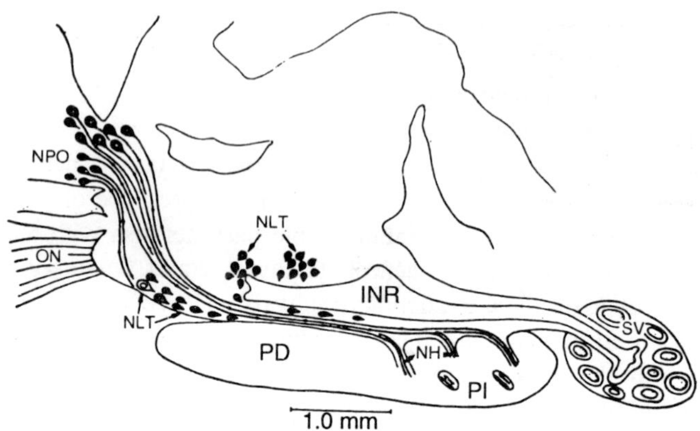

Figure 4.8 : The hypothalamus and pituitary of the garpike, Lepidosteus osseus, sagittal section. Note that the processes of the pars nervosa invade only the caudal region = pars intermedia. (ON) optic chiasma, (NPO) preoptic nucleus, (NLT) nucleus lateralis tuberis, (INR) infundibular cavity, (NH) pars nervosa, (PI) pars intermedia, (PD) pars distalis of adenohypophysis, and (SV) saccus vasculosus. From A. G. Sathyanesan and W. Chavin: Hypothalamo-hypophyseal neurosecretory system in the primitive actinopterygian fishes.

nucleus is relatively shor. In *Lepidosteus osseus* and *L. platostomus* it lies antero-dorsal to the optic chiasma, on both sides of the preoptic recess, and in a position similar to that of the teleost fish. However, in *Amia calva*, the preoptic nucleus is dorsal to the pituitary; this is because the gland is drawn forward rostrally in a remarkable manner, so that it lies beneath the optic chiasma. In *Amia calva*, the preoptic nucleus closely resembles that of the teleosts in being divided into a ventral region of small cells and a dorsal region of large neurons. However, in species of *Lepidosteus*, the distinction is less clear, since some large cells of the dorsal group intrude among the smaller variety, in a manner which is reminiscent of the sturgeon, *Acipenser fulvescens*. The preoptic cells contain variable amounts of aldehyde-fuchsin-positive neurosecretion. The well-marked preoptico-hypophysial tract which leaves the nucleus can be resolved into a number of groups of fibers, which pass caudally along the infundibular floor. These fibers run below the nucleus lateralis tuberis in *Amia calva* and through it in *Lepidosteus osseus*. The nucleus lateralis tuberis is much more extensive in the holosteans than in other preteleost fish, and it may add its own fibers to the tract. These are also bound for the pituitary. The preoptico-hypophysial tract which runs in the infundibular floor is rich in beaded droplets of neurosecretion and in Herring bodies. In *Amia calva*, the amount of stainable material is equivalent to that found in the pars nervosa-so great, in fact, that Sathyanesan and Chavin have suggested that the infundibular wall and the pars nervosa form a continuous functional unit. In *Amia calva*, *Lepidosteus osseus*, and *L. platostomus*, neurosecretory material has been seen in close association with blood vessels which penetrate the wall of the infundibulum. Frequently, Herring bodies surround these invading capillaries. The capillaries drain through the underlying connective tissue into the rostral region of the adenohypophysis. In the case of *Amia calva*, the dividing connective tissue is particularly thin, and the blood vessels which pass through it form wide sinusoids. These observations strongly suggest that the holosteans possess a simple median eminence and pituitary portal system, and this is in marked contrast to the situation in the teleost fish.

Many of the axons of the preoptico-hypophysial tract do not terminate on the infundibular blood vessels, but pass caudally, to form the pars nervosa. In *Lepidosteus* osseus and *L. platostomus*, there is no sign of a hypophysial stalk, and the ventral infundibular wall remains relatively flat throughout its length. The pars nervosa is formed by many hollow or solid neural processes, which grow downward and laterally from the caudal region of the infundibular floor. These processes penetrate the tissue of the pars intermedia, which is confined to the caudal region of the elongated pituitary. The general form is reminiscent of that seen in other preteleost groups, particularly in the sturgeons. In *Amia calva*, the general shape is different, but the essential relationships are similar. In this species, the infundibular wall folds outward to form a funnel which projects in a rostral direction below the optic chiasma and intrudes into the pars intermedia. Fingerlike processes arise from the funnel and ramify between the intermedia cells; unlike the neural intrusions of the teleosts, these processes are confined to the pars intermedia tissue. The processes of the pars nervosa contain a few large, argyrophylic ganglion cells, but they are more notable for their rich accumulation of neurosecretory granules and Herring bodies. In *Lepidosteus osseus*, the neurosecretory axons have been seen to terminate against blood vessels, presumably against those which form a plexus around the neural processes of the pars nervosa. It appears that neurosecretory materials could be liberated into these capillaries, and perhaps reach the general circulation; the presence of a pituitary portal system suggests that they might also reach the adenohypophysis.

The holostean neurohypophysis is typical of the preteleost bony fish in most of its characteristics. It is strikingly different from the teleost in its possession of a pituitary portal system. However, in a few points, such as the particularly clear division of magnocellular and parvocellular regions in the preoptic nucleus of *Amia calva*, there are hints of the evolution of the teleosts.

The Nature of the Neurohypophysial Principles of the Holostei

Pharmacological and chromatographic evidence suggests that the holosteans contain the same neurohypophysial principles as those of the teleost fish. Follett and Heller have used paper chromatography to resolve extracts from the pituitaries of *Lepidosteus platostomus* and *Amia calva* into two active componants. The slow-running principle had oxytocic, rat vasopressor, avian vasodepressor, and milk-ejection activities which were compatible with arginine vasotocin. The fast-running moiety gave oxytocic, avian depressor and milk-ejection effects which were quantitatively similar to 4 Ser, 8 Ile oxytocin. W. H. Sawyer investigated the peptides of the pituitaries of *Lepidosteus osseus* by both paper chromatography and column chromatography on carboxymethyl cellulose. He separated a principle with high frog bladder activity, apparently arginine vasotocin, from a neutral peptide which could not be distinguished from 4 Ser, 8 Ile oxytocin. The results of these two groups of workers confirm one another and suggest that the Holostei produce arginine vasotocin and 4 Ser, 8 Ile oxytocin. However, chemical evidence for these identities is still needed.

These results indicate that arginine vasotocin and 4 Ser, 8 Ile oxytocin were probably established as neurohypophysial peptides at an evolutionary stage before the modern holosteans diverged from the teleostean bony fish.

The Actions of Neurohypophysial Principles in the Holostei

There are no direct studies of the actions of arginine vasotocin or 4 Ser, 8 Ile oxytocin in the holostean bony fish. It can only be suggested that the histological evidence suggests a possible role in the control of the adenohypophysis and that systemic effects might well parallel those found in the closely related teleosts.

BONY FISH : THE TELEOSTS

The Structure of the Neurohypophysis of the Teleosts

The pioneer work of Scharrer, which was carried out on teleost fish, first showed the presence of neurosecretory cells in the hypothalamus. After Palay had traced neurosecretory material throughout the neurohypophysial system of a number of teleost species, both Bargmann and Scharrer utilised Gomori's chrome-hematoxylinphloxin stain to demonstrate the presence of a well-developed neurohypophysial system throughout a wide range of teleost fish. Since this time, the teleost pituitary has been widely studied, partly because of the easy availability of many species, and partly because of its interesting specialisations. Probably the most interesting of these is the remarkable penetration of neurosecretory axons throughout all the regions of the adenohypophysis. This is not only unique to the teleost fish, but it is also in marked contrast with the situation in the mammals. Recently, the structure of the teleost neurohypophysis and its relationship to the adenohypophysis has been investigated by means of the electron microscope, and the remarkable studies of Lederis, Knowles and Vollrath, and Leatherland have resulted in the teleost neurohypophysis becoming one of the best understood among the lower vertebrates.

Studies of a large number of teleosts have shown that their preoptic nuclei lie approximately dorsal to the optic chiasma on both sides of the third ventricle. There are variations in both the position and the length of the nucleus in different species, with a tendency for it to become located more rostrally in more advanced species. In the relatively primitive eel, *Anguilla anguilla*, each nucleus consists of a thin sheet of neurons which extends close beneath the ependyma of the third ventricle, in the shape of an inverted "L". Charlton's diagrams (1932) suggest that this pattern is relatively common, while Oztan's description of the preoptic nucleus of *Platypoecilus maculatus* as being arc-shaped may indicate slight variations from this basic theme. The cells of the nucleus are surrounded or separated by processes from the overlying ependymal cells, and by other structures which include glial cells, extracellular colloid droplets of unknown significance, and axons of unknown connections. In *Salmo irideus*, Follenius has seen synapses which contact the surface of the preoptic cells, and which may well be important in their control.

Unlike the preoptic nucleus of the elasmobranchs, that of the teleosts is divided into two contiguous regions which appear to differ mainly in the size of their cells. The ventral and more rostrally localised region is called the pars parvocellularis, and it consists of small, rounded, and closely packed cells, which are without any internal vacuoles in *Platypoecilus maculatus*. The dorsal region, which extends much further in a caudal direction, is termed the pars magnocellularis, and it consists of those large cells which first called attention to the neurosecretory activity of the hypothalamus. Its cells are loosely packed in *Anguilla anguilla*, occur in small clusters in *Carassius auratus*, but form larger groups in *Gad us morrhua*, *Cottus bubalis*, and *Salmo irideus*. Their shape has been described by a variety of terms-flask-shaped in *Carassius auratus*, fusiform or round in *Porichthys notatus*, and polygonal in *Platypoecilus maculatus*. However, Stutinsky has pointed out that in *Anguilla vulgaris*, the cells close to the ependyma are bipolar while the majority of cells, which lie deeper, are rounded, polyhedral, and multipolar.

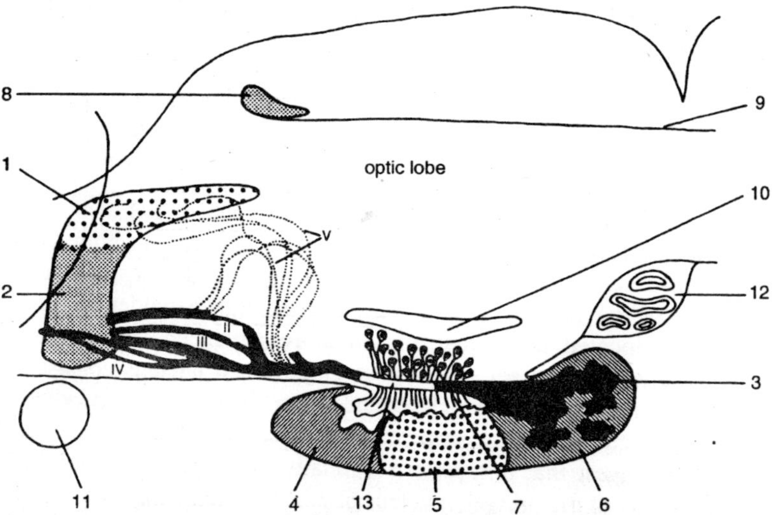

Figure 4.9 : The hypothalamus and pituitary of the teleost, Anguilla anguilla, *to show the preoptico-hypophysial tract; diagrammatic representation. (1) Preoptic nucleus, pars magnocellularis; (2) preoptic nucleus, pars parvocellularis; (3) pars nervosa; (4) rostral pars distalis or pro-adenohypophysis; (5) proximal pars distalis or mesoadenohypophysis; (6) pars intermedia or meta-adenohypophysis; (7) nucleus lateralis tuberis; (8) subcommissural organ; (9) Reissner's fiber; (10) third ventricle; (11) optic chiasma; (12) saccus vasculosus; and (13) subterminal region. I-V indicate axonal bundles (tracts) from the preoptic nucleus.*

The nuclei of the preoptic cells contain a well-marked, rounded nucleolus. These nuclei are eccentrically placed in the cells, and although sometimes round or oval in shape, they become irregular and folded in appearance, owing to deep invaginations of the cytoplasm, which may even include secretory granules. It is possible that extreme indentation of this type could account for the multinucleated cells seen in *Anguilla vulgaris by* Stutinsky, and in the carp and sazan by Polenov. Enami has observed similar indentations in *Anguilla japonica*, and in addition he has described a remarkable transformation of whole cellular nuclei into structureless, Gomori-negative colloid. This fundamental change has not been reported by any other workers. However, Oztan (*Platypoecilus maculatus*) has seen differences in nuclear structure which may have reflected different states in the activity of the preoptic cells, and Polenov (carp, sazan) has observed apparently pycnotic nuclei in cells which were densely loaded with neurosecretion.

Outside the nucleus, the cytoplasm of the neurosecretory cells can be divided into an inner and an outer zone. Usually, the outer, peripheral zone contains only the membranes and cysternae of the endoplasmic reticulum (Nissl substance), but electron-lucent vesicles have been seen in the eel, *Anguilla*. In contrast, the inner, central zone contains many inclusions, some of which do not fall into any ready classification. Nevertheless, rod-shaped mitochondria have been seen in many species, for example, in *Gadus morrhua*. Large, round vacuoles, or smooth surfaced vesicles with electron-lucent contents have been described in *Carassius auratus*, *Gadus morrhua*, and *Zoarces viviparus*. Scharrer, working with *Opsanus tau*, has reported that similar large vesicles may have openings which expose their finely granular contents to the outer cytoplasm, but this may be an artifact of fixation. In addition, Scharrer has reported the presence of cilia in the neurosecretory cells of this species. The Golgi complex is found in the central zone, and it is well developed; it has been described as consisting of narrow tubules and vesicles, or as groups of agranular membranes, in *Gadus morrhua*, *Cottus bubalis*, *Salmo irideus*, and *Perca fluviatilis*. There is almost universal agreement that the Golgi complex produces the elementary vesicles (= elementary granules) which are the neurosecretory material, as seen under the electron microscope.

The elementary vesicles, which can occur throughout the central zone of the cytoplasm, have a complex structure. The outer surface is smooth and consists of a single membrane, and immediately within it is a narrow, clear zone. The center of the vesicle is full of a fine, osmiophylic, electron-dense, and perhaps lipid-containing granular material. These elementary vesicles, which are seen in the cytoplasm of the preoptic cells, are larger than the closely similar structures found in the neighboring nucleus lateralis tuberis. Although their size may vary between different species, it is often possible to divide the elementary vesicles of a particular species into two distinct ranges.

If these correspond to the similar vesicles present in the pars nervosa, as would seem reasonable, studies in *Gadus morrhua* would suggest that arginine vasotocin is associated with those of smaller size. Leatherland and Dodd have made the important observation that the two different sizes of elementary vesicles present in the eel, *Anguilla anguilla*, can be correlated with two different types of preoptic cell. The larger elementary vesicles, 2150 $\pm$ 30 A in diameter, are associated with a more electrondense type of cell, which contains swollen and irregular cysternae in its cytoplasm. In contrast, the smaller elementary vesicles, 1627 $\pm$ 31 A across, are found in more electron-lucent preoptic cells, which possess a narrow and regular endoplasmic reticulum. These results lead to the interesting speculation that the two neurohypophysial hormones of the teleost fish may well be made by different cells within the preoptic nucleus.

Studies with the classic staining techniques suggest that neurosecretory materials, once formed, concentrate around the nucleus of the cell, and then spread out into its processes. In some species, the secretion may leave the cell in greater quantities at particular seasons, perhaps in correlation with the breeding cycles; such seasonal variations in stainable neurosecretoo, and in the cytology of the preoptic cells, have been reported in the carp, the sazan, in *Salvelinus leucomaenis pluvius*, and in *Anguilla anguilla*. Oztan has noted reduced numbers of elementary vesicles and smooth-surfaced vesicles in the preoptic cells of *Zoarces viviparus* during summer, and also in response to continuous light.

Neurosecretor products may leave the preoptic nucleus by a number of possible routes. Since the preoptic cells are intimately associated with a rich capillary plexus, the possibility of a direct passage into the blood of the hypothalamus cannot be discounted at present. Indeed, there are observations which suggest that it could occur. Oztan has observed that dendrites from the preoptic cells connect to nearby capillaries in *Platypoecilus maculatus*, and Sathyanesan has noted that section of the preoptico-hypophysial tract, followed by "activation" of the system by continuous light, will produce dilation of blood vessels around the preoptic nucleus of *Porichthys notatus*. However, in general, this direct route into the hypothalamic blood vessels is not considered to be important.

In some species, neurosecretory materials may pass to the cerebrospinal fluid. In a number of species of *Anguilla*, it is clear that thick dendrites or even whole preoptic cells may protrude into the third ventricle, where they are sometimes associated with a coagulum. Enami has suggested that the dendrites continue as Reissner's fiber and carry neurosecretory granules to the subcommisural organ for storage; however, these interesting possibilities have not yet been confirmed, and Leatherland and Dodd have demonstrated staining differences between the Gomori-positive materials of the preoptic nucleus and those of the subcommisural organ. Dendrites-sometimes with granules-have been seen to reach the ependyma, and ventricle of the carp, the sazan, and the goldfish, *Carassius auratus*. However, this may not be a universal situation in the teleosts, for Oztan has remarked on the absence of such connections in *Platypoecilus maculatus*. This observation would prevent the general acceptance of Knowles and Vollrath's suggestions that the ventricular dendrites are concerned in feedback mechanisms which control the preoptic nucleus. Nevertheless, such connections are well marked in a number of teleosts, and it is reasonable to suggest that they are concerned either with sensory functions or with the conduction of neurosecretory products to the ventricle. Despite such alternative routes, most workers consider that neurosecretory materials leave the preoptic cells down the main axons of the preoptico-hypophysial tract.

The preoptic cells give rise to broad hillocks, which contain all the structures of the main cell body, and in addition, long, thin, straight canaliculi which have led Palay to suggest that the axons which they produce are, in fact, enlarged dendrites. These fleshy, undulating, unmedullated axons form the preoptico-hypophysial tract. They pass out of the nucleus in an irregular and diffuse manner, often leaving in a lateral direction and then bending ventrally before forming a discrete tract in the infundibular floor. In the stickleback, *Gasterosteus aculeatus*, there is an extreme condition in which the axons cross much of the infundibular floor as separate, lateral tracts, and only come together to enter the pituitary itself. The tract has been particularly clearly mapped in the eel, *Anguilla anguilla*, by Leatherland *et al.*, 1966. Here, the pars parvocellularis gives off four well-defined and ventrally located tracts, which turn and run in a caudal direction for some distance, before they are joined by a diffuse curtain of axons which descends from the large, dorsally located cells of the pars magnocellularis; all these axons unite to form a single tract which converges on the midline, above the optic chiasma, and continues either past or through the nucleus lateralis tuberis toward the pituitary.

The main preoptico-hypophysial tract of many teleosts has been shown to carry rich accumulations of neurosecretory material, often in a beaded form. Oztan has remarked on the presence of Herring bodies in the tract of *Platypoecilus maculatus*, and she has added the interesting observation that there is an increase in the neurosecretion present in the tract of sterile fish produced as a back-cross between this species and *Xiphophorus helleri*. Follenius has examined the neurosecretory droplets present in the tract of *Salmo irideus* by electron microscopy, and he has shown that they consist of "packets" of elementary vesicles, essentially similar to those present in the cell body, but now grouped into discrete masses. There is no information concerning the mechanism of formation of these masses. Follenius has shown that the elementary vesicles do not increase in size as they move down the tract, and he has concluded that there is no extra synthesis by the axons themselves. In some species, such as *Salmo irideus*, similar neurosecretory elementary vesicles have been seen in the nearby nucleus lateralis tuberis. This nucleus, which is close to the preoptico-hypophysial tract, and which is sometimes penetrated by it, also connects to the pituitary. It shows marked seasonal activity. However, its vesicles are smaller and have different staining properties from those of the preoptic nucleus; the two types of vesicle are almost certainly biologically dissimilar.

Although the main preoptico-hypophysial tract is often rich in neurosecretory granules in its more caudal regions, Leatherland *et al.* and Leatherland have observed that the tract of the eel, *Anguilla anguilla*, shows a small caudal accumulation of neurosecretion, which is followed by a short empty area. This area, which was termed the "subterminal area," lies above the anterior margin of the adenohypophysis, and it may represent the terminations of certain neurosecretory axons. It was not possible to relate any nerve terminations to blood vessels, but it is known that a small number of capillaries unite the vascular network of the brain with the pituitary in this species. However, these capillaries may connect to the pars nervosa. Similar scanty capillary connections between the brain and the pituitary have been noted in *Lucioperca lucioperca* and in *Salvelinus fontinalis*, and it may be pointed out that even rare capillary connections could be important if their transmitter materials were present in high local concentrations and if they were precisely directed. However, it is only in *Phoxinus phoxinus* that there has been any claim to demonstrate definite portal vessels, with Gomori-positive granules associated with their hypothalamic capillary network. In contrast, several groups of workers have felt that the teleosts lack a conventional pituitary portal system. Henderson's recent studies of the vascular supply of the pituitary of *Salvelinus fontinalis* have emphasised the possibility that in teleosts the portal vessels are replaced by a capillary network which is completely internal to the pituitary. This capillary bed joins the rostral region of the pars nervosa directly to the adenohypophysis, and although it is anatomically different from a classic portal system, it may well be its functional equivalent. It can only be concluded that the conventional pituitary portal system of the teleosts is at least anatomically underdeveloped and that it may even be absent. In connection with these problems, it is interesting to note that Sathyanesan has shown that hypophysectomy of *Porichthys notatus* resulted in the development of a new relationship between the cut stump of the infundibulum and the local blood vessels. This was accompanied by an accumulation of neurosecretory material at the central end of the severed axons, and a similar accumulation has been seen after hypophysectomy in *Anguilla anguilla* and in *Lophius piscatorius*.

The connection of the preoptico-hypophysial tract to the pituitary is notably different in *Anguilla anguilla* and in *Lophius piscatorius*. The pituitary of *Anguilla*, and of other relatively primitive teleost fish, is closely applied to the infundibulum, in a way which is reminiscent of the preteleost groups. On the other hand, the pituitary of *Lophius*, and of many other more advanced teleosts, is suspended from the brain by a long stalk, which contains the distal preoptico-hypophysial tract. *Lophius* represents an extreme case, since its pituitary is not merely suspended and rotated, as in *Esox lucius* and *Cyprinus carpio*, but it is drawn out to lie rostral to the brain. This unusual stalk has been used in an attempt to determine whether neurosecretory cells are capable of conducting nervous impulses. Potter and Loewenstein detected impulses which were similar in conduction velocity (0.5 meter/ sec) and in refractory period (3-10 msec) to those of unmyelinated, Type C fibers from the frog. When these investigators stimulated the hypothalamus close to the base of the peduncle, the electrical responses of the stalk were variable and delayed, which suggested the presence of synapses between neurosecretory cells and other nerve cells. These results are interesting, but they are open to slight doubt since a few nonneurosecretory fibers are known to be present in the stalk. In some cases, the stalk appears to be a hollow structure; Sathyanesan has noted that the cavity of the third ventricle extends into the remarkably long peduncle (4-5 mm) of *Porichthys notatus*. In the case of *Platypoecilus maculatus*, the stalk is funnel-shaped, and it shows a decussation of nerve fibers at its narrowest point; after this the axons continue into the pars nervosa, in association with blood vessels.

The preoptico-hypophysial tract of the more primitive teleosts, such as *Anguilla*, enters the pituitary gland medially, and toward its caudal region. It forms a dorsal area of neural tissue, which constitutes the most discrete part of the pars nervosa. This pars nervosa is essentially a thickening of the infundibular wall from which processes project ventrally; in' structure, it is reminiscent of that of the preteleost bony fish. In more advanced teleosts such as *Lophius piscatorius* or *Cyprinus carpio*, the relationships may appear to be altered by variations in the direction of the incoming stalk, and by the tendency of the adenohypophysis to envelop the neural component, so that the pars nervosa becomes the central core of the gland.

However, the pars nervosa is often hard to delineate precisely in the teleosts: Early workers realised that the adult teleost pituitary had no clearly separate lobes, and that the various regions were fused into a complex, which, although relatively consistent throughout the group, was unlike that found in the higher vertebrates. Although there is a clear division between the neural and the adenohypophysial tissues in juvenile teleosts, one of the most notable features of the adult pituitary is the penetration of the neural component into all parts of the adenohypophysis. In adult teleosts, the pars nervosa forms digitate processes which branch and form rootlike structures that penetrate deeply into the adenohypophysial tissue. In the eel, *Anguilla anguilla*, the pars nervosa retains primitive characteristics, for its caudally directed neural processes are penetrated by narrow canals which extend down from the third ventricle in a situation reminiscent of the wider canals seen in preteleost groups. In a contrasting situation found in adult specimens of *Hilsa ilisha*, Sathyanesan has described processes of the pars nervosa which pass between adenohypophysial ducts; these unusual ducts connect to a patent oro-hypophysial canal.

In accordance with the early suggestions of Diepen, Leray and Stahl divided the pars nervosa of *Mugil cephalus* into two main regions- a rostral region which penetrated the meso-adenohypophysis, and a caudal region which ramified intimately into the pars intermedia. The rostral region contained less neurosecretory material than the caudal division. It passed through the dorsal areas of the pituitary, close to blood vessels and came into association with acidophyl cells of the meso-adenohypophysis. The situation in *Hippocampus guttulatus* was similar, except that the fibers passed to basophylic, gonadotropic cells present in the same region of this species. Similar processes, characterised by the absence or near absence of stainable droplets of neurosecretion, have been seen in other species such as *Gasterosteus aculeatus*, *Perca fluviatilis*, *L ebistes reticulatus*, and *Anguilla anguilla*. Stutinsky has suggested that in the eel, A. *vulgaris*, these less heavily stained tracts may have originated partly in the nucleus lateralis tuberis, and recently this suggestion has received support from the electron microscope studies of Knowles and Vollrath. Knowles and Vollrath found that in A. *anguilla*, the greater proportion of the nerve fibers which invaded the *pro*- and meso-adenohypophysis did not stain by classic neurosecretory methods; but, nevertheless, they contained irregular neurosecretory vesicles, with central electron-dense granules, about 700 A in diameter (Type B). Similar structures were found in the nucleus lateralis tuberis. The neural processes within the rostral regions of the adenohypophysis contained few pituicytes and no central canals to link them to the ventricles. They appeared to discharge neurosecretory products into the perivascular and intervascular spaces which permeated the rostral adenohypophysis. Knowles and Vollrath suggested that these spaces represented the hypophysial portal system, while the fibers themselves corresponded to an internal median eminence. This is in agreement with the suggestions of other authors, who felt that the.rostral region of the pars nervosa, as recognised by Diepen and by Leray and Stahl, should be regarded as the functional equivalent of a median eminence. However, it must be remembered that Knowles and Vollrath also recognised that the rostral neural digitations contained a smaller number of axons with classic, Type A elementary vesicles, 1400 A in diameter, and similar to those of the preoptic nucleus. The classic neurosecretory materials were rare in young specimens but became more abundant during maturity.

The great majority of the classic, Gomori-positive, nonmyelinated neurosecretory fibers (Type A) of the preoptico-hypophysial tract appear to enter the fingerlike processes of the caudal region of the pars nervosa. These penetrate extensively and intimately into the pars intermedia. The caudal neural processes contain a number of different structures. A few myelinated nerve fibers have been seen in *Perca fluviatilis* and *Anguilla anguilla*. Ganglion cells occur in *Salmo trutta*, *Gadus* sp., and *Mystus vittatus*. Pituicytes has been recognised in *Salmo irideus*, *Cymatogaster aggregata*, *Gasterosteus aculeatus*, *Lebistes reticulatus*, and Anguilla *anguilla;* in the last two species they appear to be localised mainly in the center of the neural digitations, where their processes isolate and envelop groups of neurosecretory fibers. The pituicytes have been particularly well described in *Anguilla anguilla*, by Knowles and Vollrath. Here they possess a deeply indented nucleus with cytoplasm which is either electron dense and vacuolated or electron lucent and filled with fibrilli. In this relatively primitive species, an epithelial type of pituicyte, with cilia or thin processes, lines the central ventricular canal which extends down the center of each neural process. The pituicytes extend from this central canal in a roughly radial manner and reach the outer surface of each process, where they contact the external intervascular space, which forms a boundary surrounding each neural digitation. The axons of the preoptico-hypophysial tract pass down between the radial pituicytes, which divide them into bundles. These bundles carry droplets which stain with the classic neurosecretory stain, and the caudal processes of the pars nervosa form the chief area for the storage of neurosecretion. It is possible that species living in freshwater may store larger quantities of neurosecretion than those in seawater (*Salmo irideus*). In general, neurosecretory material may occur throughout the processes, or it may be densely packed toward the boundary with the pars intermedia. It occurs as irregular granules, or as beadlike droplets, but Herring bodies, which are present in adult fish, appear to be absent in juvenile specimens. Electron microscopy has shown that the neurosecretory material consists of elementary vesicles similar in size and form to those of the preoptic nucleus. If the results of different workers are compared, there are apparent discrepancies between the sizes of these vesicles in different parts of the neurohypophysial system, even within one species. However, these differences probably result from variations in fixation procedures, since Leatherland has noted that the sizes of elementary vesicles in the preoptic nucleus and in the pars nervosa are identical, if they are compared in the same preparation. Knowles and Vollrath, working in *Anguilla anguilla*, have suggested that the two size ranges found in these elementary vesicles indicate that the caudal pars nervosa contains two types of nerve fibers; one possesses elementary vesicles of 16001800 A in diameter (Type A,), while the second contains elementary vesicles 1200 A across (Type AZ). It is tempting to speculate that these may correspond to the two hormonal peptides found in the teleost neurohypophysis. Lederis has noted that in *Gadus morrhua* the pressor activity typical of arginine vasotocin is associated with the relatively small elementary vesicles, 800-2000 A in diameter, but the source of the 4 Ser, 8 Ile oxytocin has not yet been determined. Knowles and Vollrath have recognised the presence of a few scattered fibers with small irregular vesicles, 700 A in diameter, in the caudal neural processes; these are similar to the Type B fibers of the rostral region of the pars nervosa, so that it is probable that the fibers typical of each region are not strictly segregated from one another.

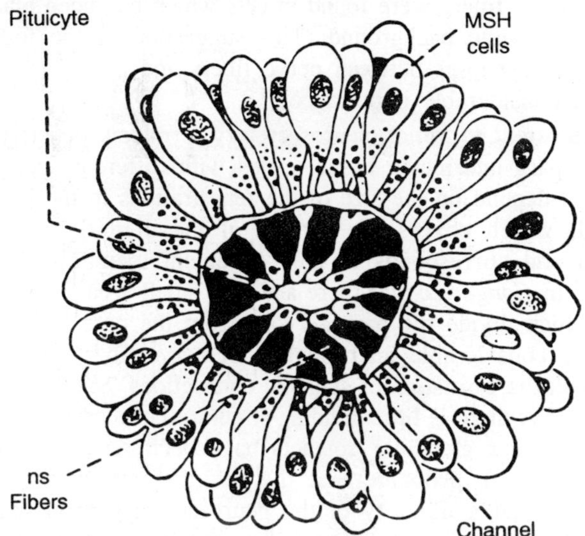

Figure 4.10 : Diagrammatic representation of the cross section of a neural process of the pars nervosa of the teleost fish, Anguilla anguilla. *The process is surrounded by cells of the pars intermedia (MSH cells).*

In general, the nerve fibers terminate in swellings, which appear under the light microscope either as varicose expansions of fine fibers or as greatly hypertrophied structures often termed "Herring bodies. Electron microscopy shows that the swellings contain mitochondria and large numbers of small, relatively electron-lucent vesicles, 500A in diameter, which are generally regarded as synaptic vesicles. In *Anguilla anguilla*, these synaptic vesicles may be clustered in the center of the fibers or they may lie along an electron-dense area next to the neural membrane. In some cases, the neural membrane is thickened at this point. These endings, which are apparently synaptic terminals, also contain elementary vesicles; and in *Anguilla anguilla* these structures appear to break down, lose their electron density, and perhaps release small particles into their surroundings.

The terminals of the neurosecretory fibers can be found in four main places: They occur on pituicytes, within the pars intermedia, on the outer intervascular channel, or around capillaries. Knowles and Vollrath have given a clear description of synapses between neurosecretory fibers and pituicytes in both *Anguilla anguilla* and *Conger conger*. The number of synapses between pituicytes and Type A_1 terminals appeared to increase when specimens of *Anguilla anguilla* were transferred to seawater, while a greater number of synapses with Type A_2 fibers were found in eels which had been placed on an illuminated white background. This suggestion of the formation of "transient" synapses between nerve fibers and pituicytes may prove to be of considerable importance.

The extent to which neurosecretory terminals penetrate directly into the pars intermedia tissue varies with different species, and it may well depend on the size of the fenestrations in the intervascular channel, which covers the surface of the digitations of the pars nervosa and corresponds to the "basement membrane" of earlier authors. In *Conger conger* there are only small areas of intervascular space which divide the pars nervosa from the pars intermedia, and one finds considerable intermingling of nerve fibers and intermedia cells, with the formation of synaptic connections. Similarly, in *Gadus* and *Phoxinus* there is little dividing structure between the two regions, and direct neuroglandular contacts have been reported. In *Lebistes reticulatus* it is probable that the dividing structures are partly developed, and a minority of the neurosecretory axons end against pars intermedia cells. In *Salmo irideus* the barrier is probably better formed, and only a few fibers pass through it, while in *Perca fluviatilis*, where the dividing structures are well developed, nerve terminations in the pars intermedia are rare. In *Anguilla anguilla* a few neurosecretory axons of type A_2 pass through gaps in the outer barrier (intervascular channel), but no synaptic terminals have been found on intermedia cells.

In *Anguilla anguilla*, by far the majority of the neurosecretory fibers (A_1, A_2, and B) terminate as a mass of rosettelike structures on the fine intervascular channel which envelops the outer surface of the neural processes. This intervascular channel is thought to correspond to the PAS-positive "basement membrane" seen by earlier optical methods. At intervals it expands to encompass capillaries within its inner space, and it would appear to be capable of transferring neurosecretory materials to either the pars intermedia cells on its outer side or to the capillaries within its framework. Jasinski has investigated the vascular supply of the pituitary of *Anguilla anguilla* and has shown that these capillaries receive their blood supply from arterioles on the dorsal surface of the pituitary; they form a fine network over the surface of the digitate processes of the pars nervosa, and although they do not penetrate the relatively avascular pars intermedia, they send ramifications into the pars nervosa tissue. Similar networks of capillaries on the surface of the neural processes, and to some extent within them, have been seen in many other species. In many teleosts these capillaries are closely associated with neurosecretory terminals, and it seems reasonable to suppose that active agents are passed into them (*Lebistes reticulatus; Fol*lenius and Porte; *Porichthys notatus;* Sathyanesan; *Salvelinus fontinalis;* Hill and Henderson; Henderson). Follenius and Port believe that most of the blood from the pars nervosa of *Lebistes reticulatus* drains directly through the hypophysial vein into the systemic circulation, so that active materials could reach many tissues. However, Hill and Henderson have suggested that in *Salvelinus f ontinalis* capillaries of the pars nervosa drain into all regions of the adenohypophysis, and that these vessels could carry neurohypophysial factors which might be important in adenohypophysial control. Recently, Henderson has greatly clarified the problem of the destination of the active agents of the neurohypophysis by a detailed analysis of the vascular supply of the pituitary of this species. She has emphasised that the pituitary is served by two largely independent arterial supplies. The first vascular route is well adapted to carry neurohypophysial products from the rostral region of the pars nervosa to the rostral and proximal divisions of the pars distalis (adenohypophysis). The second vascular circuit irrigates the caudal pars nervosa, and would appear well adapted to carry material out into the general circulation. Therefore, it is probable that active agents could be supplied both internally to the adenohypophysial cells and externally to the body tissues. However, there is no guarantee that the active agents carried in each of the two circuits are necessarily chemically similar.

It is clear that the teleost neurohypophysis has undergone considerable specialisation. Two specialisations appear particularly important: one is the absence, or profound reduction of the pituitary portal system, and the other is the development of direct neural contacts with the entire adenohypophysial complex. It is natural to wonder whether these notable modifications could be interrelated. Since pituitary portal systems appear to exist in more primitive vertebrates, and there are suggestions that they are present in the preteleost bony fish, it would seem possible that the development of direct nervous innervation in the teleosts has made a portal form of control obsolete. In this case, the portal system may have become almost nonexistent. It is possible that the median eminence is represented by the small subterminal area of Leatherland *et al.*. It is also possible, and perhaps more likely, that the median eminence became incorporated into the pituitary as the rostral region of the pars nervosa. Whatever the true relationships may prove to be, anatomical considerations suggest that the neurohypophysis may control the adenohypophysis either by direct nervous action or by a diffuse liberation of active agents into its general blood supply. Similarly, it may influence the pars intermedia either by direct neural contact or by passing active agents into the intervascular channel where this exists. Finally, the neurohypophysis may pass hormones into the general circulation and control particular organs throughout the body.

The Nature of the Neurohypophysial Principles of the Teleosts

As early as 1941, Heller had demonstrated that pituitaries from *Gadus sp.* contained far more frog water-balance activity than could be accounted for by the presence of mammalian principles. In 1959, Pickering and Heller adapted paper chromatography to the study of extracts from *Gadus sp.*, *Salmo sp.*, and *Pollachius virens*, and succeeded in separating the "frog water-balance factor" from an oxytocinlike agent. They found that almost all the pressor and antidiuretic activity of the extract was localised with the frog water-balance factor. At the same time, W. H. Sawyer *et al.* showed that crude extracts from the pituitaries of *Pollachius virens* possessed remarkably strong activity on the frog bladder and on the hen oviduct. They suggested that these actions, together with the results of other biological assays, could be accounted for by a mixture of oxytocin and arginine vasotocin-a peptide which had been known only as a synthetic material up to that time. This observation was extended by Chauvet *et al.*, who used chromatography on Amberlite IRC-50 resin to separate two active peptides from the pituitaries of *Merluccius merluccius*. Although the first principle to elute had mainly oxytocic activity, the second principle showed frog bladder, vasopressor, and natriferic activities which were consistent, once again, with the presence of arginine vasotocin. A similar separation was achieved for *Pollachius virens by* Heller and Pickering; they used column chromatography on Amberlite CG-50 and obtained a frog water-balance and pressor principle which could not be distinguished from arginine vasotocin, either by chromatography or by biological assays. More important, they obtained a tentative analysis of its amino acid composition and found a general agreement with those amino acids present in arginine vasotocin. This result was extended by Rassmussen and Craig; these workers used countercurrent distribution methods to purify the frog water-balance principle of *Urophycis tenuis* (incorrectly identified as *Merluccius vulgaris;* Sawyer and Pickford). They obtained a more precise analysis of the purified principle and showed that it contained the same amino acids as arginine vasotocin. They also noted that there was a general agreement between the biological potencies of their purified product and those of synthetic arginine vasotocin. Since this time, similar amino acid analyses have been obtained for *Gadus luscus*, *G. morrhua*, *Germo alalunga*, *Scomber scombrus*, and for the freshwater species, *Cyprinus carpio*. Recently, Wilson and Smith have achieved a complete sequence analysis of the frog water-balance factor of spawning salmon, *Oncorhynchus tschawytscha*, and its identity with arginine vasotocin has been clearly established.

During the early investigations, which concentrated on arginine vasotocin, the presence of an oxytocinlike peptide had been indicated in four species by pharmacological or chromatographic means. At first it was assumed that this principle was oxytocin itself since its chromatographic behaviour and its general spectrum of biological activities seemed similar to those of the mammalian principle. However, in 1961, Heller *et al.* made a careful comparison of the two peptides and showed that the oxytocinlike principle of *Pollachius virens* differed from oxytocin in relatively small, but significant, ways. It differed in the magnesium potentiation of its oxytocic activity, in its higher avian depressor activity, and in its lower frog water-balance potency. Further, it had almost no effect on sodium transport through the frog skin. Clearly, it was not oxytocin but a new neutral principle. In the following year, Acher *et al.* used chromatography on Amberlite IRC-50 resin to separate the oxytocinlike principle of *Pollachius virens*, *Merluccius merluccius*, and *Gadus luscus*. Analysis of its amino acid content, together with partial enzymic degradation, suggested that the peptide was 4 Ser, 8 Ile oxytocin. Later, the same group obtained a similar amino acid analysis for the neutral peptide of the marine species *Gadus morrhua*, *Germo alalunga*, and *Scomber scombrus*, and for the freshwater fish, *Cyprinus carpio*. The 4 Ser, 8 Ile oxytocin was synthesised by Guttman *et al.* and by Johl *et al.*, and the product was found to be biologically similar to oxytocin, except for its lower intrinsic pressor and antidiuretic activities; its properties were compatible with those of the teleost principle. At the sane time, W. H. Sawyer and van Dyke purified the oxytocinlike principle of

Pollachius virens and confirmed that its biological activities compared well with those of synthetic 4 Ser, 8 Ile oxytocin. Similarly, W. H. Sawyer and Pickford showed that the purified neutral principle of *Fundulus heteroclitus* was comparable to 4 Ser, 8 Ile oxytocin in its chromatographic and biological characteristics. They also found that it was preferentially lost from the pituitary of female fish during the reproductive season, while arginine vasotocin showed no seasonal changes. In 1964, Follett and Heller confirmed, yet again, the biological and chromatographic similarity of synthetic 4 Ser, 8 Ile oxytocin and the purified oxytocinlike principle of *Pollachius virens;* in turn, they showed that the pollack peptide shared many of its properties with the neutral agents of *Salmo irideus*, *Cyprinus sp.*, *Anguilla anguilla*, *Esox lucius*, and *Gadus callarias*. However, the most accurate identification of the naturally occurring 4 Ser, 8 Ile oxytocin was the complete sequence analysis achieved in spawned-out salmon, *Oncorhynchus tschawytscha*, by Wilson and Smith. The successful purification and analysis of this principle was helped by the exceptionally high activities found in these specimens, which contained a total oxytocic activity of 750-1000 mU/pituitary, as compared to values which ranged from 5 to 115 mU/pituitary (*Gadus callarias*, and *Anguilla anguilla*, respectively) in the species studied by Follett and Heller.

In the mammals, it is well known that the neurohypophysial peptides are associated with a special protein, neurophysin, which probably acts as a carrier molecule within the neurohypophysis. Little is known of the neurophysinlike carriers of the lower vertebrates, but recently Pickering has isolated such a material from the pituitary of *Gadus morrhua*.

This teleost neurophysin was acidic and rich in cystine, and it appeared to have a greater capacity for arginine vasotocin than for mammalian arginine vasopressin.

It can be concluded that the, neurohypophyses of all the teleosts investigated so far appear to contain arginine vasotocin and 4 Ser, 8 Ile oxytocin. However, the number of species studied is minute compared with the approximately 20,000 species which exist in this highly diversified group. It is possible that isolated mutations could have resulted in the production of other active peptides, but so far none of these has been demonstrated.

The Actions of the Neurohypophysial Principles in the Teleosts

Until recently, the functions of the neurohypophysial principles in the teleost fish received only sporadic attention, and the majority of studies utilised mammalian principles. Although the information available is unsatisfying, it is beginning to suggest that the peptides are concerned in at least two main fields-in osmoregulation-salt balance and in reproduction.

The earliest indications of an importance in osmoregulation came from the considerable body of histological evidence which suggested that there was a depletion of neurosecretory materials from the neurohypophysis of freshwater and marine fish when they were immersed in saline which was hypertonic to their previous environment. Lederis has shown that this is accompanied by the emptying of elementary vesicles, as seen in *Salmo irideus* with the electron microscope. Also, he carried out parallel oxytocic and pressor assays, which suggested that there was depletion of arginine vasotocin, but not of 4 Ser, 8 Ile oxytocin. This is corroborated by Carlson and Holmes, who observed that the transfer of *Salnw gairdneri* from freshwater to saltwater resulted in an initial fall in the antidiuretic activity of the pituitary, while the oxytocic content tended to increase. All hormonal levels had returned to normal within 6 hr. Not all fish show these changes clearly; *Anguilla anguilla* appears to be relatively insensitive to changes in salinity, as judged by these histological criteria. The transfer or return of fish to hypotonic conditions may result in the expected accumulation of neurosecretory material in the neurohypophysis, as in *Callionymus lyra*, *Ammodytes lanceolatus*, and *Phoxinus laevis*; but *Gasterosteus aculeatus* responds with a marked fall in its neurosecretory level. Although this anomaly may be explained by possible depletion by "stress," it may emphasize the fact that the amount of neurosecretory or biologically active material present in the neurohypophysis is always a balance between the rate of hypothalamic supply and the rate of pituitary release into the circulation. Therefore, the important point made by these experiments is not the direction of the change but the fact that a change has occurred.

Despite these responses to changes in the tonicity of the environment, it is clear that the injection of neurohypophysial peptides into teleost fish does not have any water-balance effect of the type seen in the frog. This has been confirmed by the administration of pure arginine vasotocin (36 mμ moles/kg) to *Trichogaster trichopterus*, where there was no detectable changq in body weight. Although infrequent antidiuretic effects have been seen in the teleost kidney, most attention has been focused on the strong diuretic response which usually follows injections of arginine-lysine vasopressin, oxytocin, arginine vasotocin, or teleost pituitary extracts into *Carassius auratus*. The concentration of sodium in the urine does not change, so that sodium loss is parallel to the diuresis. It is probable that this response results from an increase in the glomerular filtration rate. W. N. Holmes and McBean have shown that injections of 100 mU of either oxytocin or vasopressin will nearly double the glomerular filtration rate of *Salmo gairdneri*. Recently, this explanation has been supported by the observation that arginine vasotocin failed to cause a diuresis in the aglomerular kidney of *Opsanus tau*, even though it caused pressor effects. However, the situation may be more complicated in some circumstances, since W. N. Holmes noted that a high dose (1000 mU) of mammalian vasopressin caused a reduction in sodium output, ascribed to the kidney, in *Salmo gairdneri*, and Dlouha *et al,* have suggested that oxytocin causes reabsorption of sodium in the kidney of *Myoxocephalus scorpius*. Indeed, the physiological significance of the diuresis which follows the administration of neurohypophysial peptides is not clear, since it requires the use of high doses of hormones. Further, a number of workers have noted that there is a profound fall-not a rise-in glomerulár filtration rate when freshwater fish enter a marine environment-and this is at a time when loss of neurosecretory material from the neurohypophysis is supposed to take place. However, the diuresis, and its concomitant sodium loss are transitory, and they are outlasted by more profound affects on sodium movement in and out of the fish at extrarenal sites such as the gills. In freshwater fish, such as *'Carassius auratus*, it has been shown that oxytocin and 4 Ser, 8 Ile oxytocin stimulate influx of sodium, probably at the gills, while arginine vasotocin will enhance both influx and outflux. In the marine fish, such as *Platichthys flesus*, oxytocin increased sodium outflux, and arginine vasotocin was potent in promoting rapid sodium exchanges. Lysine vasopressin had little or no potency in these tests. These results can be summarised by saying that neurohypophysial principles (except lysine vasopressin) enhanced sodium fluxes against the prevailing osmotic gradients; they accelerated the process of adaptation to an environment of higher salinity.

It is possible that diuretic responses in teleost fish result from general vascular effects of the neurohypophysial peptides. There has been little work on possible circulatory effects of these principles in teleosts. However, recently Lahlou *et al.* have found pressor effects in *Opsanus tau*, after the injection of low, possibly physiological doses of arginine vasotocin (5 ng/ kg). They obtained a good log dose-response curve from their injections.

Maetz has pointed out the importance of avoiding shock or "stress" effects, which often cause a "laboratory" diuresis during experiments on salt-water balance in teleosts. "Stresses" such as hypoxaemia or handling are known to result in histological changes, with depletion or accumulation of neurosecretion in the neurohypophysis. Since Stevens and Randall have shown that in *Salmo gairdneri*, adrenal medullary hormones may be liberated as a physiological mechanism for the control of the circulation during moderate or strong exercise, it would be interesting to know whether neurohypophysial principles are liberated in the same circumstances. Perhaps such moderate levels of "stress" could cause the release of neurohypophysial peptides; these might help in controlling sodium fluxes across the gills during the rapid water movement which must occur during respiratory activation owing to exercise.

Adrenal cortical tissue could also be involved in conditions of "stress" and in sodium balance. It is interesting to find that Rasquin and Stoll have shown that injections of mammalian arginine-lysine vaso-pressin into *Astyanax mexicanus* led to a loss of glycoprotein from basophylic cells in the adenohypophysis, and they interpreted this as a possible release of adrenocorticotropic hormone (ACTH) into the circulation. However, it is not certain that the pituitary-adrenal axis is activated in all species. Another possible indication of effects on the adenohypophysis comes from suggestions that neurohypophysial peptides may cause the release of gonadotropins; Bacon and Ball have noted that mammalian pars nervosa extracts appear to cause the maturation of free-spawning minnows. However, despite the highly suggestive nature of the anatomical connections between the neurohypophysial system and the adenohypophysis, there is still little physiological evidence for any functional relationship between the two regions.

There are a number of lines of evidence which suggest a possible connection between the neurohypophysis and reproduction in teleosts. Most of the evidence comes from observations of seasonal changes in the histology of the system, and it is often assumed that reproductive cycles are the most likely correlate. Polenov has seen seasonal variations in the preoptic nucleus of the carp and sazan; however, the pars nervosa of this species was consistently full of material. Other workers have seen seasonal changes in the granular content of the neurohypophysis of *Salnw salar* and *Anguilla anguilla*, and these changes may have been correlated with migratory activity. Perhaps the most interesting correlation has been that with the process of mating or spawning. In *Salvelinus leucomaenis pluvius*, neurosecretion within the neurohypophysis increased in quantity until the time of spawning, when it was found concentrated around the blood vessels of the pars nervosa; after spawning, it had decreased in quantity. In *Fundulus heteroclitus*, Sokol noted a similar loss of neurosecretion in both the male and female specimens during the reproductive season, but W. H. Sawyer and Pickford, using pharmacological methods, were able to confirm this only in the case of the females. They made the notable observation that, in the females, there was a preferential depletion of 4 Ser, 8 Ile oxytocin, with no change in the arginine vasotocin content of their glands. The significance of these results has been increased by Wilson and Smith's recent observations on specimens of *Oncorhynchus tschawytscha*, which were collected, "spent out," after spawning. Although these glands contained an unusually high hormonal activity, they also showed an unusually low proportion of 4 Ser, 8 Ile oxytocin as compared to those of other teleosts. This was true of both sexes. It suggested that 4 Ser, 8 Ile oxytocin might have been lost preferentially during spawning. Unfortunately, the glands were not compared with controls from other stages of the life cycle, and no clear conclusion can be drawn. However, the results are suggestive, when combined with other data.

The role of neurohypophysial peptides in the females has been suggested by the work of Egami and Ishii; these workers found that the injection of neurohypophysial extract into gravid specimens of *Oryzias latipes* precipitated egg laying. A similar effect was obtained in *Gambusia sp.*, together with behaviour typical of spawning. Extracts from fish or frog pituitaries were said to be more effective than mammalian preparations. Although egg laying also occurred in *Rhodeus ocellatus*, species such as *Misgurnus anguillicaudatus* and certain salmonids did not respond to neurohypophysial peptides, so the effect may not be universal.

A possible function for neurohypophysial principles in the male has been suggested by the work of Wilhelmi *et al.*. It was found that injection of crude or purified teleost pituitary extracts into *Fundulus heteroclitus* produced reflex movements reminiscent of spawning behaviour, and this effect could be mimiked by high doses of synthetic oxytocin or purified arginine vasopressin (20-60 mU/g) . Similar activity has been found in extracts from *Perca fluviatilis*. It is probable that the effect is mediated by the nervous system; it does not require the presence of the gonads, and it can be elicited in both sexes.

It is too early to be certain of the precise roles of the neurohypophysial peptides in the physiology of the teleosts. If the scanty shreds of evidence available are considered, it seems possible that arginine vasotocin is concerned in osmoregulation and salt-water balance, while 4 Ser, 8 Ile oxytocin has a role in reproduction, particularly in mating and egg laying. This would be an interesting parallel with the mammal, where the basic principle (arginine vasopressin) is concerned in kidney function, and the neutral principle (oxytocin) is important in reproduction. However, further work may well prove that these speculations are unfounded.

LUNGFISH : THE DIPNOI

The Structure of the Neurohypophysis of the Dipnoi

The Dipnoi, or lungfish, are important because they may be considered to occupy a phyletic position between the ray-finned bony fish and the tetrapods. Although the amphibians have not descended directly from the lungfish, it is likely that they shared a common crossopterygian ancestor, and until the pituitaries of crossopterygian species become available for study, our picture of the evolution of the neurohypophysis from fish to tetrapod must depend on investigations of the lungfish. As early as 1926, de Beer was struck by the strong resemblance between the lungfish pituitary and that of the amphibians. Since this time, many workers have made similar observations; in fact, the diagrams of Wingstrand show that it is not easy to distinguish a sagittal section of the pituitary of *Protopterus*, the African lungfish, from that of a urodele amphibian. This resemblance is partly true of the neurohypophysis, which possesses a well-marked, dorsally located pars nervosa, similar to that of an amphibian in its gross morphology. However, it retains the more intimate intermingling of neural and intermediate elements which is typical of the bony fish.

Recently, Kerr and van Oordt have remarked that the intermingling of the pars nervosa and the pars intermedia in *Protopterus aethiopicus* is comparable to that of the elasmobranchs and primitive bony fish, so that the lungfish form a good transitional group between the fish and the tetrapods. The general anatomy of the lungfish pituitary has been clearly described by Wingstrand, and the neurohypophysial system has been carefully studied by Dorn, and by Kerr and van Oordt. The recent studies of Kerr and van Oordt are the basis of the following general description of the neurohypophysial system of the Dipnoi.

The preoptic nucleus of *Neoceratodus forsteri* lies dorsal to the optic chiasma but rostral to the average position found in teleosts. It

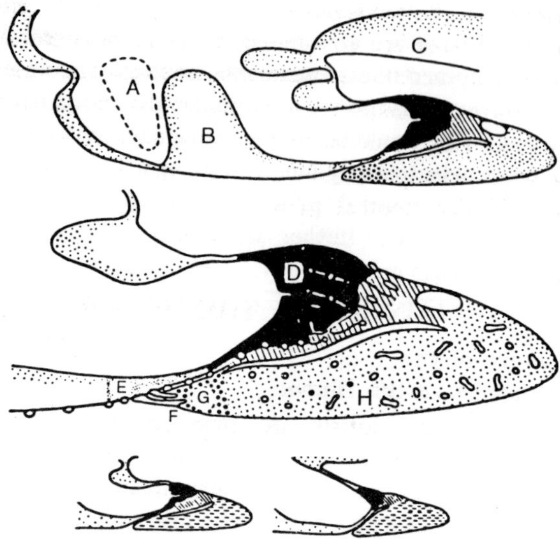

Figure 4.11: The hypothalamus and pituitary of the lungfish, Protopterus annectens, *sagittal section. Upper diagram: the relationship of the pituitary to the hypothalamus. A indicates preoptic nucleus; B, optic chiasma; and C, rhombencephalon. Middle diagram: the relationship of the pars nervosa to the adenohypophysis. D indicates pars nervosa (neural lobe) ; E, median eminence; F, portal blood vessels; G, "pars tuberalis"; H, pars distalis; and K, pars intermedia. Lower diagrams: the relationship of the pituitary of* Protopterus *(left) to that of the urodele amphibian* Ambystoma *(right).*

is one of the shortest nuclei found in any fish, and Charlton was impressed by its general similarity to that of the preteleost bony fish rather than to the more widespread teleosts. In *Protopterus annectens* the preoptic nucleus consists of notably large, bipolar cells, which appear to lie remarkably anterior, and somewhat dorsal to the optic chiasma. In both *Protopterus annectens* and *P. aethiopicus*, they contain only a few distinct neurosecretory granules, but in *Protopterus annectens* there is also a diffuse general staining of the cytoplasm. The preoptic cells send processes between the ependyma cells of the ventricle, and these terminate in small globules which are immersed in the cerebrospinal fluid. The opposite end of the cells gives rise to axons which form the preoptico-hypophysial tract, which runs first laterally and then ventrocaudally into the walls of the infundibular process.

Neoceratodus f orsteri (syn. *Epiceratodus*), the Australian lungfish, possesses a distinctive infundibular process; it grows vertically downward during development so that the pituitary lies a considerable distance ventral to the optic chiasma. In species of *Lepidosiren*, the

South American lungfish, it remains oriented in a ventrocaudal direction. In *Protopterus annectens* and *P. aethiopicus*, the African lungfish, the infundibulum is a wide, flattened, and well-delimited funnel, which projects caudally from the floor of the brain; its ventral wall is almost horizontal, and its caudal extremity is divided into two hollow pockets similiar to those found in the lampreys.

In *Protopterus annectens*, Dorn has shown that the preoptic axons are spread widely as they enter the infundibular process, and there appears to be a tendency for them to be divided into different tracts. In *Protopterus aethiopicus*, the preoptico-hypophysial tract which enters the infundibulum is notably diffuse, but its clearest component curves laterally through the walls of the infundibular process, on both sides. It subdivides into upper and lower tracts in a manner reminiscent of preteleost species such as *Polypterus senegalis*, and the sturgeon, *Acipenser fulvescens*. Many of the axons of the ventral tract contact a dense capillary network, which has been seen to run in furrows within the infundibular wall, just anterior to the rostral margin of the pituitary, in *Protopterus annectens*. This region of the infundibular wall is not thickened in any way, and is not clearly demarcated from the pars nervosa (*Protopterus annectens;* Wingstrand, Dodd and Kerr). However, it can be distinguished from the neural lobe by its lower content of neurosecretion and by the presence of the invading capillaries (*Protopterus annectens*). The capillaries connect to short portal vessels which pass into the rostral tip of the adenohypophysis.

It is clear that the lungfish possess a functional median eminence and pituitary portal system.

Although a few axons of the ventral tract carry beaded neurosecretory droplets past the median eminence into the neural lobe, most of the preoptico-hypophysial fibers reach the pituitary through the dorsal branch of the tract. These dorsal axons pass through the lateral and dorsal walls of the caudal infundibular process and form a pars nervosa at its extremity (*Protopterus aethiopicus*).

The pars nervosa originates as an outgrowth of neural tissue in the me dial walls of the two caudally directed pockets, which are formed at the caudal limit of the infundibulum (*Protopterus* sp., Wingstrand; *Protopterus aethiopicus*, Kerr and van Oordt). It is a large, conspicuous structure, which lies dorsal to the pars intermedia, and was once referred to as the "infundibular gland". It is made up of a mass of ramifying rods of neural tissue, and of intricate tubules which are penetrated by extensions from the infundibular cavity.

The tubules are lined by ependymal cells, and these cells also form a central core to many of the solid rods of neural tissue (*Neoceratodus forsteri*, Griffiths; *Protopterus annectens*, Dodd and Kerr; Wingstrand; *P. aethiopicus*, Dawson; Kerr and van Oordt). During the embryology of *Protopterus aethiopicus*, the neural tissue is invaded by solid strands of ependymal cells, which extend from the ventricle, and then open up to form a tubular lumen along much of their length. Dodd and Kerr have remarked that the pars nervosa of *Protopterus annectens* does not contain a great proportion of neural tissue, since so much of its structure is composed of ependymal cells, spaces, lumina of tubules, and blood vessels. Nevertheless, the pars nervosa is penetrated by the axons of the preoptico-hypophysial tract, and the fibrous matrix is heavily laden with neurosecretory material. Some axons spread into the ependyma of the dorsal wall of the infundibular cavity, where Herring bodies have been seen close to the cerebrospinal fluid. Kerr and van Oordt have noted the presence of occasional pituicytes in the pars nervosa of *Protopterus aethiopicus*.

The outer surfaces of the neural processes and tubules are often associated with the tissue of the pars intermedia. The intermedia tissue lies ventral to them, and it is largely separated from the rest of the adenohypophysis by a hypophysial cleft-except in the case of adult specimens of *Protopterus aethiopicus*, where this cleft is partly occluded. In species of *Protopterus*, the pars intermedia is mainly composed of tubules, like the pars nervosa; the cords of intermedia cells are divided by hollow spaces and by tubelike extensions from the hypophysial cleft. The extent to which the pars nervosa penetrates and intermingles with the pars intermedia appears to vary between different species. A comparison of different authors suggests that intermingling is least marked in species of *Lepidosiren*, where the pars intermedia is particularly thin. In *Neoceratodus forsteri* the invasion of neural elements into the pars intermedia appears to be more marked. In *Protopterus annectens* and P. *dolloi*, the intimate intermingling of the two tissues recalls that typical of fish, and not the more discrete and self-contained pattern of the tetrapods. In adult specimens of *Protopterus aethiopicus* there is a highly complicated interdigitation between columns and tubules extending from both the pars nervosa and the pars intermedia.

Despite this intimate intermixing of neural and intermediate tissues, the neural processes of species of *Protopterus* are bounded and separated from the intermedia cells by connective tissue. This connective tissue barrier has been described as a thin membrane in *Protopterus annectens* and as a strong sheet in *Protopterus aethiopicus*. In *Protopterus annectens*, a few neurosecretory fibers pass through it into the pars intermedia. The connective tissue investment of the neural processes contains a dense capillary network. If the comments of different authors are correlated, it seems probable that there is a rich blood supply through these vessels, although the pars intermedia is often notably avascular. Further, if the drainage of these vessels is comparable to that of the similarly located capillaries of the bony fish, it is possible that they could supply neurosecretory materials to the systemic circulation.

Clearly, the lungfish pituitary is transitional between that of the fish and of the tetrapods. It combines an essentially amphibian pattern with a fishlike intermingling of the pars nervosa and the pars intermedia. However, it is interesting to note that it is most closely related to the pituitary of the preteleost bony fish; it shares with them a short preoptic nucleus of comparable location (some species), a median eminence supplied by a special ventral neurosecretory tract, a pituitary portal system, and large, well-developed ventricular extensions which penetrate the pars nervosa. The anatomy of the lungfish pituitary is not only a clue to the evolution of the tetrapod neurohypophysis, but it is also a reminder of the remarkable degree of specialisation which has been developed by the teleost bony fish.

The Nature of the Neurohypophysial Principles of the Dipnoi

The close similarities between the structure of the lungfish and amphibian pituitary are borne out by the nature of their neurohypophysial peptides. The situation in the lungfish is far from clear, but their pituitaries may contain arginine vasotocin, 8 Ile oxytocin, and perhaps oxytocin itself. Follett and Heller subjected extracts from the pituitaries of *Protopterus aethiopicus* and *Neoceratodus f orsteri* to paper chromatography and separated two active moieties. The slow-running peak (R_f approximately 0.3) showed strong vasopressor activity, and estimations of its oxytocic and natriferic potencies suggested that it was arginine vasotocin. The fast-running peak (R_f approximately 0.5-0.6) was mainly oxytocic in activity and moved in a manner similar to oxytocin, or to that of a related neutral peptide. Its pharmacological properties were clearly different from those of 4 Ser, 8 Ile oxytocin, the neutral principle typical of most bony, fish. Follett and Heller noticed that the lungfish neutral peptide was similar to oxytocin in its milk-ejection activity and in its magnesium potentiation, but the relatively high avian depressor potency indicated that the two agents were not the same. Follett and Heller suggested that their results were compatable with the presence of 8 Ile oxytocin, a peptide which was known to occur in amphibia. However, it was possible that this peptide could be mixed with oxytocin or with some new principle. W. H. Sawyer, working with *Protopterus aethiopicus*, and using similar methods, was unable to find an unusually high avian depressor activity in the neutral principle and was inclined to ascribe the actions to oxytocin itself. Further, W. H. Sawyer was able to separate the active principles by column chromatography on carboxymethyl cellulose resin. He confirmed the presence of a basic peptide with vasopressor, antidiuretic, and frog bladder stimulating activities consistent with arginine vasotocin. However, he found that the majority of the oxytocic activity was eluted as a large peak in the position typical of oxytocin. It could not be distinguished from this peptide. Sawyer considered all the evidence available and concluded that probably the pituitaries of *Protopterus aethiopicus* contained oxytocin, 8 Ile oxytocin, or a mixture of the two, and that the particular combination might vary with the individual, the race, the location, the season, or the treatment of the tissues. Indeed, it was known that there were considerable variations in the hormonal content of the lungfish pituitary, since the ratio between the neutral oxytocic peptide and arginine vasotocin ranged from about 2.6 to 25.0, as judged by the oxytocic and vasopressor assays (vasopressor/ oxytocic = 0.04 to 0.38). However, the oxytocic principle was always present in far greater quantities. A possible variation in the proportion of oxytocin (?) to 8 Ile oxytocin was supported by W. H. Sawyer's studies on different batches of glands. One collection of glands appeared to differ significantly from 8 Ile oxytocin and to parallel oxytocin itself. The other three batches showed pharmacological similarities to 8 Ile oxytocin. However, the general trend has been to establish the presence of 8 Ile oxytocin more firmly, and W. H. Sawyer's most recent studies have indicated that at least some glands of *Protopterus aethiopicus* contain an overwhelming or total preponderance of 8 Ile oxytocin. Nevertheless, there is still support for the possible presence of oxytocin among other species of lungfish. In *Lepidosiren paradoxa*, thin-layer chromatography has separated a principle which appears to be arginine vasotocin from a neutral peptide with the biological properties of oxytocin. However, the authors point out that 8 Ile oxytocin might have been mixed with the oxytocin. Many of these ambiguities result from the close similarity between the properties of oxytocin and of 8 Ile oxytocin. It is possible that some individual biological assays might fail to make a clear distinction between them. For this reason, the discrepancies which occurred during assays against oxytocin standards carry the greatest weight in indicating the presence of a peptide such as 8 Ile oxytocin. Similarly, the demonstration of discrepancies between an unknown and synthetic 8 Ile oxytocin, when they were compared directly by different methods of biological assay, would be better evidence for the presence of oxytocin-or some unknown analog-in the lungfish extracts. However, many difficulties would be resolved : if the neutral peptides could be separated and analysed by chemical methods.

It appears probable that the lungfish share arginine vasotocin and 8 Ile oxytocin with the amphibia. Even if they prove to possess oxytocin, this cannot be regarded as an anomaly, since Munsick has suggested that oxytocin occurs in at least one species of amphibian, *Rana pipiens*.

The Actions of Neurohypophysial Principles in the Dipnoi

Lungfish are not readily available to most investigators. Nevertheless, their important position between the bony fish and the amphibians, which appear to utilize their neurohypophysial principles in somewhat different ways, has led to a surprising amount of work on the actions of the neurohypophysial principles in the metabolism of the dipnoans. However, the work is confined almost entirely to studies of water and salt balance in the African species, *Protopterus aethiopicus*.

There is no evidence that the injection of neurohypophysial principles into the lungfish causes a water-balance effect analogous to that seen in the frog. Large doses of oxytocin, or of a mammalian vasopressin preparation (2500 mU/100 g body weight), or of pure arginine vasotocin (36 mμ moles/kg) had no effect on the total body water of free-swimming specimens of *Protopterus aethiopicus*. Although it is possible that a water-balance effect might occur in estivating specimens taken from the mud in the dry season, no studies have been made at this time. However, present information suggests that the water absorption mechanisms of the skin do not come under neurohypophysial control until the level of the amphibians has been reached, with the hagfish the only possible exception.

Although there is no general water-balance effect, neurohypophysial principles do appear to affect the kidney. In *Protopterus aethiopicus*, the injection of arginine vasotocin by either the intraperitoneal route (500 ng/kg) or by the intravenous method (10-50 ng/kg) resulted in a marked increase in the volume of the urine. Oxytocin (4000 ng/kg) caused small, inconsistent diuretic responses, but 8 Ile oxytocin (4000 ng/ kg) had little or no diuretic effect when given into the peritoneal cavity. Again, it is not known whether an antidiuretic effect would be elicited during the dry season. In part, the diuretic effect could result from the increase in glomerular filtration rate which was observed, but unlike the situation in such teleosts as *Carassius auratus* other factors may be involved since the sodium concentration of the urine was also increased. It is possible that the rise in urinary sodium was caused by depression or sodium reabsorption by the kidney tubules. Arginine vasotocin (500 ng/ kg) was particularly potent in increasing sodium excretion in the urine, which may rise several hundredfold under its influence; in contrast, oxytocin (4000 ng/ kg) had a slight effect, and 8 Ile oxytocin (4000 ng/ kg) was ineffective. When arginine vasotocin was given by the intravenous route, doses as low as 10-50 ng/kg caused loss of sodium in the urine. This partly answers the criticism that the doses required to produce natriuretic responses by the intraperitoneal route are so great that the pituitary hormone content is barely sufficient for a single response. Problems of this type have been discussed during considerations of the lampreys. Although an effect on the sodium content of the urine has been observed a number of times, it is not clear whether the principles affect sodium loss by other extrarenal routes. However, their action causes an increase in the net loss of sodium from the body. The physiological significance of this enforced sodium loss is difficult to see since it would appear to be detrimental to a totally freshwater animal. Perhaps its effects are normally in a delicate state of balance with opposing influences, similar to the antagonism seen between insulin and glucagon in the control of mammalian blood sugar. Further work is needed.

The diuretic response of the kidney to neurohypophysial peptides may be partly related to the increase in glomerular filtration rate which also occurs. W. H. Sawyer has considered the possibility that this rise in filtration rate could be mediated by direct vascular effects. The injection of doses as low as 2.5 ng/kg of arginine vasotocin into *Protopterus aethiopicus* resulted in sharp increases in the blood pressure in the dorsal aorta. However, the pressor response was outlasted by the increase in glomerular filtration rate, so that it was unlikely that the increased filtration was totally dependent on a raised systemic blood pressure. Nevertheless, it is possible that more subtle effects on the glomerular arterioles might persist after the gross change of aortic blood pressure.

There is no information concerning the role of the neurohypophysial principles in the control of the pars intermedia or of the adenohypophysis of the lungfish; all that can be said is that the anatomical relationships are suggestive of a close association. There is no information on the possible effects of the principles on reproduction. It is tempting to speculate that the neutral peptides of the Dipnoi could be important in this sphere; as W. H. Sawyer has pointed out, they are stored in greater quantities than arginine vasotocin, but their functions are as mysterious as those of neutral principles in the nonmammalian tetrapods and in the male mammals.

CONCLUSION

Many of the data presented above are summarised in Tables II and III. However, these tables are no substitute for the facts presented in the review, since the observations listed vary from well-established facts to solitary, and often brief, comments. It is dangerous to try to generalize in a group as varied as the fish. Indeed, there are more variations, both fundamental and individual, between the different classes and species of fish than exist throughout the neurohypophyses of the tetrapods. However, a few generalisations may be useful, and they will undoubtedly provide a rich field for future critics.

The complete neurohypophysial system exists from the cyclostomes up through the vertebrate scale. Presumably, its primitive origins are lost in extinct species. In the hagfish it is indistinct, and in both the hagfish and the lampreys the pars nervosa, although present, appears to be poorly developed. The pars nervosa becomes more marked as one progresses to more advanced groups of fish. Above the evolutionary level of the cyclostomes, the pars nervosa becomes closely intermingled with the pars intermedia, although the degree of this association may be somewhat reduced in some species (e.g., *Squalus acanthias*). This notable fusion between the neural and intermediate elements persists into the lungfish, even though the Dipnoi develop the more localised, dorsal pars nervosa typical of the tetrapods. The median eminence and pituitary portal system is a common feature of the neurohypophysial system from the level of the primitive hagfish up through the vertebrate tree. In its early stages of development it appears to possess connections to the neurointermediate lobe (Myxiniformes, some Elasmobranchii). We are faced less with the problem of how and why it developed in the vertebrates, and more with the question of why it is inconspicuous in the lampreys and absent in the teleosts. The lampreys do not have a marked portal system, but the evidence presented here favors the presence of a small neural and neurovascular connection between the preoptic nucleus and the adenohypophysis; its effects may well be aided by the large sinuses which lie between the pars nervosa and the rest of the pituitary. The teleosts have developed direct neural contacts with the adenohypophysis, and it is possible that the portal system became obsolete and was lost. However, the importance of both neural and vascular connections between the neurohypophysis and the adenohypophysis suggests that the neurohypophysis may well mediate a control over the pars intermedia and over other adenohypophysial regions as well.

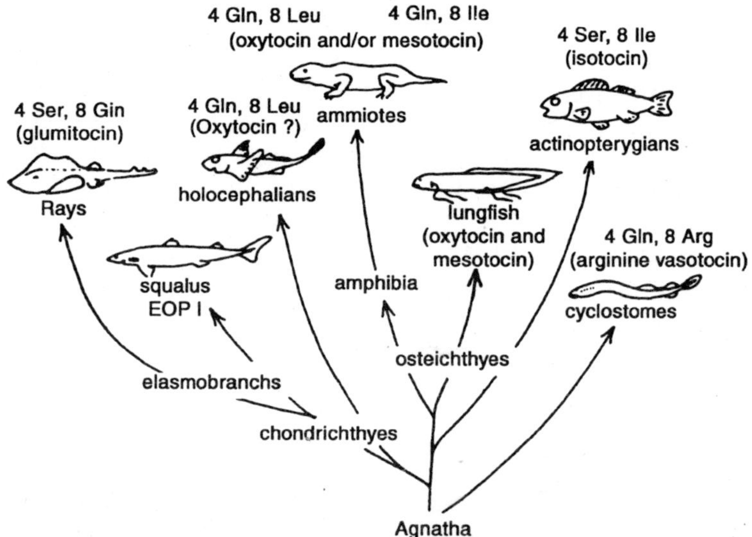

Figure 4.12 : The distribution of neutral neurohypophysial principles throughout different classes of fish. The cyclostomes are exceptional in possessing only the basic principle, arginine vasotocin.

Arginine vasotocin has been demonstrated in most fish which have been examined, and it is strongly suspected in most others. Although it may be present in trace quantities in cartilaginous fish, its constant presence is a remarkable example of evolutionary stability. By contrast, the neutral neurohypophysial peptides show more variation. They may be absent in the cyclostomes, although this is not well established. The elasmobranchs may contain 4 Ser, 8 Gln oxytocin and other unknown principles, but the bony fish appear to utilize 4 Ser, 8 Ile oxytocin. The lungfish contain 8 Ile oxytocin, and perhaps oxytocin, and are effectively amphibians in terms of their neurohypophysial principles.

Although there have been clear advances in the study of the morphology of the fish pituitary, and the nature of the neurohypophysial principles is becoming better established, there is more confusion and uncertainty with regard to the actions and functions of the hormones in fish metabolism. At present there are strong suggestions that the neurohypophysial peptides are concerned in salt and water balance, most particularly on sodium metabolism. The loss of water by the kidney may be controlled by effects on glomerular filtration rate, perhaps through actions on the circulatory system. There are suggestions that the neurohypophysial peptides might be important in reproduction and influence spawning and egg laying. It is possible that the basic principles are more important in osmoregulation and salt balance, while the neutral peptides find a function in reproduction: This would parallel the situation in the mammal, but at the present time this is almost pure speculation. Although a basic pattern of function may emerge, the marked differences between the metabolism of such diverse fish as elasmobranchs and teleosts may well imply that neurohypophysial principles have widely different uses in different fish.

5

Hormones of Metabolic Gland

In higher vertebrates, prolactin has physiological actions on biological features that are missing from fishes: stimulation of the pigeon crop and the mammary gland and of the postovulatory corpus luteum in some mammals, and induction of the water-drive and land-water integumentary changes in the urodele amphibians. In addition, prolactin is generally considered to be a causal agent in the induction of maternal behavior and broodiness in mammals and birds, although it is not certain whether this is a direct action or one mediated by gonadal steroids.

For many years, on the basis of a misunderstanding of the work of Leblond and Noble, it was accepted that the teleostean pituitary contained a pigeon crop-stimulating factor, i.e., a typical prolactin; and overenthusiastic interpretation of the results of Noble *et al.* led to the erroneous view that fish prolactin had been shown to be involved in the causation of parental care behavior in teleosts. Reexamination of the response of the pigeon crop to fish pituitary material led Nicoll and Bern to conclude that a typical prolactin (i.e., a factor with full effect on the pigeon crop) does not occur in fishes, with the significant exception of the lungfish *Protopterus*, although typical prolactin reactions were obtained with amphibian glands. Study of the mammary gland response showed that typical responses could be elicited with amphibian and reptilean material, but not with fish pituitary material, and in this case *Protopterus* pituitary was ineffective. However, more recently, Chadwick, Nicoll *et al.*, and Nicoll and Bern have come to consider not that there is no pigeon crop-stimulating factor in the fish pituitary but rather that there is present an incompletely or minimally effective stimulating factor. The essence of the present situation is that definitive prolactin appears only at the tetrapod stage of vertebrate organisation; the position of the dipnoans requires further elucidation, but it is probable that these animals stand closer to amphibians than to other fishes in this respect, as in so many others.

Apart from its historically primary pigeon crop and mammary gland activities, purified mammalian prolactin has a specific action in inducing the return to water ("water drive") in hypophysectomised newts or efts, *Diemyctilus viridescens*, which normally prefer to stay on land. Grant and Pickford were able to induce water drive in hypophysectomised efts with fish pituitary material, and so provided the first unequivocal demonstration that the fish pituitary possesses in full a property characteristic of tetrapod prolactin. From that time on, interest in the presence and possible physiological actions of a kind of prolactin in the fish pituitary has grown, and in one field, concerned with effects of prolactin on electrolyte regulation in certain teleosts, evidence has been obtained which substantiates the presence of a prolactinlike hormone possessing some, but not all, of the properties of mammalian prolactin.

Many different effects have been described following administration of mammalian prolactin (usually the ovine hormone) to teleosts. In this review, these effects will be discussed in turn, roughly in the order of the volume of research they have attracted.

Prolactin and Osmoregulation in Teleosts

The reader is referred to recent reviews for details that are beyond the scope of this chapter. As with so much of modern fish endocrinology, this burgeoning research field stems from the pioneering investigations of Dr. Grace E. Pickford. Essentially, the situation at present is that certain euryhaline teleosts, mainly cyprinodonts and other Atheriniformes, but including at least two nonatherine species, *Betta splendens* and *Tilapia mossambica*, are unable to survive in freshwater after hypophysectomy for more than a limited period (varying from 1 to 12 days) ; but they can live for much longer, perhaps indefinitely, in seawater or dilute (1:3) seawater, or in a fish Ringer solution. In some cases (*Fundulus heteroclitus*, *Poecilia latipinna*, *Xiphophorus maculatus*, *Gambusia sp.*, and *Tilapia mossambica)*, injections of mammalian prolactin have been shown to prevent this failure and to maintain hypophysectomised fish for long periods in freshwater.

Other teleosts, euryhaline or stenohaline, have been found able to live in freshwater after hypophysectomy, which perhaps suggests that the prolactin-mimicked pituitary mechanism is unusual and of only limited taxonomic distribution. However, investigations concerned with electrolyte regulation, though limited in extent, point to a more widespread mechanism, one that may not be essential for life in freshwater but which is necessary for normal electrolyte regulation in this medium. The data on cyprinodonts indicates that the main deficiency after hypophysectomy concerns conservation of electrolytes rather than failure to excrete water; thus, in failing *F. heteroclitus*, the mean decrease in serum osmolality or chloride content was about 40-60%, but the average weight increase (presumably owing to water accumulation) was only about 6-7%, and when hypophysectomised *P. latipinna* fail in freshwater the 25% or greater fall in plasma sodium occurs in the presence of normal potassium levels, indicating that hemodilution is not a major factor in failure.

Failing hypophysectomised *F. heteroclitus* exhibit a marked fall in plasma chloride and total serum osmolality, the latter being preventable by ovine prolactin. Similarly, failing *P. latipinna* in freshwater display a fall in plasma sodium, but not of potassium, which is corrected by a single injection of ovine prolactin. It is in the work on *F. heteroclitus* and *P. latipinna* that there is to be found the most considerable evidence for regarding fish prolactin as a physiologically important part of the pituitary equipment of teleosts. In evaluating the possible physiological significance of the effect of a mammalian pituitary hormone on a teleost, a first question must be whether the effect is specific, or if it can be produced by other pituitary factors. Dr. Pickford and her collaborators have addressed themselves to this question in the case of the killifish, *F. heteroclitus*, and have shown in this fish that the ability to promote freshwater survival after hypophysectomy is unique to prolactin; it is not shared by thyroxine, ACTH, TSH, growth hormone, ACTH-TSH-growth hormone combination, posterior lobe extract, arginine vasotocin, isotocin, urophyseal extracts, DOC, cortisol, aldosterone, corpuscles of Stannius extracts, hog renin, and parathyroid hormone. Effective substances were ovine and bovine prolactins (the latter less effective than the former in terms of its known pigeon crop activity), and two primate growth hormones (monkey and human) known to contain prolactin activity. Pituitary brei of rat, freshwater perch, and *F. heteroclitus* were effective, but not homogenates or extracts of pituitaries from the marine pollack or the freshwater carp. Furthermore, the η cells, source of fish prolactin, are very much more active and more numerous in killifish adapted to freshwater than to seawater, suggesting that fish prolactin really is physiologically involved in maintaining the fish in freshwater, but it is not so important in seawater.

Following the finding that ovine prolactin prevents the rapid fall in plasma sodium in hypophysectomised *P. latipinna* in freshwater, this action was shown to be a specific property of prolactin. Ineffective hormones, each tested at a low and a high dose, were oxytocin, vasopressin, arginine vasotocin, isotocin, ACTH, growth hormone, TSH, and α-MSH. Gonadotropins were excluded, since natural or experimental alterations in the pituitary-ovary axis were found not to affect survival in freshwater. The response to prolactin was dosedepe-ndent, and in a slightly modified protocol frog pituitary homogenates produced a response paralleling that to ovine prolactin, indicating that the presence of other hormones in the homogenates did not distort the effects of the frog prolactin in any essential way. Parallel responses to homogenates of *P. latipinna* pituitary were also obtained, which permitted the bioassay of the activity in the fish gland. The question then arose whether the activity should be ascribed to a single hormone in the fish gland, which could be designated fish prolactin, or whether there were any reasons for thinking that mammalian prolactin in the experiments was mimicking the physiological action of two or more fish hormones in combination, or of some quite different fish hormone such as ACTH.

By bioassay, it was shown that there is approximately double the "prolactin" activity in pituitary glands taken from P. *latipinna* adapted to freshwater compared with glands taken from fish adapted to dilute sea water, this increased activity being already established 72 hr after the fish enters freshwater, coincident with restoration to normal of the initially declining plasma sodium. In this respect, the activity behaves more as though it were the property of a single hormone rather than of a combination of hormones. The parallelism between the responses to fish and frog pituitary homogenates supports this idea, since there is no reason to attribute the activity in the amphibian gland to anything other than a single hormone. Additional evidence comes from ectopic pituitary transplants in *P. formosa*. In these preparations some functions pass into total or partial abeyance, but the sodium-conserving activity persists, together with TSH secretion. One would not expect these results if the sodiumconserving activity were owing to ACTH, gonadotropins, MSH, or growth hormone, alone or in combination; and the involvement of endogenous

TSH is unlikely, since treatment with thiourea or thyroxine does not impair survival in freshwater.

Considering now the work on pituitary histophysiology in *P. latipinna* in relation to freshwater survival, we find a situation that parallels the specificity screening of the mammalian pituitary hormones. The origin of the fish prolactinlike activity has been experimentally ascribed to one cell type in the rostral pars distalis, the η cell. This is distinct from the cell type that secretes growth hormone, just as the prolactin activity is distinct from growth promotion, and is distinct from the ACTH cells, the TSH cells, and the gonadotrophs and pars intermedia cells. Furthermore, the η cells, but no others, become activated during the first 72 hr after transfer to freshwater, in correlation with the increase in prolactinlike activity in the gland and with the restoration of the initially falling plasma sodium levels and the cells remain active during life in freshwater. In summary, it appears that the prolactinlike activity originates from a single distinct cell type which behaves as if this activity were involved physiologically in sodium conservation in freshwater.

There is some evidence from *Gambusia* and *Fundulus kansae* that prolactin in these species may act like ACTH in stimulating the interrenal, but considerable experimental evidence has been adduced which shows that ovine prolactin does *not* stimulate the interrenal in *Fundulus heteroclitus* or *Poecilia latipinna*, and of course in both species ACTH did not mimic prolactin in promoting electrolyte conservation. This is additional evidence, adding to that from mammalian hormone specificity screening and pituitary histophysiology, that ovine prolactin really does imitate a specific and distinct fish hormone in the conditions of the work on these species. This fish hormone is physiologically involved in electrolyte conservation, and its kinship to tetrapod prolactin is attested by the unique ability of prolactin to mimic its action. It may appropriately be called "fish prolactin" or "paralactin".

Thus far, we have considered only teleosts which require the pituitary gland for survival in freshwater, from which the paralactin-based mechanism could appear to be of only limited distribution within these fishes. However, evidence from other teleosts which tolerate hypophysectomy in freshwater points to a more widespread osmoregulatory role for fish prolactin.

Although the eel, *Anguilla anguilla*, does not fail in freshwater after hypophysectomy; it nevertheless suffers a slow reduction in plasma sodium, potassium, and calcium, which can be retarded by maintenance therapy with ovine prolactin. Chan *et al.* found that while sodium and calcium decreased slightly following hypophysectomy of the eel in freshwater, plasma potassium *increased* slightly, recalling the more pronounced increase in deionised water found by Olivereau and Chartier-Baraduc. Chan *et al.* were not able to influence the decline in sodium and calcium with bovine prolactin treatment, although this hormone did reduce the elevation of plasma potassium. Butler also could not demonstrate an effect of a very low dose of ovine prolactin on electrolyte levels in the hypophysectomised eel. However, Chan *et al.* found that bovine prolactin prevented the excessive hydration of eel muscle that occurs after hypophysectomy in freshwater, and they suggested that fish prolactin might act in the intact animal to maintain a relative impermeability of the body surface (particularly the gills) to water. This hypothesis runs counter to Stanley and Fleming's suggestion that in *Fundulus kansae* prolactin acts in both seawater and freshwater to *increase* the water permeability of the integument.

Another role for paralactin in the eel, and one that would explain Olivereau and Chartier-Baraduc's results, is suggested by the findings that hypophysectomy of freshwater eels leads to an increased outflux of sodium from the body, which is corrected by ovine prolactin. The failure of Chan *et al.* to alter plasma sodium values in the eel with bovine prolactin may lie in the fact that sodium outflux is of minor importance relative to the total electrolyte economy of the eel, and it is this component alone that is influenced by prolactin. Another point is that bovine prolactin appears to be less effective than the ovine hormone in sodium conservation in *P. latipinna*, and in inducing freshwater survival in *F. heteroclitus*, comparing pigeon crop units of the two preparations; and in the work on the eel the French workers used ovine prolactin while Chan *et al.* used bovine prolactin. At any rate, the eel offers an example of a teleost in which hypophysectomy does not abolish freshwater tolerance, but in which, nevertheless,the operation entrains some disturbance in mineral metabolism which can be partially corrected by prolactin. Adrenocorticotropic hormone and corticosteroids appear to be more important than paralactin in freshwater osmoregulation in this species.

The goldfish, *Carassius auratus*, is another teleost which survives for long periods in freshwater after hypophysectomy; nevertheless, 3 weeks after the operation, goldfish in freshwater have very low plasma levels of sodium and chloride, which apparently results from enhanced extrarenal loss of ions (probably across the gills), renal sodium loss being unaltered by hypophysectomy. Correspondingly, hypophysectomised goldfish on being transferred from saline to freshwater displayed a lower than normal plasma osmolarity, and this defect could be corrected by pretreatment with ovine prolactin or with salmon acid-acetone-extracted pituitary powder; presumably in this case, as in *F. heteroclitus* and *P. latipinna* (below), the prolactin acted to limit the enhanced sodium outflux of the hypophysectomised fish.

The eel and the goldfish appear to be able to live in freshwater after hypophysectomy because of two rather different factors. In the eel the prolactin-activated restriction of sodium outflux is of only minor importance to the animal compared with other, perhaps largely nonendocrine mechanisms that operate when it moves to freshwater, whereas in the goldfish, in which the fall in plasma electrolytes after hypophysectomy is very marked (from about 135-87 mEq/liter plasma sodium), it would seem that the paralactin-based mechanism is important for normal electrolyte control, but that the fish tissues are able to tolerate large changes in the electrolyte composition of the body fluids.

The examples of the eel and the goldfish serve to warn us against being overly impressed by the notion of failure in freshwater after hypophysectomy as being a necessary index or sign that a prolactin-based mechanism normally operates. It so happens, because of ease of observation, that the facts of failure in freshwater after hypophysectomy, and its correction by prolactin, were the salient points observed in the earlier work; and some authors have been misled by this into postulating a unitary pituitary mechanism that operates to guarantee survival in freshwater (or hypotonic media) with the implication that this mechanism must be absent in species in which freshwater survival is not impaired by hypophysectomy. It has even been suggested that paralactin can play no role in a fish in seawater, and that the η cells must turn to the secretion of some other factor in seawater. It is now apparent that we are actually dealing with something more subtle than a kind of all-or-nothing survival mechanism, and we need to think not in terms of life and death but of modulation of electrolyte (especially sodium) movements across the body surface. As will be seen, hypophysectomy and prolactin influence these sodium movements in the same manner whether *P. latipinna* is in freshwater (where the pituitary is essential for survival) or in dilute seawater (where the gland is not essential for survival).

It is probable that the major site of action of prolactin in relation to sodium conservation in freshwater is on the passive outflux of sodium from the body, presumably mainly at the gill. In *F. heteroclitus* and *P. latipinna*, hypophysectomy results in a marked increase in sodium outflux in freshwater which is corrected by prolactin, but not by cortisol in the latter species. These changes are largely extrarenal in the killifish, and we assume that this is true also in *Poecilia.* In the eel, too, we have seen that it is the sodium outflux that is affected by hypophysectomy and prolactin, and the same probably applies to the goldfish. In *P. latipinna* living in a dilute seawater with approximately the same sodium content as the plasma, the sodium turnover rate (i.e., percentage of exchangeable sodium leaving the body in unit time) is strongly enhanced by hypophysectomy and is restored to normal by chronic prolactin treatment, although not by an otherwise effective ACTH therapy. It is noteworthy that the prolactin cells in the pituitary of this species are not totally regressed in this dilute seawater as they are in full-strength seawater, and the data suggest that in both dilute seawater and freshwater the physiological action of fish prolactin lies in reducing the passive permeability of the integument (presumably especially the gills) to sodium. It would seem that the importance of this "impermeabilisation" action of fish prolactin in relation to other components of the osmoregulatory machinery varies from one species to another, and that this may be the determinant of, for example, the drastically rapid loss of sodium in hypophysectomised *P. latipinna* in freshwater compared to the slow sodium loss suffered by the eel in like circumstances, with *Xiphophorus* and the goldfish coming somewhere between these two extremes.

The pituitary gland of the marine form of the stickleback, *Gasterosteus aculeatus (trachurus)*, seems unable during the winter to secrete paralactin in amounts adequate for survival in freshwater, although competent to do so in spring and summer, a seasonal difference probably triggered by photoperiodic changes. Thus, intact winter fish may be regarded as being "physiologically hypophysectomised," at least where paralactin is concerned. In nature, the fish lives in seawater or brackish water during the winter, and migrates to freshwater to breed in spring or early summer. Winter fish transferred from seawater to freshwater suffer a high mortality which can be reduced by prolactin treatment, and they also display a greater fall in plasma osmolality and a smaller fall in urine osmolarity than spring fish transferred to freshwater in the same way. The fall in plasma osmolality in winter fish after transfer to freshwater was paralleled by a rapid drop in plasma sodium and chloride, which could be corrected by a single injection of prolactin given 24 hr before the transfer. It will be seen that although not surgically hypophysectomised, the winter stickleback behaves remarkably like hypophysectomised P. *latipinna*. In the stickleback, the loss of electrolytes after transfer to freshwater, and the prevention of this loss by prolactin, is paralleled by the behavior of the gill mucous cells, which were increased in density by prolactin treatment. Lam quotes preliminary studies showing that prolactin reduces extrarenal outflux of sodium in sticklebacks transferred to freshwater, as in *F. heteroclitus* and *P. latipinna;* and renal effects of prolactin also seem operative in this species.

A striking illustration of the fact that even within the same genus different species may vary greatly in their physiological mechanisms is seen in the example of the plains killifish, *Fundulus kansae.* Unlike *F. heteroclitus*, *F. kansae is* able to survive without its pituitary in freshwater, and even in deionised water provided sufficient calcium is added; whereas calcium will not protect *F. heteroclitus* from failure in deionised water or even in freshwater. *Fundulus kansae* will survive for weeks in a calcium-rich freshwater after hypophysectomy, although in negative sodium balance; such fish display a reduced serum and total body sodium despite normal or higher than normal extrarenal sodium influx, because of enhanced sodium loss via the kidneys. These workers concluded that hypophysectomy, which reduced urine flow, must actually reduce the permeability of the integument to water. A low dose of prolactin increased urine flow in hypophysectomised fish in freshwater, but ACTH was not effective. The renal loss of sodium after hypophysectomy in freshwater resulted from impairment of renal sodium resorption which was not compensated by reduced urine flow. Prolactin reduced the sodium content of the urine of hypophysectomised *F. kansae*, but because of the associated increase in urine flow the actual sodium loss remained elevated. Thus, although prolactin probably enhanced the resorption of sodium by the renal tubule in this fish, it had no net result in terms of the total sodium economy. In intact *F. kansae*, however, prolactin did reduce sodium loss via the kidney. These workers concluded that prolactin acts in both freshwater and seawater to increase the permeability of the integument, resulting in freshwater in increased passage of water into the body and hence in increased urine flow. The effects they had demonstrated on sodium metabolism were minor or equivocal. However, more recent work has indicated that prolactin at higher doses does have extrarenal effects in *F. kansae*, in that it stimulates the active uptake of sodium in hypophysectomised fish in freshwater, while having no effect on the outflux. These findings are very different from the effects of prolactin on sodium outflux in *F. heteroclitus* and *P. latipinna* and may be related to the fact that in *F. kansae*, but not in the other two species, prolactin has an ACTH-like action on the interrenal.

It is difficult to generalise about the site of action of prolactin in these fishes. We have seen that the evidence points largely to an extrarenal action of prolactin in the eel and *F. heteroclitus*, although it is not impossible that some sodium loss could take place across the eel skin and be reduced by prolactin. An extrarenal action of prolactin is probably paramount in *P. latipinna*, although not actually demonstrated, and in the goldfish. On the other hand, prolactin appears to have distinct renal actions in *F. kansae*, possibly normally in synergism with other pituitary factors, and in addition probably stimulates active sodium uptake at the gills even in hypophysectomised *F. kansae*. At least part of the action of prolactin in osmoregulation in the winter stickleback transferred to freshwater appears to be on the kidney. Possibly in teleosts in general prolactin potentially acts on both the kidney and the integument (especially the gills), and its main locus of action differs with species. The epidermal mucous cells have been put forward as important factors in fish osmoregulation, in maintaining a layer of nonstirred mucus over the surface of the skin, and especially of the gills, and effects of prolactin on these cells have been described.

Fish prolactin, or paralactin, is certainly only one of several hormones implicated in electrolyte regulation in teleosts, and its importance in the maintenance of homeostasis appears to vary with species. The physiological status of this fish hormone in teleosts is well established by the work on *F. heteroclitus* and *P. latipinna*. In the only reported effect of prolactin on electrolyte regulation in a nonteleostean "fish," Chester Jones *et al.* found that prolactin, when injected daily for 8 days, lowered the muscle sodium concentration in the cyclostome *Myxine* kept in 60% seawater. A similar effect could be produced by corticosteroids or ACTH, and the authors were inclined to think that the injected ovine prolactin acted as a mimic of endogenous ACTH and stimulated interrenal secretion. We have seen that this possibility is in line with the effects of prolactin in *F. kansae* and *Gambusia*, but that in other teleosts prolactin appears to have no ACTH-like properties. Implications of these differences have been briefly discussed elsewhere.

PROLACTIN AND MELANOGENESIS IN TELEOSTS

Work on this aspect of effects of prolactin in fishes is virtually confined to one species, *Fundulus heteroclitus*. After hypophysectomy this fish becomes noticeably pale, owing to loss of melanin pigmentation, the effect being particularly marked on the normally dark dorsal surface ("dorsal paling"). Treatment of pale hypophysectomised fish with a variety of purified mammalian pituitary preparations demonstrated that darkening of the dorsal surface, with the reappearance of melanin in the depigmented melanophores, was elicited by prolactin, but not by MSH or by ACTH (which appears to promote melanogenesis in the goldfish). Prolactin did not cause proliferation of new melanophores, but intermedin (MSH) did cause melanophore proliferation, and its action was potentiated by prolactin. Thus melanogenesis and melanophore proliferation are separable in *F. heteroclitus* on the basis of their responses to prolactin and MSH. In extending this work, it was found that dopa tyrosinase activity in fin tissue *in vitro* decreased after hypophysectomy (this being an essential enzyme in the pathway of melanin synthesis) and that enzymic activity was restored by prolactin, which also restored melanin pigmentation *in vivo;* however, *in vitro* restoration of dopa tyrosinase activity is not necessarily correlated with restoration of *in vivo* pigmentation, since ACTH and MSH also restored the *in vitro* activity without affecting *in vivo* melanogenesis. In interpreting their finding, Kosto *et al.* suggested that the unique ability of prolactin to restore the melanin content of faded melanophores may lie in an action that makes available some melanin precursor, the supply of which is deficient after hypophysectomy. They also emphasised that prolactin appeared to increase the dopa tyrosinase activity of preexisting melanophores, whereas ACTH and MSH primarily increased the number of melanophores, each probably with only a low level of enzymic activity.

These results demonstrate definitely an effect of ovine prolactin in promoting melanogenesis in *Fundulus* and differentiate this action from the proliferative effects of MSH (and, to a minor extent, ACTH). The distribution of this prolactin effect in teleosts is not known. Work on the goldfish indicates that in this species it is ACTH, not prolactin, that stimulates melanin synthesis. It seems likely that impairment of melanin formation after hypophysectomy is general in teleosts, although sometimes obscured by the persistence of neural mechanisms for background adaptation, and sometimes not apparent because of too short a period of observation after the operation. In the author's laboratory *P. latipinna* has repeatedly been observed to develop pallor after hypophysectomy, associated with depletion of measurable melanin in the dorsal skin; the pituitary factors involved are still under investigation, but it can be said that ovine prolactin restores melanin after hypophysectomy in this species as in *F. heteroclitus*. Curiously, beef prolactin did not induce melanogenesis in hypophysectomised *F. heteroclitus*, although primate growth hormone, with intrinsic prolactin activity, did exhibit melanogenic potency. Ovine prolactin caused melanodispersion in the eel, even after hypophysectomy, and may also stimulate melanogenesis.

It is not certain that stimulation of melanogenesis is a physiological function of the native fish prolactin. An observation suggesting that this may be so is that the prolactin (η) cells were extremely active in hybrid *Xiphophorus-bearing* melanomes. One postulate that should be checked is that if fish prolactin is indeed a stimulator of melanogenesis, the skin of *Poecilia* or *F. heteroclitus* in freshwater might be expected to contain more melanin than the skin of seawater fish, since fish prolactin is probably secreted at a higher rate in the former medium.

PROLACTIN AND EPIDERMAL MUCOUS CELLS IN TELEOSTS

Evidence is accumulating that suggests the possibility that fish prolactin might maintain the epidermal mucous cells in at least some teleosts. Burden showed that the activity and number of mucous cells in the gills of *F. heteroclitus* were reduced by hypophysectomy, a change which he suggested might be correlated with the inability of the operated fish to live in freshwater. Removal of the pituitary was subsequently shown to reduce the mucous cells in the skin of *Betta splendens*, again associated with impaired ability to tolerate freshwater, and prolactin was shown to cause proliferation of mucous cells in the skin of certain intact cichlids. Mucous cells are reduced in number and size in the goldfish skin following hypophysectomy and are maintained by ectopic pituitary transplants, possibly by fish prolactin secreted by the transplants. A special case is the discus fish, a cichlid *Symphysodon discus*, in which the skin of brooding fish hypertrophies and secretes mucus ("discus milk") which is eaten by the fry. It has been claimed that skin proliferation and skin claminess (= excessive mucus secretion) was induced in young nonbreeding *Symphysodon* by repeated injections of prolactin on alternate days for 10 days. The dose of prolactin was extremely high (each injection, 1 mg of prolactin, probably about 20 IU in fish weighing about 4-7 g). However, much lower doses (0.03-2.0 IU) caused proliferation of mucous cells in *Symphysodon* when given over a period of 5 weeks. Prolactin has also been shown to increase the density of mucous cells in the gills of winter *Gasterosteus* transferred to freshwater.

The above works suggest an influence of prolactin on mucous cells, but not all workers have been able to obtain this effect. It should be noted that Bliim and Fiedler were not able to demonstrate an effect on the mucous cells of *Pterophyllum* because the number of cells vary greatly from one part of the body to another. Attempts to demonstrate a reduction in mucous cells of *Tilapia mossambica* after hypophysectomy, and stimulation of the cells by prolactin, have so far been unsuccessful. To date it has been impossible to detect any consistent effects of hypophysectomy or prolactin treatment on the gill mucous cells in *P. latipinna*. On the other hand, as in the case of the goldfish, we have repeatedly observed that hypophysectomised *Poecilia* secrete less mucus than intact fish, in that they feel distinctly less slimy to the hand. Furthermore, not only prolactin treatment but also (and more noticeably) TSH and thyroxine restore the normal sliminess to the skin. These rather subjective data (which nevertheless are consistently borne out by our records of the condition of the skin on fish in many different experiments) require backing by more objective methods of assessment, but they do suggest that in this species the effects of prolactin on sodium movements are not mediated by alterations in general skin mucus secretion, since neither thyroxine nor TSH has the same effects on sodium metabolism as prolactin. Similarly, our failure to detect effects of hypophysectomy or prolactin on gill mucous cells may cast doubt on the hypothesis that these cells mediate the action of prolactin on sodium movements. However, it remains true that increased mucus secretion need not necessarily be accompanied by histologically detectable increases in the number of mucous cells in the skin or gills, and it is quite possible that the secretion of mucus on the gills of *P. latipinna* may be altered by experimentation even though the number of mucous cells showed no definite changes.

One important and curious point is that none of the workers who have demonstrated an effect of prolactin on mucous cells has employed hy pophysectomised animals; thus, contributions from the *in situ* pituitary cannot be excluded until more rigorous experimentation has been reported.

MISCELLANEOUS EFFECTS OF PROLACTIN IN FISHES

Several different categories of prolactin effects have been described following the administration of mammalian prolactin to (usually intact) fishes. It should be said at once that the physiological status of most of these effects is uncertain. Some are related to reproductive phenomena and may be considered at this point. It is well known that in birds prolactin generally seems to act to promote care of the young, being impli cated in brood patch production, pigeon crop sac development (for feeding the young), promotion of broodiness, and suppression of gonadal activity; in general, it seems to promote parental care and behavior in mammals, including, of course, lactation. Because of this background information from work on higher vertebrates, many investigators have been predisposed to expect that prolactin might promote parental behavior in lower vertebrates. One result of this predisposition was the incorrect interpretation that Noble *et al.* had shown prolactin to promote parental behavior in a cichlid mouthbrooding fish, *Hemichromys*. More recent investigators have reported that prolactin protects the embryos of the viviparous cyprinodont *Gambusia* from the deleterious effects of injected estradiol benzoate. Contrary to the statement by Egami and Ishii, Ishii did not demonstrate the importance of prolactin in the maintenance of gestation in the viviparous embiotocid fish, *Neoditrema;* this is just one out of several possible interpretations of his data. It is premature to argue from these observations on intact fish that fish prolactin plays some physiological role in gestation in viviparous teleosts, since it is possible that the results actually reveal peripheral interactions between estrogen and exogenous prolactin rather than pointing up any normal role of fish prolactin. The pituitaryy is certainly not necessary to maintain gestation in *P. latipinna* and *P. formosa*, nor, apparently, in *Gambusia*, despite earlier claims to the contrary.

A distinct stimulation of a component of parental behavior by prolactin has been reported by Fiedler and Blum and Fiedler. Certain cichlid fishes exhibit characteristic parental care behavior in which they fan their eggs by movements of the pectoral fins; during this phase, the tendency to fight is depressed and the appetite reduced. Injections of ovine prolactin at low doses, in the absence of eggs, elicited fanning behavior in intact fish, directed toward a definite point as though toward eggs. Prolactin simultaneously inhibited fighting behavior and feeding. These effects of prolactin were opposed by gonadotropins (FSH, LH, and HCG), and higher doses of prolactin actually inhibited fanning behavior. In contrast to this work on cichlids, the similar parental egg fanning in *Gasterosteus* was not induced by prolactin treatment. However, R. J. F. Smith and Hoar emphasise the possibility raised by work from their laboratory on prolactin and ionic regulation (Section I, B, *Gasterosteus)* that fish prolactin could be a causal in promoting the movement of this fish from the sea to fresh-water for spawning, and they compare this postulated action of prolactin with the newt water-drive effect (Section I, A). In both cichlids and *asterosteus*, the critical experiments involving hypophysectomy and treatment with prolactin remain to be done before we may feel certain that fish prolactin is physiologically involved in inducing egg fanning or migratory behaviour.

By far the most complete evidence of the participation of fish prolactin in aspects of parental care comes from an extensive series of investigations on the sea horse, *Hippocampus*, by Boisseau. The male *Hippocampus* incubates developing eggs in a ventral pouch (marsupium), the development of which is under testicular and gonadotropic control. The maintenance of the connective tissue structure of the marsupium during incubation appears to depend upon ACTH and corticosteroids. During the incubation, the epithelial lining of the marsupium proliferates and secretes a protease which breaks down proteins that are present M· the marsupial fluid and derive from the yolk of early developing eggs; the amino acids released are probably absorbed by the embryos. As it were, the embryo sits in a nutrient broth which is predigested for its consumption by a paternal enzyme. Secretion of the protease is arrested by hypophysectomy of the father, and the operation also leads to histological regression of the marsupial epithelium. Prolactin treatment of both intact and hypophysectomised fish leads to strong histological stimulation of the marsupial epithelium, and in the intact fish it was shown to accelerate secretion of the protease. The physiological role of fish prolactin implied by these findings was confirmed by Boisseau, who showed by partial hypophysectomy that the endogenous hormone responsible for maintenance of the marsupial epithelium originated from the rostral pars distalis, known to be the part of the fish pituitary concerned with paralactin secretion. Boisseau also demonstrated that the $_{q}$ (paralactin) cells in this region of the gland displayed cyclical variations in activity which correlated closely with the development and secretory activity of the marsupial epithelium. The importance of this most interesting work on *Hippocampus* cannot be overemphasised, and it constitutes the most satisfactory demonstration of a parental role for fish prolactin. It may be significant that this parental role is in a marine teleost, in which possibly the hormone is not concerned with regulation of electrolyte exchanges across the body surface.

Other actions of prolactin in the sexual rather than the parental phase of fish reproduction have been described. The male Indian catfish, *Heteropneustes fossilis*, has well-developed seminal vesicles that change seasonally in rhythm with the testicular cycle. The seminal vesicles regressed more rapidly after hypophysectomy than after castration, suggesting that they may be partially maintained by some pituitary hormone (s) other than gonadotropins. In a detailed investigation, Sundararaj and Goswami showed that prolactin by itself did not stimulate the atropic seminal vesicles of intact (regressed), castrated, or hypophysectomised fish, but they showed that prolactin did stimulate growth and secretory activity of the seminal vesicles of intact fish primed with HCG (human chorionic gonadotropin). Similarly, androgen priming of the vesicles of castrated or hypophysectomised catfish allowed prolactin to exert a stimulatory effect on the growth and secretory activity of these structures. Growth hormone also synergised in the same way after androgen priming of hypophysectomised fish, and the maximum response was obtained by simultaneous treatment with androgen, prolactin, and growth hormone. These results suggest that endogenous fish prolactin may participate in the maintenance and activity of the seminal vesicles, although the physiological status of the observed effects of mammalian prolactin is not completely certain. Similarly, it is not clear that the various recorded actions of exogenous prolactin on male accessory structures in mammals have any physiological validity.

In the only report of an effect of prolactin on reproduction in fish other than teleosts, Carlisle reported that spermiation, which was interrupted by starvation, was resumed in dogfish following treatment with prolactin or with gonadotropins. As usual, the physiological meaning of this observation is obscure, although the presence of fish prolactin in the elasmobranch pituitary is attested by a positive response in the red eft water-drive test and the negative performance of elasmobranch material in pigeon crop and mammary gland tests.

Several effects of prolactin have been demonstrated in fishes that seem to have no relation to osmoregulation or reproductive processes. Thus, Olivereau found that chronic injections of ovine prolactin resulted in marked histological activation of the thyroid in intact eels, *Anguilla*

anguilla, but not in hypophysectomised animals, the TSH cells in the pituitary becoming highly active after prolactin treatment. As Olivereau suggested, these results could be explained either by prolactin acting as a goitrogen at the thyroid level or by prolactin somehow stimulating TSH output by a hypothalamic or pituitary action in the presence of a normally functioning thyroid. In amphibians, exogenous prolactin appears to act as a goitrogen in *Rana catesbiana*, and there is some evidence from pituitary transplantation experiments that endogenous prolactin may have an antithyroid action in this frog. However, in the urodele *Triturus*, the prolactin thyroidal effect is apparently one of true thyroidal activation (i.e., elevated secretion of TSH). More recently, Olivereau has found that in the eel, as in *Triturus*, prolactin stimulates ^{131}I uptake by the thyroid, as well as producing morphological activation. Thus prolactin does not act as a goitrogen in *Anguilla* but presumably activates the TSH cells directly or via the hypothalamus, perhaps by resetting the thyroid hormone-TSH homeostat to a higher level.

A metabolic action of mammalian prolactin in an intact teleost was reported by Lee and Meier, who found that daily injections of prolactin (or LH) induced fattening (increase in the lipid fraction of the total body) in sexually mature golden topminnows, *Fundulus chrysotus*. Prolactin had little or no such effect in hypophysectomised fish, and there is a marked diurnal variation in the responsiveness of intact fish to the hormones since injections of prolactin or LH early in the photoperiod actually caused a loss in lipid in contrast to the gain produced by injections in the middle of the photoperiod. It would appear that the fattening effect of prolactin requires the presence of some other pituitary factor(s) which probably undergo a diurnal fluctuation in secretion rate. In addition, differences in the responses of hypophysectomised *F. chrysotus* depending on the time of day they were injected suggest the presence of some nonpituitary factor which conditions the tissue response to prolactin and which varies during the photoperiod. It is interesting to recall that prolactin and gonadotropins probably synergise in inducing fat deposition resembling premigratory fattening in the bird, *Zonotrichia*.

Another possible metabolic role for fish prolactin has been suggested by Johansen. He found that the resistance of goldfish to heat stress was promoted by long photoperiod and was mediated by some hypophysial mechanism that was still active in the ectopically transplanted gland. Although exogenous prolactin failed to increase the impaired heat resistance of hypophysectomised goldfish, Johansen compared the behavior of the transplanted goldfish pituitary with that in *Poecilia* and suggested that endogenous fish prolactin is essential for normal resistance to thermal stress, possibly by alleviating the presumed osmoregulatory disturbances induced by thermal stress. This is an interesting possibility but requires more direct experimental backing. Another miscellaneous effect of the hormone is seen in the maintenance of normal levels of circulating erythrocytes and thrombocytes by chronic prolactin injections in hypophysectomised P. *latipinna*; this is not a specific action of prolactin since erythrocytes were also maintained by ACTH and thrombocytes partially by thyroxine and growth hormone, and a physiological role for fish prolactin in hemotopoiesis is far from proven. In an earlier investigation Slicher had obtained no effects of prolactin on erythrocytes or thrombocytes in hypophysectomised *F. heteroclitus*, but in one experiment out of two she found that prolactin elevated the leukocyte count. This last effect has not been duplicated in our work on *Poecilia*.

Currently there is considerable interest in demonstrations of growth promotion by prolactin in various vertebrates. Evidence exists of overlap of properties between growth hormone and prolactin in various vertebrates, and in particular there are several demonstrations of growth promotion by exogenous prolactin in amphibians, reptiles, birds, and mammals. Little information is available for teleosts, perhaps partly because of the difficulty of providing the ideal conditions necessary for experimental fish to exhibit vigorous growth. Prolactin has never been found to promote growth in hypophysectomised *F. heteroclitus*, and endogenous fish prolactin and growth hormone are certainly separate factors in *Poecilia*. Prolactin treatment of hypophysectomised *P. latipinna* has not induced any significant growth in length, although like TSH it somewhat alleviates the shrinkage of hypophysectomised fish.

Extraction of Paralactin and Hypothalamic Control

There have been no published studies on the isolation and physicochemical properties of paralactin. Acid-acetone extraction of salmon pituitaries yielded a fraction which probably contained fish prolactin, inasmuch as it opposed the drop in serum osmolarity when hypophysectomised goldfish were transferred to freshwater. Application of a modified Bates and Riddle procedure to carp pituitary material produced a fraction that was presumed to be fish prolactin, but which had no biological activity when tested on hypophysectomised *F. heteroclitus*. Material from carp and pollack, but prepared by different procedures, proved able to cross-react *in vitro* with a rabbit antibody to ovine prolactin in precipitin tests.

In mammals, prolactin is the only pars distalis hormone which is secreted at a high rate when the gland is separated from the ypothalamus, and prolactin is apparently under inhibitory control of the hypothalamus in this group. In fish, the ectopically trans-planted pituitary proved capable of secreting sufficient paralactin to maintain freshwater survival in *Poecilia formosa*, and *Xiphophorus*. In *P. latipinna* the paralactin from ectopic transplants maintains the normal low rate of sodium outflux in freshwater and the low rate of sodium turnover in isotonic dilute seawater. Although it seems certain that in cyprinodonts at least, hypothalamic connections are not essential for secretion of paralactin at normal levels, the absence of rigorously quantitative data on rates of secretion of the hormone in grafted fish compared with normal fish precludes the assumption that a hypothalamic paralactin-inhibiting factor exists in teleosts, particularly since prolactin in some birds appear to be under stimulatory hypothalamic control.

GROWTH HORMONE

Introduction

Growth hormone (GH), or somatotropin (STH), is best known and most readily characterised as the pituitary factor which promotes overall growth of the body, even when acting in the absence of the other pituitary hormones, for example, when injected into hypophysectomised animals. Many complex actions of the hormone on metabolic processes underlie this growth promotion, and recent accounts of these metabolic actions of GH in mammals will be found in the reviews by Knobil and Hotchkiss and Evans *et al.* and in the symposium edited by Pecile and Miiller. Growth hormone is known from studies in higher vertebrates to be particularly important as a synergist with other hormones, including among its properties for instance the power to enhance the effects of tropic pituitary hormones on their target glands.

Nearly all studies concerned with the physiology of GH in fishes relate to stimulation of body growth, and there is little published data on other aspects of its action. The concept of growth is notoriously vague and difficult to define, but it would probably be generally agreed that the idea of growth of the whole body, as commonly used, implies the permanent addition of differentiated tissues but excludes temporary additions to bulk and weight, such as result from the seasonal development of the gonads, and also excludes the accumulation of food reserves such as fat or glycogen. In practical terms, it follows that increase in linear dimensions offers the safest criteria for the detection of body growth, certainly over the short periods involved in experimental studies.

Effects of Hypophysectomy on Growth in Fishes

Among teleosts, the basic observation that growth permanently ceases after removal of the pituitary seems to have been made on only two genera. Pickford was the first to demonstrate conclusively that a hypophysectomised teleost does not grow, when she reported that hypophysectomised killifish, *Fundulus heteroclitus*, do not grow in length and show only irregular changes in weight. Sometimes hypophysectomised fish actually decreased in length, a change which was partially corrected by TSH therapy. Pickford has emphasised how misleading weight changes can be in respect to growth, pointing out for example that some hypophysectomised killifish develop oedema because of renal calculi forming in the kidney ducts, and this may lead to weight increases without any linear growth. Similarly, the liver in hypophysectomised fish is enlarged and full of fat and glycogen reserves, which sometimes leads to weight increments although the length of the fish is unaltered.

In agreement with this pioneer work on *F. heteroclitus*, hypophysectomy arrests growth in length in *Poecilia formosa* and *P. latipinna*, even under conditions in which intact or sham-operated controls grow vigorously. More noticeably than *Fundulus*, hypophysectomised *Poecilia* usually decrease in length. In an experiment in which 46 *P. latipinna* were hypophysectomised and then observed for 14 days, all the fish decreased in length during this period (mean 1.5 $\pm$ 0.09%), while intact controls were growing under the same conditions at a mean rate of 0.7 $\pm$ 0.17% in 14 days.

For elasmobranchs, Vivien, in the only published work on this topic, showed that hypophysectomised dogfish ceased to grow.

Effects of GH on Growth in Hypophysectomised Fishes

Purified mammalian GH will stimulate linear growth in intact teleost, and an extensive series of investigations has been directed to the analysis of the more significant fact that purified beef GH will also promote growth in hypophysectomised male *F. heteroclitus*. One important point that emerges from these experiments is the necessity of providing optimal living conditions in order to obtain a good response to GH, and also the necessity of avoiding too frequent handling or other stressful procedures which oppose the growth response. The response of hypophysectomised killifish to a standard dose of beef GH is temperature-dependent, little or no significant length increment being obtained below 15°C, with an optimum plateau for the response between 20° and 25°C. Standard beef GH causes a linear log dose response in the length increase of hypophysectomised killifish over a dose range from 3 to 30 μg/ g body weight, a finding which permits the bioassay of fish GH in hypophysectomised *F. heteroclitus*. The slope of the log dose growth response is influenced by time after hypophysectomy among other factors. Hypophysectomi-:ed killifish also respond with linear growth to pig, sheep, monkey, and human GH.

In hypophysectomised *P. latipinna* and *P. formosa* pig and beef GH, respectively, can induce linear growth; but the responses have not yet been studied systematically.

Hypophysectomised *F. heteroclitis* are also reponsive to a purified crystalline fish GH isolated from glands of pollack, hake, and cod by Wilhelmi. In demonstrating this response, Pickford found that even the most highly purified fish GH was less potent that the equivalent dose of beef GH. This may be because the beef GH contained traces of TSH, which maintained the thyroid, whereas the fish GH preparation had no stimulatory effect on the regressed thyroid of the hypophysectomised recipient fish. This thyroid stimulation probably potentiated the growth-promoting action of the beef GH, since although TSH alone has no growth-promoting activity, the addition of small amounts of TSH to hake GH significantly augmented the growth response in hypophysectomised killifish.

By bioassay on hypophysectomised *F. heteroclitus*, Swift and Pickford have estimated the GH content of the pituitary of the perch, *Perca fluviatilis*, during the annual cycle in Windermere. They found that during the winter, when the perch do not grow appreciably, the pituitary contained a low resting level of GH. The GH content of the gland increased in the spring and reached a maximum in June, at about the time of onset of the natural growth period in Windermere, and then fell to a very low level in August. This depletion of pituitary GH store corresponded to the period of most rapid growth in the natural cycle and presumably reflects the fact that at this time rapid secretion of the hormone is only just balanced by synthesis, and little is stored.

A point of special interest to fisheries biologists is that this work on *F. heteroclitus* showed that removal of the pituitary led to failure of formation of new circuli (growth rings) on the scales, correspo-nding to cessation of body growth. Administration of beef GH to hypophysectomised fish elicits a resumption of scale growth which starts with formation of an irregular zone resembling the year mark, and which shows an approximate correlation between the body length increment and the member of new circuli formed on the scales. Furthermore, Swift and Pickford showed that although a body length increment of 3% was required before any scale growth was detectable, length increments greater than 3% were associated with proportional increases in scale width. These experimental findings thus support some of the main assumptions on which a great deal of fishery biology work is based, that is, the existence of a direct correlation between body growth and scale growth, and a correspondence between spacing of circuli and rate of body growth. As Pickford pointed out, since temperature affects the response of hypophysectomised killifish to GH, temperature is probably an environmental factor involved in the establishment of the usual seasonal pattern of scale and body growth in temperate zone fishes. In addition, food supply (which may be controlled by temperature) may determine the growth pattern in some species in some environments. The importance of the annual changes in day length as an environmental governor of the annual growth cycle was suggested by a field study of a population of brown trout, *Salmo trutta*, but experimental studies on the effects of day length on fish GH secretion have not been undertaken.

Metabolic Effects of GH in Fishes

Beef GH rapidly improves the appetite of hypophysectomised *F. heteroclitus*, which suggests some stimulation of metabolism preceding the growth response. Unfortunately, there are no reported detailed studies of this aspect of GH physiology in teleosts. After hypophyse-ctomy, the liver in killifish increases in size and is laden with fat and glycogen, and GH did not reduce liver size or stores in hypophysec-tomised fish. Hypophysectomised *Poecilia* also have enlarged livers with increased glycogen stores, and GH did not reduce liver size or glycogen stores, although ACTH was effective on both counts. Matty has reported nitrogen retention following injection of GH into intact *Cottus*, which is in line with the protein anabolic effects of GH in mammals. Preliminary work by Enomoto suggests that chronic treatm-ent of intact rainbow trout with beef GH caused growth and increased the crude protein content of the carcass.

For elasmobranchs, Orias and Abramowitz *et al.* reported amelioration of the diabetes of pancreatectomised dogfish following hypophysectomy. This recalls the Houssay phenomenon in higher vertebrates which hinges principally on the action of GH in inhibiting peripheral glucose utilisation. Thus in dogfish, as in tetrapods, CH appears to be diabetogenic. Comparable studies on teleosts or other fishes have not been reported, but GH injections elevated plasma glucose in intact *Cottus* ; and in preliminary experiments, Enomoto found that a single injection of beef GH caused transient glucosuria, but no elevation of blood glucose, which suggests a renal effect of the GH rather than an action on tissue utilisation of glucose. Despite some dissenting claims in the earlier literature, it appears that fish pituitary extracts and purified fish GH are not diabetogenic in higher vertebrates and have no effect on rat cardiac glycogen. Purified fish GH also failed to stimulate nitrogen retention in the rat, which is in line with its failure to promote growth of rats.

MISCELLANEOUS EFFECTS OF GH IN FISHES

Some possible effects of GH on electrolyte metabolism have been adumbrated in the literature, although in each case the data are only suggestive and require extension. D. C. W. Smith found that GH increased the tolerance of trout to high salinities, and Hoar cites results of J. E. McInerney as demonstrating changes in the salinity preference of young coho salmon after a long period of GH injections. These j two findings suggest a possible role for GH in the complex of physiological changes at smoltification in salmonids. Although GH had no marked effects on plasma electrolytes in intact trout, it did increase the potassium content of the muscles, an observation to be placed alongside the extreme activation of the GH cells in eels kept in deionised water. It is possible that in the intact eel in deionised water an increased GH secretion opposes the movement of potassium from muscles to plasma, a movement that is more marked in the absence of the pituitary. It should be noted that pig GH has, no sodium-conserving activity in P. *latipinna*, and beef GH did not promote tolerance of freshwater in hypophysectomised killifish.

Like prolactin, GH induced weight increase and stimulated secretory activity in the androgen-primed seminal vesicles of the hypophysectomised Indian catfish, *Heteropneustes fossilis*, and synergised with androgen and prolactin to produce a maximal response. This recalls the well-documented ability of GH to potentiate the tropic effects of gonadotropins and androgens on the testis and accessary structures in mammals.

Porcine GH slightly elevated the depressed thrombocyte count of hypophysectomised P. *latipinna* without affecting any of the other blood elements. In hypophysectomised *F. heteroclitus*, GH had no action on any hematological parameter.

EFFECTS OF FISH PITUITARY MATERIAL ON GROWTH IN OTHER VERTEBRATES

Although hypophysectomised killifish respond with linear growth to GH from pig, beef, sheep, monkey, and man, teleostean pituitary material will not promote growth in the rat or in the tadpole *of Rana temporaria*. It has been suggested that induction of growth in *Rana temporaria* tadpoles is an indicator of prolactin rather than of GH, but according to Enemar tadpoles of the size used in his laboratory respond to GH but not to prolactin. Cyclostome pituitary material is also without action on tadpole growth. Furthermore, crystalline fish GH, active in *F. heteroclitus*, is without effect on growth in hypophysectomised rats. However, Geschwind has recently reported that the pituitary of the lungfish, *Protopterus aethiopicus*, showed good growth-promoting activity in the rat, but that glands from the dogfish, *Squalus*, were inactive. Once again, the dipnoan allies itself with tetrapods rather than with other fishes. These data presumably indicate differences in the molecular structure of fish and tetrapod GH, about which regrettably little is known. G. Extraction of Fish GH Wilhelmi showed that a highly purified crystalline growth hormone (identified by its action on hypophysectomised killifish; Pickford) could be prepared from pituitary glands of hake and pollack by application of extraction procedures designed to be effective with mammalian glands. Wilhelmi gives the procedure in detail, and Pickford has summarised the physicochemical properties of the material, as determined by Dr. Wilhelmi. In general terms, and as far as this preliminary data goes, fish GH and beef GH are quite similar, but already differences between hake GH and pollack GH are indicated. Hake GH and beef GH have similar isoelectric points, but pollack GH, like that of pig and horse, is a more acid protein. On the other hand, Wilhelmi estimated the molecular weight of beef GH by ultracentrifugation as 44,000-47,000 and that of fish GH as 22,000-26,000. More recently, Dr. Wilhelmi's laboratory have revised their figures for the beef GH, now finding a molecular weight of 22,400, but revised figures are not available for the fish hormone.

HYPOTHALAMIC CONTROL OF FISH GH SECRETION

Growth continued in hypophysectomised *P. formosa* with ectopically transplanted pituitaries (homotransplants), but at an extremely low rate, indicating a partial failure of GH secretion by the transplanted gland, correlated with a marked reduction of typical GH cells in the transplants. Growth hormone secretion appears to continue at a low rate from the autotransplanted pituitary in *P. latipinna*, although not in all cases, and the matter is still being studied. These limited data suggest that as in higher vertebrates the hypothalamus may exert a stimulatory influence on GH secretion by the fish pituitary, but much more information will be needed to substantiate this proposition.

6

Hormones of Heterocrine Gland

It has not been possible to make this chapter an exhaustive review of thyroidal function in fishes, nor is such a review really desirable. Much of the research on piscine thyroid physiology has merely established that in the broad sense it is not too different from thyroid function in other vertebrates. What makes the study of controlled thyroid function in fish exciting to the comparative endocrinologist is that some differences do exist. Furthermore, since these characteristic piscine differences are found in the most primitive of living vertebrates, they offer the only clues available concerning the evolution of thyroid function.

The general vertebrate pattern of thyroid function is diagrammed in Figure elsewhere in the book. The four components of this functional system are the following:

(1) The brain, and more precisely a complex of nervous afferents to a center (or centers), presumably in the hypothalamus, which regulate pituitary thyrotropic (TSH) secretion. This part of the system links changes in the extrinsic environment (e.g., temperature and photoperiod) to eventual changes in thyroid secretion rate. The brain may contain thyroxin-sensitive centers, and possibly these are the same as the hypothalamic centers just referred to; they are part of a feedback mechanism

(2) The pars distalis of the pituitary gland, and more precisely the TSH secreting cells in it. Secretion of TSH, a protein hormone,

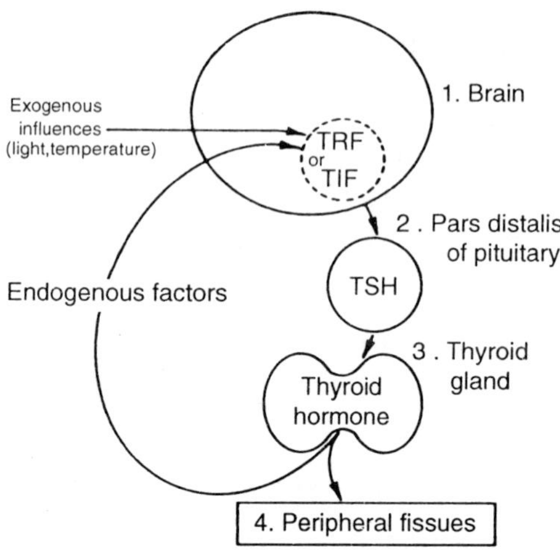

Figure 6.1 : (thyroid to pituitary), also regulating TSH secretion, but through intrinsic influences.

is controlled in higher vertebrates in part by direct feedback responsiveness of the TSH cells to levels of circulating thyroid hormones.

(3) The thyroid gland, whose structural and metabolic properties depend upon the circulating level of TSH. The characteristic thyroid function is accumulation of iodine and the formation of the tyrosinederived hormone, thyroxine.

(4) The peripheral tissues of vertebrates, particularly in growing and differentiating animals, respond in a variety of ways to circulating thyroxine levels. In part these responses are basic metabolic phenomena (nitrogen turnover, respiratory oxygen utilisation) ; in part they are structural and secretory responses.

HYPOTHALAMIC AND OTHER REGULATION OF TSH SECRETION

Anatomically there are enough differences in the relationship between ; the brain and pituitary of fishes compared to other vertebrates, that we may suspect some functional differences as well. Unfortunately, there is little or no *direct* evidence from fishes of hypothalamic regulation of TSH or any other tropic hormone. In Agnatha, for example, there appears to be no vascular or nervous relationship between brain and pars distalis. In elasmobranchs, a large part of the TSH is said to be formed in the ventral lobe, an extension of the pars distalis that appears to have no anatomical connection with the brain. In teleosts, on the other hand, there is an interdigitation of the neurohypophysis and adenohypophysis. Accordingly, in teleosts there seems to exist the possibility for the brain to influence the adenohypophysis either through a vascular route or by direct innervation. Which of these is utilised is not known.

Although an anatomical means for hypothalamic control over TSH (or other tropic hormones) appears to be absent in cyclostomes, there is some evidence that it may still occur, possibly through the secretion of hypothalamic releasing factors into the systemic circulation. For example, continuous illumination of ammocoetes larvae of lampreys affects stainable hypothalamic neurosecretion. Purely circumstantial evidence resides in the fact that thyroid gland metamorphosis and sexual maturation of larval lampreys are closely linked to season. Finally, there is dubious evidence based on the hyperplasia of ammocoetes endostyle (larval thyroid) epithelium following treatment with goitrogenic drugs. In higher vertebrates, at least, goitrogens act by blocking thyroid hormone synthesis, and the resulting lowered level of thyroid hormone in blood evokes TSH secretion. Unfortunately, the possibility that ammocoete endostylar hypertrophy is evoked in this way seems denied by recent experiments of Barrington and Sage. They found that hypophysectomy will not prevent the "goitrogenic" effect of thiourea, thus this hyperplasia in lampreys cannot result from evoked TSH secretion.

For elasmobranchs, also, information concerning control of TSH secretion is almost completely absent. In these forms there is a well-developed, frequently double median eminence and a portal system of vessels connecting this part of the hypothalamus with the pars distalis. Yet, as has been mentioned, Dodd *et al.* have reported that a special adenohypophysial structure of elasmobranchs that produces TSH (the ventral lobe of the pars distalis) is not related to this portal system. Olivereau has reported that goitrogens do not alter thyroidal histology in sharks, so that feedback control through the hypothalamus or direct pituitary sensitivity to blood thyroxine may also be lacking in these forms. The TSH secretory mechanism remains unstimulated in goitrogenised *Scyliorhinus* despite the fact that it can be shown that the goitrogen is in fact blocking iodine metabolism in the thyroid. In contrast to these results, Pritchard and Gorbman by repeated injections of thiouracil into late embryos of *Squalus suckleyi* produced a slight histological stimulation of the thyroid, suggestive of an activated TSH secretion. The entire question of possible hypothalamic control over TSH secretion in cartilaginous fish obviously requires attention. Although there is no conclusive evidence for it, the possibility that it exists in some form is reasonable since several workers have described annual histology cycles in elasmobranch thyroids that are closely linked with season.

Evidence for the existence and the character of hypothalamic control over teleostean TSH is much fuller than for cyclostomes or elasmobranchs, but it is still relatively incomplete. Many descriptions have appeared characterizing cyclic structural changes in teleost thyroids correlated with annual environmental cycles. To supplement this there have been studies of the specific influence of temperature and photoperiod upon thyroid structure and function. Thus, there would appear to be sensory-evoked nervous afferents to the hypothalamus of teleosts that can eventually alter TSH secretion by the pars distalis. Lesioning experiments to show the positions of these afferents, or of the putative TRF-forming loci in the hypothalamus have not yet been done. Some very interesting experiments bearing on the question were done by Ball *et al.* by transplanting the pituitary gland of *Poecilia formosa* to nonhypophysial loci in hypophysectomised specimens. Away from any hypothalamic innervation these pituitaries secreted as much TSH as the normal pituitary, or more, to judge from thyroid histology. Gonadotropic or adrenotropic function in such transplants was reduced. Thus, it would appear that in this fish, the hypothalamic influence in regulating TSH secretion is normally inhibitory, not stimulatory as described in higher vertebrates.

Further evidence comes from experiments with goitrogens and with low-iodine diets or low-iodine environmental waters. Limited availability of exogenous iodine would be expected to lower circulating thyroxine levels and, through negative feedback, to evoke hypothalamic (or direct pituitary) activation for eventual increased TSH secretion. The classical work of Marine and Lenhart showed that low-iodine waters promote overwhelming thyroid hyperplasia in trout, and that this thyroid condition can be reduced or prevented by addition of iodine to the water. Since then many similar observations have been made in other species of fish. Olivereau *et al.* have described an increase in pituitary TSH cells in trout kept for months on a low-iodine diet or radiothyroidectomised. Similarly, goitrogenic drugs given to a variety of teleosts, unlike the reported unresponsiveness of elasmobranchs, activate the pituitary-thyroid axis, evoke cytological changes in the TSH cells of the pars distalis, hyperplasia of pharyngeal thyroid tissue, and at the same time at least partially block thyroxine synthesis.

Conversely, administration of thyroxine to several species of teleosts has been shown to be inhibitory for the pituitary-thyroid system, both morphologically and physiologically (thyroid uptake and organification of radioiodine).

Several workers have shown parallel variations in stainable hypothalamic neurosecretion in teleosts that accompany cyclic changes in thyroid function, or follow treatment with goitrogens or with thyroxine. However, at this time it is difficult to interpret changes in stainable hypothalamic neurosecretion in terms of thyrotropic activation by thyrotropin releasing factors (TRF).

In summary it may be said that reasonably good evidence of diencephalic control over TSH secretion exists only for teleosts, and even this is not well characterised. Because of the different and characteristic anatomical relations between the brain and pars distalis in different groups of fishes, it would seem that this would be a rewarding and significant area for further research.

THYROTROPIC FUNCTION OF THE FISH PITUITARY GLAND

The evidence for secretion of a thyrotropin by the cyclostome pituitary is very limited and for the reasons given below must still be considered equivocal. In the only extraction experiments reported, Dodd *et al.*, in a review, tabulate data not otherwise published showing that hagfish pituitary contains a very low concentration of TSH activity and lamprey pituitary a slightly higher level. In the same table it is reported by Dodd *et al.* that frog neurohypophysis contains as much TSH activity as the lamprey whole pituitary, fish neurohypophysis three times as much, bird neurohypophysis four times, and mouse neurohypophysis 60 times as much. In these tests, using the McKenzie procedure as the bioassay, the high levels of activity in neurohypophysis tissue must be considered indicative of the background or meaningless level of the assay technique; hence, the results could be interpreted to show that no TSH activity was extractable from cyclostome pituitary, and the question remains open.

Two other direct approaches to the question of TSH in cyclostome pituitary have been made. Knowles hypophysectomised larval lampreys (ammocoetes), and Larsen has hypophysectomised young adult lampreys; in neither set of experiments was there any observable effect upon endostylar or thyroid histology, or upon rate of ^{131}I accumulation by the thyroid. On the other hand, although TSH injections by Knowles into larval *Lampetra planeri* were without effect, both Olivereau and Clements-Merlini, using other species of ammocoetes, found that mammalian TSH evoked morphological changes in the endostyle and a slight stimulation of $^{13'I}$ metabolism.

Under these circumstances, as long as hypothalamic control of the cyclostome pituitary remains dubious, and as long as definitive proof of TSH production still is lacking, it is difficult to interpret claims of thyrotropic cells in the cyclostome pars distalis.

The use of adult temperate zone selachians in laboratory experiments, particularly when this requires keeping them for extended periods of time, presents numerous practical difficulties and high mortality rates. For this reason there are relatively few studies of the effects of hypophysectomy, or of the effects of repeated injections of TSH. Furthermore, these studies are conducted at marine laboratories where available facilities and techniques modify experimental design, and somewhat less than definitive experiments result. These difficulties are mentioned at this point in explanation of the fragmentary and imperfect experiments to be described in the following paragraphs. This also explains why several of the studies have utilised embryonic or newly hatched sharks; these are much more easily used in laboratory experiments.

If we examine the evidence based on pituitary tissue extraction, and on hypophysectomy, we must admit that definitive proof of a TSH-like principle in elasmobranch pituitaries also remains to be provided. The pituitary extraction and bioassay experiments of Ferguson, Dodd, and Hunter show by the use of the McKenzie TSH bioassay that the ventral lobe of the *Scyliorhinus canicula* pars distalis contains a concentration of TSH activity about equal to that of a salamander and higher than that in a teleost, *Pleuronectes*, pituitary. This important observation opens so many questions, including that of hypothalamic control raised in the previous section of this chapter, that it deserves confirmation and publication in full detail for better evaluation. This appears to confirm earlier data of Olivereau and Vivien who observed histological stimulation of *Scyliorhinus caniculus* thyroids after injection of crude suspensions of *Scyliorhinus* pituitary tissue.

Eight days after hypophysectomy of late-embryo sharks by decapitation,. thyroidal ^{111}I uptake was apparently reduced in experiments by Vivien and Rechenmann However, these authors judged the ^{131}I uptake by the use of radioautography of tissue sections, and this is not a technique that yields reliable quantitative data. Dodd and Goddard hypophysectomised adult *Scyliorhinus caniculus*, and in specimens that survived as long as 2 years they found no histological changes in the thyroid. In earlier work Waring *et al.* found no histological changes in the thyroids of *Scyliorhinus* 1 month after hypophysectomy. More recently, Vivien found that in younger *Scyliorhinus* embryos hypophysectomised by decapitation, thyroidal differentiation is incomplete.

Unlike the cyclostomes and elasmobranchs, for the teleosts there is an abundance of evidence which indicates that the pars distalis contains a thyroid stimulating hormone and that surgical removal of this gland leads to involutional changes in the thyroid and reduction of its function. The earliest literature is exhaustively reviewed by Pickford and Atz as well as by Dodd *et al.* and Dodd and Matty. Emphasis of the newer phases of this field will be made here. It may be concluded from the earlier and more recent work that the pars distalis of numerous teleosts contains cells which are responsive to and along with changes in thyroid state of the fish, and these may be considered TSH cells. B. I. Baker even finds these cells directly responsive to thyroxine in tissue cultures (of *Salmo gairdneri* pituitary), raising again the question of direct feedback responsiveness versus hypothalamic control of TSH secretion. In this context reference should be repeated to the papers of Ball *et al.* these workers found that pituitary glands transplanted into a site away from the hypothalamus can secrete more TSH than in the normal site, while gonadotropic and ACTH secretion are reduced. The obvious inference from this work is that TSH secretion is under *inhibitory* control of the brain when the normal anatomical relationships of the *Poecilia* pituitary obtain.

It is quite possible that in fish, as in mammals, both levels of feedback control of TSH secretion, hypothalamic and direct TSH-cell responsiveness are effective. It is of interest that bioassay procedures reveal an annual pituitary cycle of TSH concentration in temperate zone species. It would seem that such cycles must be under hypothalamic control since they depend eventually upon sensory afferent information.

The properties and chemical characteristics of teleostean TSH have 'been a continuing interest of Y. A. Fontaine and have been the subject of a number of publications from his laboratory. Y. A. Fontaine and Condliffe purified TSH of the eel to a high level of specific activity, 15 "units" per milligram, and this as well as similar purified hormone has served for further work. A part of such further study has been concerned with the degree of molecular differentiation or evolution that has occurred during vertebrate evolution. Gorbman showed earlier that the chemical variation in TSHs is sufficiently great that it can be revealed in physiological tests by reciprocal bioassay. That is, if crude pituitary preparations from a series of vertebrates are tested for their thyrotropic potency according to their effects on thyroids of a series of vertebrates, the tendency is for greater potency in recipient species most closely related (phylogenetically) to the animal source of the pituitary.

Y. A. Fontaine and Dellerue-Lebelle have employed immunological techniques to reveal some of these relationships in a different way. They found, for example, that the mouse thyroid can be stimulated by beef or mouse TSH, but not by that from teleost pituitary. Lungfish TSH will stimulate mouse thyroid in the doses used. Anti-beef TSH antibody will neutralise mouse TSH but not that of teleosts or lungfish. Thus, the molecular differences between mouse, lungfish, and teleost TSH appear to be such that lungfish is intermediate between the other two, distinguishable by immunological properties from the mammal, but not clearly differentiated by bioassay.

In the converse sense, it would seem that the fish thyroid is more responsive to mammalian TSH, than mammalian thyroid to fish TSH. For example, goldfish thyroid is sufficiently responsive to mammalian TSH to permit use of these fish for bioassay. Mammalian thyroids clearly are not as suitable for bioassay of teleostean TSH, although some slight responsiveness can be shown.

One reason for the apparently greater responsiveness to various pituitary preparations by the fish thyroid than by the mammalian thyroid is that fish thyroid may not discriminate closely among pituitary hormones. Y. A. Fontaine has found, for example, that a substance in mammalian pituitary preparations other than TSH will stimulate the teleost thyroid. Until it is definitively identified, this substance is tentatively called the "heterothyrotropic factor" (HTF). Rodesch and Fontaine found that although HTF is strongly thyrotropic in teleosts, it is a relatively weak thyroid stimulant in mammals. In an *in vitro* test for thyrotropic activity based on ^{131}I metabolism by cultured sheep thyroid cells, Rodesch and Fontaine found mammalian TSH most active, mammalian HTF less active, and carp TSH least active. Y. A. Fontaine and Burzawa-Gerard fractionated the TSH and HTF activities in rat and beef pituitaries and found that the HTF activity could not be isolated from gonadotropic activity. This suggests strongly that the teleost thyroid is responding to mammalian gonadotropin. Pickford and Grant have confirmed the HTF activity of mammalian pituitary material in hypophysectomised *Fundulus* and in starved trout, showing that it is referable to a substance other than TSH.

It would appear, then, that the teleost thyroid responds to several glycoprotein hormones, but in the evolution both of the pituitary hormones and the thyroid mechanisms responsive to them, that further differentiation and specificity may have occurred. There is a basis here for some important further research.

An interesting parameter was added to the question of species responsiveness to TSHs of different sources by M. Fontaine and Fontaine. They found in assays for TSH activity in trout at different temepratures, that the rat pituitary is relatively. more active at 20°C than at 10°C. Eel pituitary, on the other hand, is more active at 10°C than at 20°C. This remarkable finding deserves confirmation and further study.

THYROID FUNCTION AND THYROID HORMONE SYNTHESIS

Stated most simply, the function of thyroid gland is to manufacture thyroid hormone. As will be made clear presently, the thyroid hormones of fish are relatively small molecules and they are the same in all vertebrates, namely, thyroxine (Tx) and triiodothyronine (T_3). Other iodocompounds may be manufactured by the thyroid, and even released from the thyroid tissue, but in amounts and activities that have no apparent physiological value.

The process of thyroid hormone formation has been studied very closely because it can be followed to a large extent by the use of radioiodine as a characteristic label. The first phase of the process involves concentration of iodine (iodide) within the secretory units (follicles) by means of active transport of the ion. The next phase involves oxidation of the iodide and iodination of tyrosine in a specific thyroidal globulin. Synthesis of the thyroglobulin itself has been subjected to recent study and has been found to proceed by formation of smaller protein units which are then used as monomers in buildup of a polymeric molecule. The iodinated thyroglobulin must be hydrolysed within the follicle, or within the follicular epithelium, to free the iodinated tyrosines (monoand diiodotyrosine) and their simple condensation products, thyroxine (tetraiodothyronine) and triiodothyronine. Most of the freed iodotyrosines are destroyed by enzymic dehalogen-ation, but the iodothyronines diffuse into the blood where, in loose combination with certain serum proteins, they are distributed to the tissues.

This brief synopsis has been given to serve as a background against which the general discussion of thyroxinogenesis in various fishes can be conducted. Although the process has been found generally similar in most vertebrates, there are some features that are peculiar to the fishes. Comparative descriptions of the process of thyroxinogenesis in vertebrates may be found in reviews by Berg *et al.* Dodd and Matty, and Gorbman.

An anatomical feature must be considered at the outset since it confers a unique characteristic to the fish thyroid. In the hagfish, and in a large majority of the teleosts, the thyroid follicles are not assembled into a single gland, but instead they are scattered generally in connective tissue throughout the subpharyngeal and parapharyngeal area. K. F. Baker *et al.* discovered that in several species of small teleosts thyroid follicles migrate extensively from the pharyngeal region forming a large concentration in the head kidney, but occurring also at many other places, including even the eye. Chavin confirmed this phenomenon for the goldfish and showed further that head kidney thyroid is more active than pharyngeal thyroid in the quantitative sense and responds differentially to goitrogens. However, Frisen and Frisen believe that the two kinds of thyroid are equivalent in function, and quantitative differences can be explained on the basis of differences in total numbers of follicles in the pharynx and kidney.

The cyclostome thyroid gland has additional special morphological features. In addition to widespread scattering of thyroid follicles, some of the largest known thyroid follicles, about 2 mm in diameter, occur in *Myxine glutinosa*, and they are almost completely avascular. The poor blood supply of the hagfish thyroid may indicate an extremely low physiological value of thyroid secretion to the adult hagfish. The adult lamprey thyroid resembles that of teleosts in that the follicles are concentrated in the ventral pharyngeal region and they are well vascularised. However, much interest attaches to the iodine metabolism in the embryonic precursor of the thyroid, the endostyle or subpharyngeal gland of the larval lamprey. Such studies have served as the basis for much speculation concerning evolution of thyroid function. The endostyle is a complex tubular organ which opens by a duct to the mid-ventral pharynx, and certain cells of the lining epithelium of this organ function in an adult thyroidlike manner.

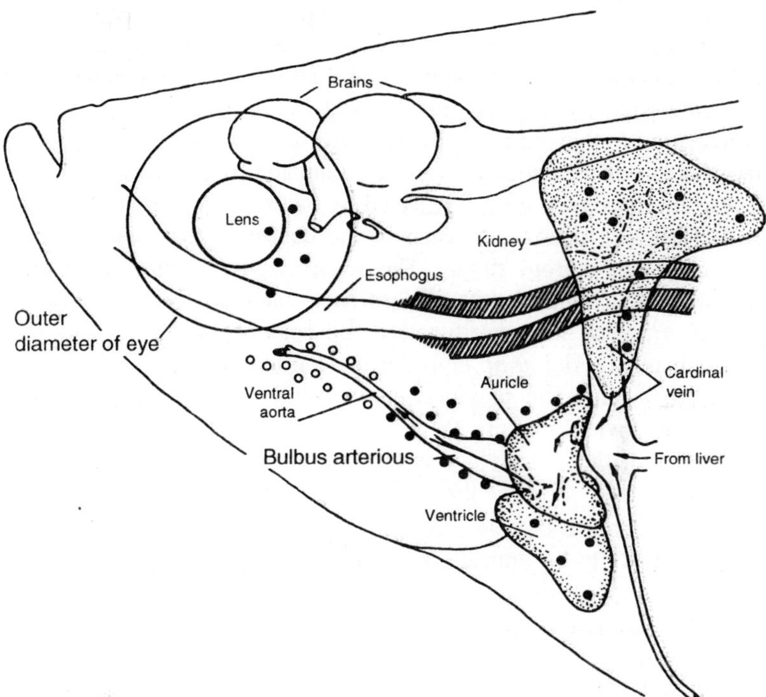

Figure 6.2 : Diagram of the anterior part of a platyfish showing the distribution of thyroid follicles: (O) Normal distribution and (●) heterotopic distribution of thyroid follicles.

Functionally, the thyroid follicles of hagfish appear typically vertebrate, but relatively slow. An injected dose of ^{131}1, as iodide, is accumulated by the pharyngeal thyroid tissue to a maximum of 2.1% of the injected dose in 4 days in *Eptatretus stouti* or 3.4% in 6 days in *Myxine glutinosa*. Tong *et al.* found that radioiodine-labeled thyroxine was eventually detectable, but only 2% of the accumulated thyroidal radioiodine was in this form after 4 days. Furthermore, *no* radiothyroxine was found in the blood until 4 days after injection of the ^{131}1. Waterman and Gorbman observed that the lumen of the hagfish thyroid follicle does not contain the characteristic thyroglobulin-containing "colloid" material seen in all other vertebrates. Instead, there are numerous colloidlike droplets within the lining epithelium of the follicles. This is of particular interest since Waterman and Gorbman and Tong *et al.* made radioautographs of tissue sections of ^{131}I-injected hagfish thyroid and located all of the radioiodoprotein in the cells and none in the colloid, as in other vertebrates.

In the lamprey larval endostyle, Gorbman and Creaser, using radioautography found that radioiodine localises in only certain of the cell types lining the cylinders, and Leloup and Berg and Leloup reported that radioiodine metabolised in the endostyle is in part converted to thyroxine. Clements-Merlini found that radioiodine metabolism in the endostyle can be stimulated by pituitary hormones. These findings have been confirmed by others, and further, Clements and Gorbman found a protease in the endostyle that resembles the one found in thyroid tissue. The iodoprotein of *Petromyzon* adult thyroid tissue, labeled with ^{125}I has been analysed by Aloj *et al.* by density gradient ultracentrifugation and fractions at 5, 12, and 17 S were obtained. Aloj *et al.* consider these fractions to be similar to the elementary units and polymerised molecules characteristic of other vertebrate thyroglobulins. With so much interest in the thyroidlike properties of the endostyle it is difficult to understand why there has been almost no study of the adult lamprey thyroid.

Elasmobranch thyroid function has received relatively little study, but it appears to be typical. In *Scyliorhinus*, for example, maximum thyroidal uptakes in excess of 20% of an injected dose of tracer ^{131}I can be measured at about 17 hr after injection. The proportion of labeled organic iodine that is found in the form of thyroxine continues to increase gradually for at least 4 days or more. Detailed study of the dynamics of thyroxinogenesis, of the enzymic activities aiding the process, or of the character of the thyroid protein has not been done.

Teleost thyroid function, as might be expected, has received the most attention, and also as might be expected, the goldfish and several salmonids have been most studied. One obvious reason for this is that analysis of iodine metabolism is a project for the chemical laboratory, and the fishes most frequently stocked in or near the laboratory are most readily studied. However, despite this, a sufficient variety of piscine species has been studied to permit some generalisation, and an excellent and detailed study was made by Hickman on a flounder, *Platichthys stellatus*. Thyroid function in the European eel, *Anguilla anguilla*, has been a continuing subject for study in Fontaine's laboratory, particularly with colleagues Leloup and Olivereau.

Apparently the small freshwater platyfish, *Xiphophorus maculatus*, was the first teleost whose thyroidal utilisation of radioiodine was studied by chromatographic methods. Qualitatively, it was typical in showing an eventual synthesis of thyroxine. However, new thyroxine synthesis, up to 6 days after injection of tracer ^{131}I was never very high in proportion. Since this work many similar experiments have been conducted with generally similar results but with some interesting differences.

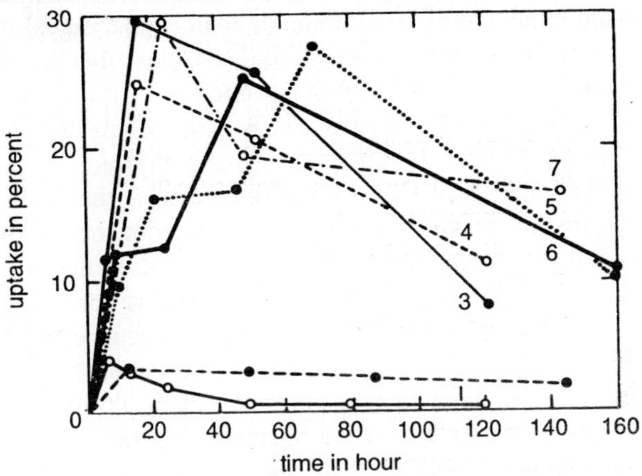

Figure 6.3. Thyroidal radioiodine uptake and loss of "'I over a period of time in seven species of freshwater teleosts maintained at 21°C.

In a purely quantitative sense, there are great differences in the avidity of the thyroids of various species of fish for administered iodine (as revealed by uptake of carrier-free radioiodine), as well as in the rate of loss of organically bound iodine from the thyroid gland. As Hickman and others have pointed out, the rates of uptake and discharge (= turnover) of iodine by the thyroid is subject not only to intrinsic properties of the fish thyroid (including regulatory influences from the hypothalamo-hypophysial system) but also to other factors like environmental salinity, temperature, and iodine concentration. Fishes differ also in the proportion of thyroxine to triiodothyronine formed in the thyroid and released into the blood. An extreme instance is the mud minnow, *Umbra limi*, which under certain natural circumstances makes considerable amounts of triiodothyronine but no thyroxine. Ordinarily, however, triiodothyronine is relatively difficult to demonstrate on the fish thyroid.

The goldfish, *Carassius auratus*, and "Pumpkinseed," *Lepomis gibbosus*, have been found by Berg and Gorbman, Berg *et al.*, and Fortune to represent those fish with very "slow" thyroids. That is, they accumulate in the pharyngeal thyroid only a small fraction of an administered dose of radioiodine and convert this to thyroxine to a very small degree, and only after a period of more than a week. Part of the explanation of the low pharyngeal iodine uptake is provided by Chavin, and Chavin and Bouwman who show that the thyroid tissue in the head kidney accumulates about twice as much radioiodine as does the pharyngeal thyroid. Goldfish head kidney thyroid is more sensitive to the radioactivity of radioiodine than is pharyngeal thyroid. This fits well with the observation of Baker-Cohen that in platyfish whose thyroid tissue is hyperplastic because of a low iodine environment, the pattern of hyperplasia is quite frequently different between pharynx and head kidney or splenic thyroid. A puzzling datum in the publication of Chavin and Bouwman is the finding of a maximum thyroidal uptake at 60 min after injection of radioiodine. This differs from the experience of others and raises the possibility that variations in experimental design or conditions for maintaining teleost fish may be the source of important discrepancies in work from different laboratories. Another remarkable datum from Chavin and Bouwman is the claim that in goldfish kidney thyroid, , which has a maximum uptake of 7% of the injected ^{131I} at 60 min and a fall from this maximum to 1.7% at 24 hr, the ^{131}I "uptake" rises again to 4.4% at 84 days. This would imply an extraordinary retention of iodine by the goldfish and a slow recycling to the thyroid after long periods in the tissues.

In salmonids special interest has been attached to correlations of thyroid function with phases of the life cycle. Hoffert and Fromm and Hunn and Fromm for example, estimated thyroxine secretion rate in trout and found a 5-10 times greater rate between the immature young and the 1- to 2-year-old. Jacoby and Hickman, using an isotope dilution method, were able to estimate absolute concentration of iodocompounds in adult rainbow trout. They found totals of 1.1 ug of thyroxine, 1.3 μg of triiodothyronine, and 1.1 $_{\mu}$g of iodotyrosines per 100 ml of blood. It would be interesting to reconcile these data with Hoffert and Fromm's to note the relation between thyroid secretion rate and blood levels of hormone. The presence of so large a proportion of iodotyrosine in the blood is surprising, and it needs further study.

Another remarkable claim that requires additional study and confirmation is made by La Roche *et al.* who have attempted radio thyroidectomy of trout and of chinook salmon by repeated injection of ^{131}I. While thyroidectomised trout underwent characteristic thyroidectomy changes, the salmon did not. Nine months after destruction of thyroid follicles, radiothyroxine still was produced by the "radiothyroidectomised" salmon from a tracer ^{131}I dose. The explana-

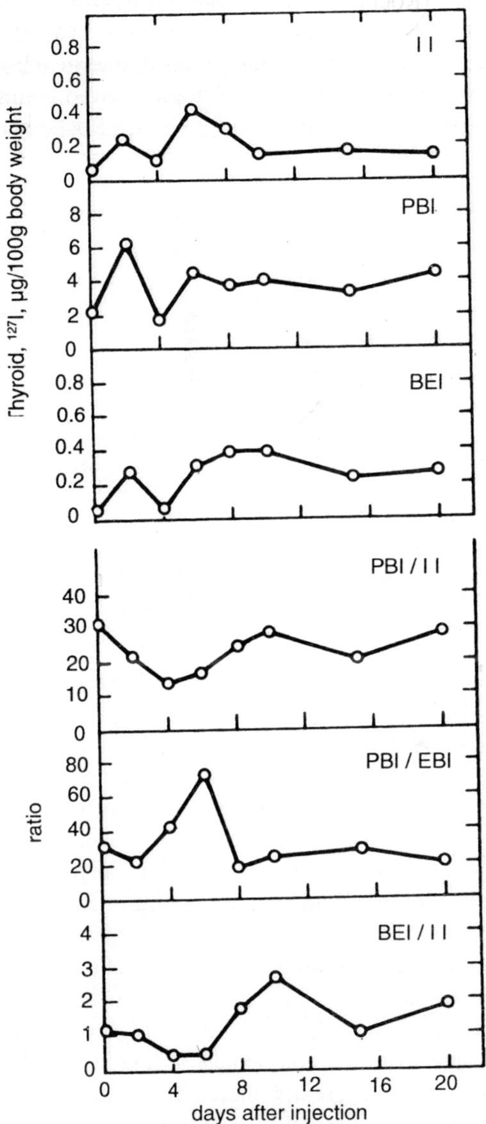

Figure 6.4 Effect of a single injection of thyrotropin (0.001 USP units/g body weight) on the thyroid organic and inorganic ^{127}I of Platichthys stellatus *and on the ratios of these fractions. Four to seven flounder were sampled at each time. The upper three curves represent the variations in inorganic iodide (II), protein bound iodine (PBI), and butanol extractable iodine (BEI), respectively.*

tion offered by La Roche *et al.* is that disorganised nonfollicular thyroid cells may be performing this function. An equally good possibility is the survival of follicular tissue in one of the many sites mentioned by Baker-Cohen where heterotopic thyroid could be found in platyfish. The intriguing questions raised by La Roche *et al.* require further study.

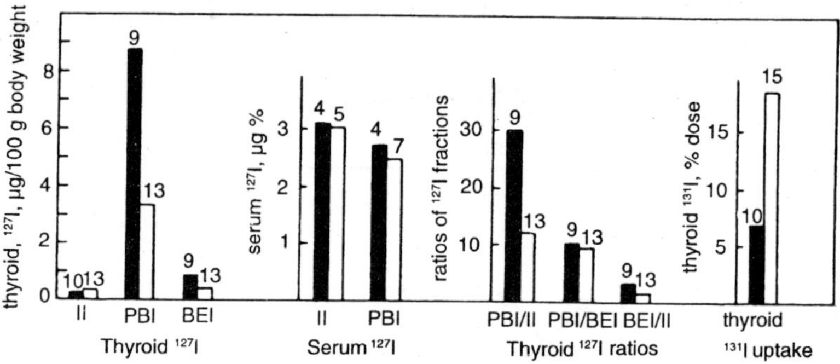

Figure 6.5 : Effect of prolonged thyrotropin treatment (six injections at 5-day intervals 0.001 USP units/g body weight) on thyroid and serum iodine content of Platichthys stellatus. Black bars represent controls, shaded bars thyrotropin-injected fish. Numbers above bars indicate sample size. Experimental temperature 15° ± 1°C. Salinity 20 ± 4%.

The influences of fluctuations of environmental factors, and of TSH, on thyroid metabolism have been studied in fragmentary fashion by many authors. However, probably no study of biochemical changes in iodine metabolism has been made in as extensive a manner in fishes as that by Hickman. The particular value of this study is that it provides data on stable iodine (^{127}I) changes as well as the shifts of tracer ^{131}I. There is a complex of changes in iodide content, PBI (protein bound iodine, considered all organified iodine), and BEI (butanol extractable iodine, considered to be iodine in hormonal form) over a period as long as 10-12 days after a single injection of TSH into a flounder, *Platichthys stellatus*. The effect of repeated injections of TSH on the same species. It is most interesting that in *Platichthys* there is a remarkable loss of stable (total) iodine at a time when the 48-hr radioiodine uptake is greatly increased in the thyroid. In the blood of *Platichthys* the most sensitive indicator of TSH action appears to be the ratio of PBI to iodide, a fact that underlines the usefulness of "conversion ratio" in the blood (PB^{131}I/total ^{131}I) to easily indicate thyroid state in fish. Use of conversion ratio (CR) for this purpose has been recommended by Hickman and by Eales. The extension of these principles to the rainbow trout has been made also by Hickman where the oppositeness in response of stable iodine content and ^{131}I uptake in thyroid also can be shown after TSH treatment. This is, of course, an indication of the fact that thyroidal iodine turnover and hormone secretion are activated. Sound advice to all who are looking for a single reliable and easy method of estimating thyroid function in fish is found in Hickman's valuable commentary: "No parameter of thyroid activity is any better than its potential as an index of hormone secretion rate." Unfortunately, for any particular species this would require a prior study, using coordinated measurements of both ^{127}I and ^{131}I, before the dependability of a single index could be judged.

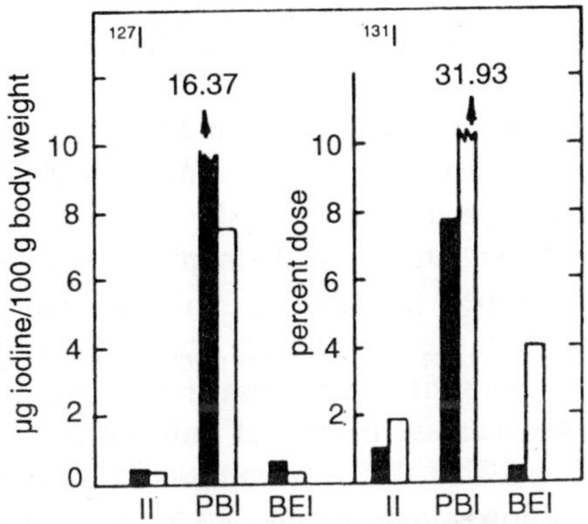

Figure 6.6 : Effect of three injections of thyrotropin (0.002 USP units/g body weight at 2-day intervals) on the thyroid ^{111}I and ^{131}I concentration of Salmo gairdneri. *Black bars represent controls, shaded bars thyrotropin-injected trout. Height of bar indicates average of three samples.*

Variation of thyroid function with respect to temperature has been a continuing and puzzling subject for research with fishes. Using various indices of function (histological and radiochemical) lists have been made of species whose thyroid function parallels, is inverse to, or is indifferent to changes in environmental temperature. One of the clear indications of this tabulation is that the criteria for judging thyroid function with respect to temperature may vary in an opposite way, contributing to some of the confusion that this topic enjoys. As Drury and Eales point out, thyroid activity judged by histology may be activated, depressed, or remain unchanged with respect to temperature variation, but radioiodine metabolic measures generally change in a direction parallel with temperature. There are two apparent exceptions, however, even to this generalisation: the mud minnow and the brown trout. It would seem that the dissociation of the histological response and the iodine metabolic response of teleost thyroids when the temperature is changed means that they are under separate and different control. A part of the reaction may be in response to a temperature actuated hypothalamohypophysial-TSH mechanism, and the rest to a direct effect of temperature on the thyroid. Further work, particularly with hypophysectomised fish, may help to clarify this problem. Leloup and Fontaine have reported that in hypophysectomised eels radioiodine levels are similar and low at both 6.5° and 25°C. At 6.5°C TSH has little or no effect upon thyroidal radioiodine uptake either in hypophysectomised or normal eels. These data indicate that part of the temperature effect may be explained on the basis of difference of kinetics of action of TSH at higher versus lower temperatures.

There is some special interest in the function of the lungfish thyroid in the humid and in the dry or cocoon state. This interest derives from the fact that an environmental factor other than temperature or light appears to be regulating thyroid function in this species, and it may be a derivative of the state of hydration of the animal. Both Godet and Dupe and Leloup, as well as Leloup and Fontaine, have described the changes in thyroidal ^{1311} uptake in *Protopterus annectens* in the wet and dry states, but kept at the same temperature. Both groups find that passage from the wet to the dry phase causes a sharp decrease in thyroid radioiodine uptake. Leloup, in addition, has shown that this is accompanied by a fall in plasma organic iodine and a lowered secretion rate by the thyroid. Returning a dry lungfish to a wet environment causes a prompt activation of the thyroid, revealed by a prompt loss of previously stored thyroidal radioiodine.

Analysis of the thyroid iodoproteins by density gradient centrifugation has been done by Lachiver *et al.* Three fractions were obtained from eel thyroid (11.1 S, 17.8 S, 25.4 S). Of these the 11.1 S protein is very feebly iodinated, but the others are highly iodinated. In trout, similarly, the principal thyroglobulin fraction (about 16 S) is iodinated. In moist or dry lungfish, *Protopterus annectens*, most radioiodine activity is associated with a fraction whose sedimentation rate is 20.8 S, but radio-Much has been written concerning peripheral tissue responsiveness in fish to thyroid hormone, but only a few of the observations have established a firm causal relationship between thyroid hormone and the particular functional feature discussed. It is useful also to remember in the following discussion that certain of the thyroxine effects claimed are relatively modest ones and that other known mechanisms may affect the same phenomena in a much more potent manner. This last comment applies in particular to the relationship of thyroid hormone to carbohydrate metabolism, osmoregulation, and perhaps also to growth in fish.

Practically nothing is known of the action of thyroid hormones in cyclostomes or in elasmobranchs. Some interest exists in the question whether thyroxine can precipitate metamorphic changes in ammocoetes larvae of lampreys. Stokes long ago showed that thyroxine does not have this action. This finding has been confirmed by a number of workers but such confirmations usually remain unpublished because of their negative character. Furthermore, Leach found that thyroid hormone treatment had no effect on oxygen consumption either in ammocoetes or in adult lampreys.

Elasmobranchs likewise have no clear respiratory response (oxygen consumption) to thyroidectomy or administered thyroid hormone. Pritchard and Gorbman treated late embryos (pups) of the shark, *Squalus suckleyi*, with repeated injections of triiodothyroacetic acid, thyroxine, or propylthiouracil over periods of several weeks and measured respiratory oxygen consumption during this time. There was a rise in oxygen consumption for 8-10 days, and then a fall despite continued hormone treatment. The goitrogen had no apparent effect. Dogfish pups of the same species were injected with thyroxine by Gorbman and Ishii who found that hypothalamic neurosecretory material was caused to appear precociously. This may be an expression of thyroxine sensitivity of the central nervous system discussed below more extensively for the teleosts.

There seems to be a general parallelism of thyroid and reproductive functions in elasmobranchs. This is more fully reviewed by Pickford and Atz, Dodd and Matty, and Woodhead.

There are so many physiological phenomena in teleosts in which thyroid appears to play some role, or has been claimed to, that it is necessary to present them in some logical order. First, will be considered the metabolic actions of thyroxine; second, structural effects; and third, effects on central nervous system and behaviour.

Respiratory stimulation by thyroxine is its best known action, yet the bulk of evidence, frequently reviewed indicates that thyroxine has no such action in teleosts. Thyroidectomy of parrot fish or of rainbow trout does not alter oxygen consumption. Antithyroid drugs had no effect in *Fundulus;* however, Osborn claims that thiouracil lowers oxygen consumption by *Campostoma*, a minnow, and makes it thereby more viable in water low in oxygen. To add confusion to this picture, Sage has found that thiourea depresses oxygen consumption of the guppy, *Lebistes reticulatus*. Thyroxine reverses the respiratory depression of thiourea, but when administered alone (without thiourea) thyroxine then *depresses* oxygen consumption.

Thyroxine injections frequently produce different respiratory effects in the hands of different investigators. Most commonly, thyroxine has evoked no respiratory response in *Carassius auratus*, *Lebistes reticulatus*, *Opsanus tau*, *Rhodeus amarus*, or *Salmo gairdneri* in doses that may produce other physiological effects (see review in Pickford and Atz,). Miiller stimulated oxygen consumption in goldfish with dosages of 0.5-1.0 mg-a tremendous dose in an animal that may produce less than 1 μg of thyroxine per day in its own thyroid. Chavin and Rossmore using the same species, the goldfish, were unable to produce a change in 0, consumption using 0.5 mg doses of thyroxine, and in their tests thiouracil also was without action.

It is of particular interest that although thyroxine has no consistent effect on teleost oxygen consumption it produces at 10^{-1} M hormone concentrations, mitochondrial swelling *in vitro* in fish tissues just as it does in mammals. According to some interpretations of the cellular action of thyroxine, such hormone-evoked mitochondrial changes may alter enzymic relations sufficiently on the mitochondrial membrane to uncouple oxidative phosphorylation. This would lead to an increased oxygen consumption in order to yield a particular amount of useful chemical energy. The fact that mitochondrial changes can be evoked without respiratory consequences in fish may be a good argument against this interpretation of thyroxine's cellular action.

A direct intervention of thyroxine in carbohydrate metabolism of fish tissues has been claimed by Hochachka. He finds that thyroxine or T, *in vitro*, acting upon brook trout liver slices or homogenates, stimulates the conversion of ^{14}C-labeled gluconate to CO_2 by as much as 125%. He has concluded that this *in vitro* action represents a possible activation of the pentose cycle.

Increased ammonia excretion by goldfish is another metabolic function of thyroxine that has been claimed. Thornburn and Matty have shown that thyroxine affects nitrogen metabolism in the brown trout as well as in goldfish and it stimulates incorporation of 14C_ labeled leucine into protein. Numerous authors have observed the silvering of the skin, especially in salmonids, that results from thyroxine treatment. It seems clear that thyroxine may therefore affect integumentary guanine metabolism, and some evidence to this effect was furnished recently by Matty and Sheltawy who report that prolonged thyroxine treatment increases the guanine content of teleost skin and also increased incorporation of ^{14}C-labeled glycine into newly synthesised guanine.

A relation to salt and water movement in teleost tissues, and therefore to osmoregulation, also has been claimed frequently for thyroxine. There appears to be little question that environmental salinity may have some effect on thyroid function in fishes. However, whether thyroid hormone has an important influence in osmoregulation remains to be proved. Some of the older research showing apparently reduced thyroid function in seawater merely showed the influence of increased iodide in the environment. Research such as that cited by M. Fontaine from his laboratory, which showed that thyroxine has a protective action for eels transferred from saltwater to freshwater, will have to be reevaluated in terms of simultaneous action of prolactin, which has an important role under these circumstances in teleosts. Other hormones that may play a larger role in water-electrolyte regulation than thyroxine are the corticosteroids and the neurohypophysial peptides.

Srivastava has studied patterns of influence on isotopically labeled phosphate uptake by goldfish by hypophysectomy, and by thiourea administration. He concluded that such uptake of ^{32}P from the surrounding aquatic medium is favored by TSH since it is reduced by hypophysectomy, but increased by thiourea.

Conversion of retinene to vitamin A in young fish has been found apparently inhibited by triiodothyronine.

Endocrine control of growth of teleosts has concerned several investigations but has not resulted in a clear definition the role of thyroid hor mone in the process. La Roche *et al.* who studied trout in particular, concluded: "It remains to be demonstrated that thyroid hormone administrations will stimulate overall growth in the presence of adequate iodine intake and thyroid function." This statement is made in the face of claims by Smith and Everett, Smith *et al.* Barrington *et al.* Piggins, Hopper, Gross *et al.*, and Bjorklund who administered thyroid hormones in various forms to rainbow trout, guppies, sunfish, goldfish, and young salmon and found some growth stimulation. Hoar's review also expresses uncertainty concerning the role of thyroid hormone in teleostean growth, and the earlier data are summarised by Pickford and Atz, Baker-Cohen, and Olivereau. It should be mentioned that thyroid treatment of some fish retards growth rate. It may be that thyroid hormone plays a permissive role in growth regulation of fishes, as it does in mammals. If so, the analysis of its role in growth may require more complex experimental design for final definition. Furthermore, as La Roche *et al.* point out, "chemical thyroidectomy" with goitrogens and radiothyroidectomy, which inhibit growth, have toxic effects which have not been evaluated adequately in interpreting their action.

If the complex character of general body growth resists a definition. of thyroxine's role in it, a somewhat more satisfactory situation obtains when we consider structural changes in individual systems or organs. It is clear that skeletal elements, for example, are responsive to thyroid state in teleosts. Gerbilsky and Saks have observed accelerated scale and bone plate formation in thyroxinised sturgeon. La Roche *et al.*, and in earlier papers of La Roche, specific effects of both thyroidectomy and thyroxine administration on phases of skeletal growth, calcification, and differentiation have been described in salmonids. Baker-Cohen also observed skeletal changes in radiothyroidectomised platyfish. In this context, it is of interest that Barrington and Rawdon found that thyroxine favors the uptake and incorporation of radiosulfur into the skeleton of trout.

The positive relationship of thyroid state to integumentary silvering has been mentioned above. La Roche *et al.* have noted the increased melanin pigmentation of the skin of radiothyroidectomised trout and have found that it is the result of an increased number of melanophores per unit of skin area. Sembrat, using thiouracil, found increased pigmentation in skin of the carp. Belsare *et al.* showed that in larval *Channa punctatus*, thiourea not only favors increased integumentary pigmentation but also melanophore granule dispersion. Thyroxin had the opposite effect. However, in young *Misgurnus fossilis* and salmon, goitrogens did not affect melanin pigmentation. La Roche *et al.*(and in earlier publications by La Roche, especially La Roche and Leblond) have found in both rainbow trout and salmon, that radiothyroidectomy affects the morphology of skin (thinning of both epidermal and dermal elements) which is corrected by thyroxine administration.

There is now an extensive literature that indicates a role for thyroid hormone in central nervous function and behaviour of teleosts. Much of this has been gathered by Hoar and Baggerman. The eventual significance of this work is that it may form a basis for explanation of important events like population migration. One thing is certain, and it is that if thyroid hormones affect migratory behaviour, they are not the only hormones involved; this may account for the imperfect correlation between migratory behaviour of salmon and thyroid state, summarised by Hoar.

Among the provocative publications in this field are two of Hoar *et al.* who found that thyroid and sex hormones affect the pattern of motor behaviour and orientation of goldfish and young salmon, and that thyroxin stimulates the degree of motor activity. Sage similarly found that thyroxinised guppies have an altered swimming and increased jumping behavioural pattern, without increasing oxygen consumption. Thiourea has an opposite effect on motor behaviour in guppies. Baggerman, working with sticklebacks and juvenile salmon, has provided evidence that thyroxine and goitrogen treatment alter "salinity preference." When thyroxinised young salmon, for example, are presented a choice between freshwater and saltwater, they "prefer" saltwater to a greater degree than do controls. Thiourea treatment favors freshwater preference. These data, although they demonstrate an interesting hormone-altered behaviour, do not actually explain migratory tendency since the salmon do not begin their downstream migration in a salinity gradient. In an imperfect and difficult experiment, Norris recently planted radiothyroidectomised and intact steelhead trout, *Salmo gairdneri*, in a small lake. He observed the tendency of these young fish to migrate out of the lake downstream and found no clear difference. Unfortunately, mortality and the difficulty of recovering specimens in such a "natural" situation reduced the numbers of specimens to a relatively small value, making a firm conclusion difficult.

Effects of thyroxine upon behaviour point out the need for more precise study of the effects of changes in thyroid state upon particular phases 'of central nervous function. A beginning has been made in studies by Hara *et al.* and by Oshima and Gorbman. These workers have selected particular sensory-evoked (optic, olfactory) electrical events in the goldfish brain and have analysed them in some detail, in terms of pattern, amplitude, threshold, and other properties. Optically evoked (by light flashes) midbrain potentials are sensitised by prior thyroxine treatment of the fish, and recovery time between stimuli is shortened. Electrical potential patterns in the olfactory bulb evoked by chemical stimulation of the olfactory organ of goldfish are affected by thyroxinisation in a more complex way. Apparently, reactivity of neurons in the olfactory bulb is potentiated by thyroxin, but centrifugal impulses from the posterior brain to the bulb are inhibited by thyroxine treatment. Thyroxinisation of goldfish also produces changes in spontaneous and injury-evoked electrical activity in the forebrain and diencephalon. It seems remarkable so far in this analysis that the influences of thyroxine and other hormones on central nervous functions in fishes are so widespread, and the implications of these influences are many. Unfortunately, the significance of these thyroxine-influenced neuroelectrophysiological changes in terms of particular behavioural or neurosecretory events remains to be defined. One suggestion has been made by Godet and Dupe, who have studied the relation of thyroid state to cerebral electrogenesis, particularly with respect to olfaction, in the lungfish, *Protopterus annectens*. On the basis of their work they have developed a neuroendocrine hypothesis which explains the awakening of the lungfish from the dry state when moistened, causing him to excyst from the cocoon to resume aquatic phase activity. Their data, at least in part, support a hypothesis requiring that increasing humidity activates the thyroid, presumably through a hypothalamic mechanism. Thyroid hormones, acting upon cerebral centers, increase sensitivity of an olfactory center, and this, in turn, evokes normal feeding and other behaviour. Data supporting this interesting idea are incomplete, but at least they illustrate how an eventual piecing together of nervous and endocrine phenomena will serve to explain endocrine and neuroendocrine regulation of behaviour.

7

Hormones of Adrenal Gland

The current revival of interest in the general physiology of fishes has emphasised our ignorance of autonomic functions in these animals. Since Nicol reviewed the anatomy and function of the autonomic nervous system of fish in 1952, there have been many new investigations; these have often been concerned with details of the innervation of systems already examined in some detail, namely, the heart, the gastrointestinal tract musculature, and chromatophores. It is rather surprising that there has not been greater interest in the innervation of, for example, the vasculature, the internal genitalia, and the digestive glands. It is possible that this will be the last of the old-style reviews of this subject, for new techniques, including the now widespread use of delicate mechanoelectric transducer recording systems and the introduction of the histochemical technique for visualizing tissue catecholamines, are already starting to revise our opinions on the autonomic nervous system of fishes.

CYCLOSTOMES

Cranial Autonomic Nerves

The oculomotor nerve is absent in hagfish, in which the lateral eyes are degenerate. The eyes of lampreys are also small, but oculomotor nerves are found. Anatomical evidence for an autonomic component of the oculomotor in lampreys is controversial, and there would appear to be no function for such a component since the eyes are probably completely devoid of intrinsic musculature.

An autonomic component of the facial nerve has been claimed for *Myxine*, largely on the basis of brainstem analysis. Lindstrom has suggested that the autonomic fibers innervate the lingual artery, but he does not exclude the possibility that the fibers are afferent. No physiological investigations of this nerve have been made.

In both hagfish and lampreys, the two vagus nerve trunks unite to form a cardiac plexus, lying on the gut wall, containing numerous nerve cell bodies and a few ganglionic aggregations. From this plexus, autonomic fibers are distributed to the region of the heart, major veins, gall bladder, and, via an unpaired nerve trunk, to the entire length of the intestine. The intestinal wall contains a nervous plexus with numerous cell bodies, some of which may be sensory neurons. The neurons become sparse in the posterior region of the gut, although the fiber plexus persists. Many of these enteric neurons are probably innervated by vagal fibers.

Of the vagal innervation systems, that of the heart has been studied in greatest detail. Greene first reported that the heart of the hagfish *Polistotrema* (=*Bdellostoma*) *stouti* was not affected by stimulation of the vagus nerve, or indeed of any other nerve. This was confirmed for the same species by Carlson and for another hagfish, *Myxine glutinosa*, by Augustinsson *et al.*. In agreement with this observation, no nerve fibers were found during an electron microscopic investigation of *Myxine* heart. Normal hearts of both species are virtually unaffected by acetylcholine and by catecholamines. In contrast, the lamprey heart is innervated by the vagus nerves. The main response to stimulation of the medulla oblongata is an acceleration of the heart followed by a period of deceleration after periods of intense stimulation, although a period of inhibition preceding the acceleration has been reported. The heart of *Lampetra* does not contain adrenergic nerve fibers demonstrable by fluorescence histochemistry, and catecholamines are variously stated to cause a slight depression or an acceleration and augmentation of the cardiac beat. It would therefore seem that the vagal innervation is cholinergic, since acetylcholine accelerates the heartbeat, at the same time depressing the force of contractions. Similar actions are exerted by other choline esters, including succinylcholine, and nicotine, but not by muscarine; the response to acetylcholine is inhibited by the nicotinic blocking drugs hexamethonium and tubocurarine but is unaffected by atropine. The cholinergic receptors are therefore of the nicotinic type. Because of this, Augustinsson *et al.* suggested that acetylcholine was acting on ganglion cells in the heart. It seems a simpler view that the lamprey heart muscle has a closer affinity to the skeletal muscle than to the cardiac muscle of higher vertebrates and that acetylcholine is acting directly on nicotinic receptors in the heart muscle itself. In this case, one would expect treatment with curare to prevent vagal transmission to the heart, but Zwaardemaker observed cardiac responses to medulla oblongata stimulation in lampreys immobilised with curare. Perhaps the cardiac cholinergic receptors are relatively resistant to blockade by curare. The vagus does not appear to innervate the heart in larval lampreys.

Stimulation of either vagal trunk in *Myxine* causes a contraction of the strongly muscular gallbladder. This pathway involves a ganglionic synapse in the hepatic plexus since topical application of nicotine to the plexus abolishes the responses. The terminal neurons are probably cholinergic, for acetylcholine causes a contraction of the gallbladder, whereas epinephrine appears to relax it.

The vagal innervation of the intestine is poorly understood. The muscularis of the gut wall is poorly developed, but it seems that all regions of the intestine are contracted by acetylcholine and relaxed by catecholamines. In addition, the isolated intestine is reported to undergo peristaltic movements. It is therefore clear that the gut musculature is functional. However, Fange and Fange and Johnels could not detect any intestinal response to vagal stimulation in *Myxine*. Previously, Patterson and Fair had found that stimulation of the vagus in *Bdellostoma* caused a very slight relaxation of the gut, but their records were obtained *in situ*, and in view of the thinness of the gut wall the response could have been artifactual. On the other hand, Fane and Johnels pointed out that they had allowed the animal to warm up to room temperature (18'C), and they suggested that the vagus might be ineffective at this high temperature. The only independent support for an inhibitory vagal innervation of the gut is the observation that nicotine causes a relaxation of previously contracted *Mijxine* intestine *in vitro*, which may indicate the presence of inhibitory neurons in the gut wall. It has been postulated elsewhere that the vagus nerve is primitively inhibitory to the vertebrate gastrointestinal tract and that the inhibitory nerve fibers are of a nonadrenergic type recently demonstrated in the mammalian gastric vagus.

The possibility has been raised that the vagus nerves contain adrenergic nerve fibers since the vagus of *Myxine* contains considerable concentrations of catecholamines, predominantly norepinephrine. Adrenergic fibers have not been found in the cranial autonomic outflow of any higher vertebrate, but this does not necessarily mean that they will be absent in cyclostomes. Alternatively, sympathetic adrenergic fibers could enter the vagi from the spinal outflow, as claimed by Marcus, since the vagi run extremely close to the mixed ventral rami of the anterior spinal nerves. Finally, the catecholamines could be stored in chromaffin cells; the vagus contains many ganglion cells, and one must remember that the "ganglion cells" found in cyclostome hearts by Augustinsson *et al.* were later shown to be specialised chromaffin cells. A fluorescence histochemical study of this problem would be most welcome.

Spinal Autonomic Nerves

There is still much confusion about the distribution of autonomic nerve fibers from the spinal cord. Some of this confusion arises from the fact that many peripheral nerve cell bodies have been observed in cyclostomes, but it is not always clear which of these neurons are autonomic and which sensory. In addition, there are reports of fibers leaving the spinal cord of *Lampetra* in both dorsal and ventral spinal nerves and running directly, without an intervening ganglionic synapse, to visceral structure. Whether all or any of these fibers are autonomic, i.e., efferent, is unknown and will probably remain so until physiological observations have been made.

There is, however, general agreement that there are neither organised sympathetic chains nor strictly segmental sympathetic ganglia in cyclostomes. There are collections of ganglion cells along the cardinal veins in the abdominal cavity of lampreys which may represent sympathetic ganglia. In *Lampetra*, visceral branches of the spinal nerves run directly to the region of blood vessels, the kidneys, and the gonads. In posterior regions there is a spinal outflow to the cloaca, rectum, and ureters, with many ganglion cells in the pathway. A similar segmental visceral outflow, possibly vasomotor, has been demonstrated in *Myxine*, but in this animal the outflow to the posterior gut appears to be mainly somatic motor, supplying the striated muscles of the cloacal sphincter, with a sensory component. It is highly likely that there is no spinal autonomic innervation of the gut in hagfish but that the smooth muscles of at least the posterior intestine of lampreys is innervated from the spinal cord. Support for a vasomotor function of the sympathetic has come from the observation of Leont'eva, who showed histochemically that there is an adrenergic innervation of blood vessels in cyclostomes.

Both lampreys and hagfish possess a rich subcutaneous nerve plexus containing uni- and bipolar nerve cells, a feature unique among the vertebrates. The ganglion cells are innervated by spinal nerve

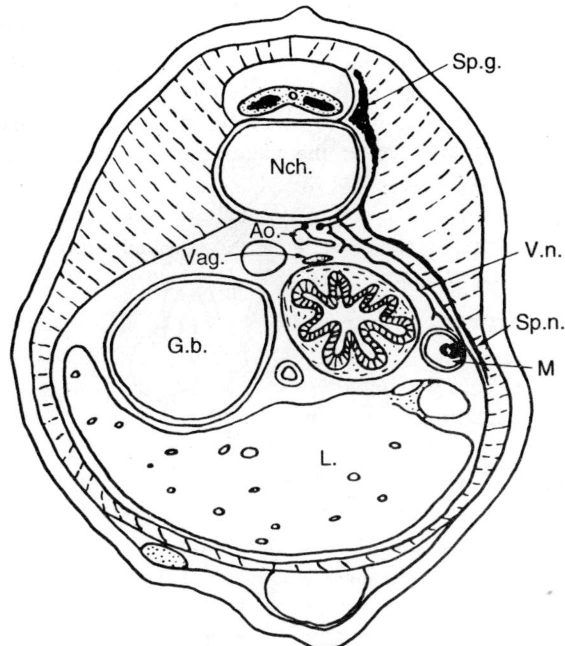

Figure 7.1: Spinal autonomic fibers in Myxine. Diagrammatic illustration of the course of the visceral (spinal sympathetic) fibers as seen in a projection of a transverse section.

fibers and are almost certainly autonomic. A number of functions could be subserved by the fibers. Bone has found that the plexus is particularly concentrated at the orifices of the slime glands in *Myxine*, associated with bundles of smooth muscle which form sphincters for the glands. In addition, the plexus is continuous about the walls of the gland and may regulate the activity of smooth muscle cells around the gland. Bone has also noted that cell bodies in the plexus often lie beside the small blood vessels of the subcutis, and he has suggested that the plexus may control blood flow in the subcutaneous blood sinuses. One further function may be proposed for the subcutaneous plexus, that of chromatophore control. Colour changes in most cyclostomes appear to be extremely slow if they occur at all. However, Wild has reported that *Petromyzon marinus* shows dramatic and rapid colour changes, indicating a nervous control of the chromatophores in this species, presumably via the plexus.

GNATHOSTOMATOUS FISH

Physiological observations on the autonomic nervous system of gnathostome fishes have been virtually restricted to selachians and

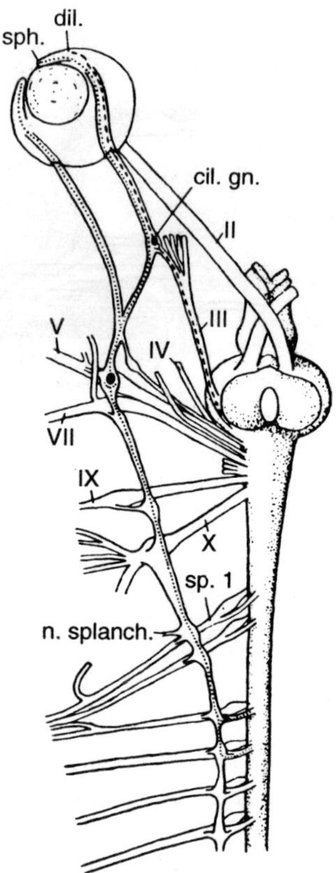

Figure 7.2 : Diagram of a ventral view of the anterior sympathetic system in Uranoscopus, *showing the sympathetic and parasympathetic innervation of the iris. cil. gn., ciliary ganglion; dil., pupil dilator muscle; n. splanch., anterior splanchnic nerve; sph., pupil sphincter muscle; sp. 1, first spinal nerve; cranial nerves numbered II, III, IV, V, VII, IX, and X. (. . .) Sympathetic pathways and (—) parasympathetic pathways; positions of ganglionic synapses shown by filled circles.*

teleosts. Although there are conspicuous anatomical differences between the autonomic systems of these two groups, especially in the arrangement of the spinal outflow, it seems likely that there are no gross physiological differences.

Cranial Autonomic Nerves

Autonomic fibers leave the brainstem of selachians in the oculomotor, facial, glossopharyngeal, and vagus nerves. In teleosts, the outflow is restricted to the oculomotor and vagus nerves. There have been no physiological investigations of the autonomic functions mediated by the selachian VIIth and IXth nerves, but it has been suggested that both outflows provide vasomotor fibers to the pharyngeal region. Presumably this function has been taken over completely by the vagus in teleosts.

Oculomotor Nerve

Oculomotor nerves are absent from forms with reduced eyes, but where present they supply preganglionic fibers to the ciliary ganglion, from which postganglionic fibers pass to the eyeball. Stimulation of the oculomotor nerves causes pupillary dilatation in the teleost *Uranoscopus scaber* and in one selachian, *Scyllium*, although this was not seen in two others. This response is in marked contrast to the pupillary constriction caused by oculomotor nerve stimulation in mammals. There is no direct evidence to show whether the response represents an inhibition of activity in the sphincter or an excitation of the dilator muscle. In *Uranoscopus*, both epinephrine and acetylcholine can cause both dilatation and constriction of the pupil; in the selachians studied both epinephrine and acetylcholinc caused predominantly dilator reactions, and it has been found that acetylcholine causes contraction of the isolated iris dilator muscle of *Scyllium*. Atropine inhibits constrictions of *Uranoscopus pupil* caused both by sympathetic nerve stimulation and by acetylcholine, but the effects of blocking drugs on responses to oculomotor nerve stimulation have not been tested. The most simple, but not necessarily the correct interpretation of these limited observations, is that both the sympathetic and the oculomotor are cholinergic and that the sympathetic provides an excitatory innervation to the sphincter (in teleosts only) while oculomotor stimulation excites the dilator muscle.

Vagus Nerve

The posttrematic rami of the vagus nerve in teleosts (as well as of the glossopharyngeal nerve in selachians) almost certainly innervate the branchial vascular bed. Direct evidence of efferent innervation is wanting, but elevations of branchial vascular resistance are known to occur during, for instance, periods of anoxia. Determination of whether this phenomenon is nerve mediated is made difficult by indications that the gill vasculature can constrict in direct response to the anoxic conditions. It is well established that catecholamines dilate and acetylcholine constricts the branchial vasculature in teleosts, although no vasomotor responses could be obtained from the isolated gills of the selachian *Squalus acanthias*. Similar vasoconstrictor actions of acetylcholine have been observed in the amphibian pulmonary vascular bed, a system derived from branchial arch vasculature. It is then reasonable to suggest that the vagi provide cholinergic vasoconstrictor fibers to the gills, if only because of the analogy with the established cholinergic vasoconstriction mediated by the vagi in amphibian lungs.

The visceral vagal rami distribute efferent fibers to the heart, the stomach, and usually no more than the most oral region of the intestine, the swim bladder in teleosts, and possibly to accessory digestive organs and to blood vessels in this region. The innervation of the heart has been reviewed by Nicol and more recently by Johansen and Martin. The evidence for a cholinergic vagal supply to the heart, causing a reduction in the rate of beating, is so overwhelmingly strong that no further comment is needed. The consensus has been that the vagi do not contain cardiac augmentor or accelerator fibers, whether of sympathetic or parasympathetic origin, and it is well established that there are no distinct sympathetic nerves innervating the heart directly. Jullien and Ripplinger found that after degenerative sections made high in the vagal trunk stimulation of the vagal cardiac branches still caused negative chronotropic and inotropic responses in teleosts, indicating that the ganglionic synapses in the vagal pathway occur within the vagal trunk itself. They also observed that the heart rate was slow and that arrhythmias were common in bivagotomised tench and that the heart returned to a fast, regular beat after atropine treatment. They suggested that the vagal postganglionic cell bodies are tonically active, holding the heart muscle under an inhibitory drive, and that the activity of the neurons is normally inhibited by fibers from the central nervous system running in the vagi, a suggestion which deserves further consideration.

A further observation made by Jullien and Ripplinger is also worth reconsidering. In a series of papers around (summarised in Jullien and Ripplinger), they provided evidence for a unique system of noncholinergic "negative tonotropic" nerve fibers in the vagal supply to the teleost heart. The crux of their argument is that in nonbeating hearts, e.g., after treatment with acetylcholine, stimulation of the vagus causes elongation of the heart; on the other hand, in hearts treated with atropine, stimulation of the vagi causes elongation of the heart when all signs of negative inotropic and chronotropic responses have vanished. Since their recordings have been made by attaching the heart to a kymograph lever *in situ*, leaving the heart attached to the body by the large veins alone, one might suggest that the " tonotropic' responses represent inhibition of venous smooth muscle tone. This suggestion is striking enough but is perhaps more acceptable than the alternative that the cardiac muscle cells are in a condition of tonus which is not related to cardiac action potential activity.

The vagal innervation of the gut of teleosts and selachians has been reviewed by Nicol, Barrington, and Campbell and Burnstock. All reviewers agree that stimulation of the vagi causes contraction of the stomach, but not of the intestine, in those animals possessing a stomach. When there is no stomach, the vagi may cause contraction of part or all of the intestine. However, cautious interpretation of this result is needed for two reasons. First, part or all of the responses observed might result from stimulation of sympathetic nerve fibers running with the vagus, at least in teleosts. Young described connections between the cranial sympathetic chain and the vagus nerve in *Uranoscopus*, and Fange has already raised this question with respect to the innervation of the swim bladder. It is pertinent to raise this possibility since all of the excitatory nerve fibers in the vagosympathetic trunk supplying the gastric musculature of an amphibian, *Bufomarinus*, are of sympathetic origin, the vagus supplying only nonadrenergic inhibitory fibers to the stomach. The presence of these nonadrenergic inhibitory fibers in the gastric vagi of both mammals and amphibians provides the second reason for caution, for it is not yet clear whether such fibers exist in fish. Campbell and Burnstock have pointed out that many of the excitatory responses to vagal stimulation recorded by earlier workers on both selachian and teleostean gut do not start until some time after the period of stimulation of the nerves is over. In this respect the contractions appear more like the "rebound" or recovery contractions which follow responses to inhibitory nerve stimulation in mammalian gut than like the virtually immediate primary contractions caused by stimulation of the excitatory innervation. In other words, there is still considerable doubt as to whether the contractions seen in fish stomach preparations following vagus nerve stimulation are mediated by excitatory or inhibitory nerves. The absence of records of actual inhibition could be easily explained under the conditions of experimentation for the nerves have been stimulated for relatively short periods, the spontaneous contractions occur at a very low rate, if at all, and the musculature has little or no tonus. Pharmacological investigations have been of little help in answering this question so far. For instance, vagal excitation of the stomach is not inhibited by atropine in the brown trout or in *Raja*, whereas vagal excitation of the smooth muscle of the intestine of the tench is prevented by atropine; in all three preparations, atropine prevents the excitatory action of acetylcholine. Only further experimentation can settle this question.

There have been no further investigations of neural control of gastric, hepatic, or pancreatic secretion in any fish since the subject was reviewed by Barrington in 1957. One can only agree with Barrington that there is no evidence for such a nervous control, while remarking that there have been too few investigations to allow any conclusion to be reached.

The processes causing filling and emptying of the swim bladder of teleosts are complex. At least three processes seem to be implicated in determining whether a net secretion or absorption of gases occurs. First, the relative amounts of secretory and absorptive epithelium exposed to the lumen of the swim bladder can be varied by differential contraction of the muscularis mucosae or by variable closure of the oval. Second, the rates of blood perfusion of the secretory epithelium, and therefore of the retia mirabilia, and of the resorptive epithelium are probably independently controlled. Third, the activity of the secretory cells themselves, which appear to act by adding metabolites to the blood leaving the secretory epithelium, is probably under nervous control.

Bohr first showed that following section of the vagi, experimentally deflated swim bladders were no longer refilled. This observation shows that the vagi contain nerve fibers which are in some way responsible for secretion into the swim bladder. However, Fange was unable to show any filling responses to vagal stimulation. In fact, the only clear response to vagal stimulation that he observed was an increase in the relative area of resorptive epithelium exposed to the enclosed gases brought about in a number of ways by contraction of the muscularis mucosae, and a vasodilation in the resorptive epithelium. This response would favor emptying of the swim bladder, and it probably masked completely those effects of vagal stimulation favoring filling. Fange showed that these "emptying" responses of the muscularis mucosae and of the vasculature were probably mediated by adrenergic nerves which, he suggested, contracted the muscularis mucosae of the secretory portion but relaxed that of the resorptive portion of the swim bladder and dilated the blood vessels in the resorptive region. In view of their adrenergic nature, Fange postulated that these fibers were sympathetic in origin, joining the vagus intracranially. In the absence of conflicting evidence, this is still the safest interpretation. Fange's pharmacological arguments that the nerves are adrenergic have received support from studies using the histochemical technique for demonstrating tissue monoamines. Fahlen *et al.*, using this technique, have shown that there is an adrenergic (norepinephrine) innervation of blood vessels throughout the swim bladder and of the muscularis mucosae only in the secretory region in cod and trout; Fahlen found that the muscularis of only the postulated resorptive region in *Argentina* was adrenergically innervated. Many or all of these adrenergic neurons have their cell bodies in or near the swim bladder, a fact which would explain why vagotomy does not alter catecholamine levels in this organ .

It is highly likely that the vagal nerve fibers which cause secretion are cholinergic, for Fange has shown that atropine treatment, like vagotomy, inhibits gas secretion into a deflated swim bladder. The problem is which structure or structures are being affected by the choline-rgic fibers. Fange has suggested that the vagi provide cholinergic fibers to contract the muscularis mucosae of resorptive regions, a response which would tend to cause filling by decreasing the area of resorptive epithelium exposed. He reached this conclusion because he found that isolated preparations of the tissue were contracted by acetylcholine and relaxed by atropine. However, he did not observe muscular contractions of this type in response to vagal stimulation, nor did atropine treatment affect the position of the swim bladder diaphragm in *Ctenolabrus*. Fange has provided evidence for vagal cholinergic fibers causing vasodilatation in the secretory portion of the swim bladder. He noted that during secretion there was a marked vasodilatation in the gas gland, whereas after vagotomy, and especially after atropine treatment, there was considerable vasoconstriction in the gland. But Fange rightly raised the still unresolved question of whether the vasodilatation occurring during activity is a primary effect caused by specifically vasodilator nerves or a secondary effect caused by the release of metabolites from cholinergically activated secretory cells.

Spinal Autonomic Nerves

The sympathetic nervous system of selachians consists of a series of paravertebral ganglia, one or more occurring per spinal nerve, linked together at most by a loose plexus of nerve bundles. There is no compact sympathetic chain. The ganglia are connected to the spinal nerves by white rami communicantes. Young claims that gray rami are absent, i.e., that there is no entry of postganglionic fibers into the spinal nerves to reach dermal structures, but there is some evidence for sympathetic nervous control of dermal melanophores. The ganglia extend no further caudally than the posterior end of the mesonephros in adults. There are no cranial sympathetic ganglia nor is there any semblance of a sympathetic chain entering the head region; however, this does not preclude a sympathetic innervation of cranial structures via perivascular nerve plexuses. There are no prevertebral ganglia, and fibers from the paravertebral ganglia are collected into anterior, middle, and posterior splanchnic nerves extending to the alimentary canal via plexuses on the arterial supply. Other fibers run directly to the genital and urinary ducts.

In teleosts, the sympathetic system has a well-defined structure resembling that found in tetrapods. Paravertebral sympathetic ganglia, usually two per spinal segment, arc linked together regularly to form two sympathetic chains which fuse to a greater or lesser extent in different species. The sympathetic chains are connected to spinal nerves by both gray and white rami communicantes. The chains extend caudally to the end of the tail and also extend rostrally into the cranium where connections are usually made with the IIIrd, Vth, VIIth, IXth, and Xth cranial nerves. The gastrointestinal tract is mainly innervated by an anterior splanchnic nerve, leaving the right sympathetic chain, but a posterior splanchnic nerve has been found in trout. Genital and vesical nerves leave the posterior regions of the chains, supplying the urinary andd genital ducts via periarterial nerve plexuses.

A sympathetic innervation of the iris occurs in teleosts, but it has never been observed in selachians. The fibers involved leave the spinal cord in the anterior spinal nerves and run rostrally up the chain to form synapses in the sympathetic ganglion in the trigeminal nerve. Stimulation of the sympathetic chain causes pupilloconstriction. The response is prevented by treatment with atropine, indicating that the nerves are cholinergic. The sympathetic also appears to provide a motor innervation to the striated extraocular muscles in teleosts because sympathetic section causes abnormal protrusion of the eyeball. There is no evidence concerning the transmitter substance mediating this effect, but it has been found that sympathetic adrenergic fibers can cause contraction of the superior rectus muscle in a mammal.

It has long been denied that there is a sympathetic innervation of the heart in fish. However, fluorescence histochemical studies of the heart of teleosts have now shown that in spite of one report to the contrary there is an adrenergic innervation of the cardiac chambers. Nerve fibers containing small granular vesicles, typical of adrenergic nerves in other animals, have been seen during an electron microscopic investigation of the trout heart. The adrenergic nerve fibers appear to come both from the vagi and along the coronary arteries and occur most densely in the sinus venosus and auricle. Gannon and Burnstock have confirmed Kulaev's report that stimulation of the vagi can cause ardioacceleration in teleosts. They have also shown that this response is prevented by treatment with the adrenergic neuron-blocking drugs, bretylium and guanethidine, or with an epinephrine-blocking drug, propranolol, a congener of which inhibits cardiac responses to catecholamines in plaice. As in the case of the vagal adrenergic innervation of the swim bladder, it is probably safest to assume that these adrenergic cardioaccelerator nerve fibers enter the vagi from the sympathetic chain via the intracranial connections. In selachians, considering the absence of sympathetic nerves running directly to the heart and the absence of intracranial connections between the sympathetic and the vagus, it seems unlikely that a sympathetic adrenergic innervation of the heart will be found.

In both selachians and teleosts we are temptingly close to being able to say that there is sympathetic nervous control of the vascular system. On the one hand are reports that catecholamines injected into the circulation cause pressor responses in selachians and teleosts. On the other hand are the many reports of dense nervous plexuses investing blood vessels and, recently, a number of histochemical studies which show that there are adrenergic nerve fibers in some of these plexuses, at least in teleosts. Kirby and Burnstock have shown that adrenergic and cholinergic nerves, both mediating contraction of isolated spiral strips, occur in the ventral aorta of trout and eels. And yet there is still no report of a direct effect of autonomic nerve stimulation on the resistance of any vascular bed in fishes. What little evidence there is goes against belief in a sympathetic vasomotor tone. Wilber and Sudak found that there were no compensatory cardiovascular responses to hemorrhage in *Mustelus* and *Squalus*, while Burger and Bradley showed that a ganglionblocking drug, tetraethylammonium, and an epinephrine-blocking drug, dibenamine, did not cause any alteration of ventral or dorsal aortic pressure in *Squalus*. Randall and Stevens found that the epinephrine-blocker, phenoxybenzamine, caused a fall of only 1-2 mm Hg in dorsal aortic pressure in coho and sockeye salmon. At the moment it seems likely that if there is a sympathetic vasoconstrictor innervation of systemic vascular beds, it is used to regulate blood flow to individual organs rather than to sustain a certain level of blood pressure. Any generalised control of systemic vascular resistance is likely to be mediated by circulating epinephrine from chromaffin cells.

The sympathetic innervation of the gastrointestinal tract has been reviewed recently. Interpretation of the available results is again made difficult by the possibility that the observed excitatory responses to sympathetic nerve stimulation are in reality rebound contractions caused by the stimulation of inhibitory nerves. However, these authors concluded that a number of workers had shown true primary contractions mediated by sympathetic nerves. Other excitatory responses could be interpreted as rebound contractions, and there is one clear example of an inhibitory response to ' posterior splanchnic nerve stimulation in trout. In summary, these authors suggested that the sympathetic supply to all regions of the gut in both selachians and teleosts contains a mixture of excitatory and inhibitory nerve fibers. One may suggest that the excitatory nerve fibers are cholincrgic, for at least some of the excitatory responses are abolished by atropine treatment. The inhibitory nerve fibers may be adrenergic since catecholamines cause relaxation of the gut of a number of species and an adrenergic innervation of gastrointestinal muscle has been demonstrated histochemically in trout, eel, and tench, but direct evidence is lacking. In addition to the adrenergic, presumably sympathetic, nerves reaching the swim bladder in the vagus trunk of teleosts, there is a direct sympathetic innervation by a branch of the splanchnicnerve. Fange showed that if this branch is cut there is a slight rise in the oxygen content of the swim bladder gases. Fange suggested that this might occur because of a loss of sympathetic vasoconstrictor tone in the gas gland. However, Fahlen *et al.* have shown that adrenergic nerve terminals, deriving from cell bodies in the coeliac sympathetic ganglion, form pericellular networks about the adrenergic neurons of the gas gland ganglion in cod. This raises the possibility that bysectioning the splanchnic nerve one removes an inhibitory influence on synaptic transmission in the gas gland ganglion, thus enhancing a secretory activity being mediated by that neural pathway.

A few passing observations have been made on the sympathetic innervation of the urinogenital system. Stimulation of the sympathetic ganglia in selachians causes contractions of the oviducts, as does the application of acetylcholine. Similar responses of the ovaries occur in the teleosts *Lophius* and *Uranoscopus*. In addition, stimulation of the sympathetic innervation of the urinary bladder in these teleosts causes a contraction and the expulsion of urine, a response also mimicked by acetylcholine. Young found that the excitatory responses of the bladder to nerve stimulation were reduced by atropine, suggesting that the nerves are cholinergic.

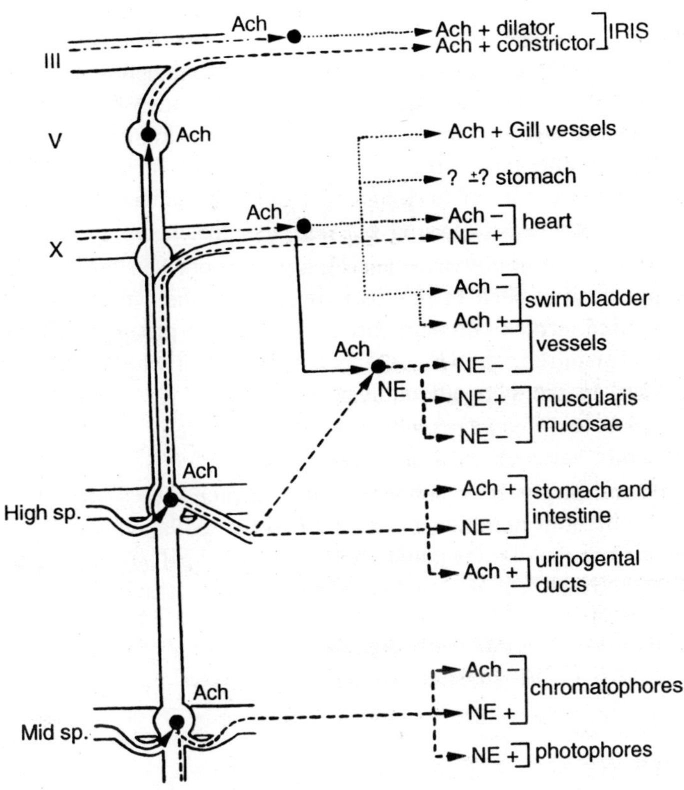

Figure 7.3 : Diagram to illustrate some principles of the arrangement of the autonomic nervous system in teleosts.

The number of investigations of nervous control of dermal chromatophores in fish is so great that no detailed discussion will be given here. The main conclusion reached has been that the melanophores in many if not all teleosts are under the control of sympathetic adrenergic nerve fibers, stimulation of which causes concentration of the melanophore pigment. A comparable sympathetic adrenergic innervation of dermal photophores has been postulated for the teleost *Porichthys*. Although it has been suggested that there are cholinergic sympathetic fibers which mediate pigment dispersal in melanophores, it seems to this reviewer that there is no definitive evidence for their existence. In selachians, colour changes are generally slow or absent, but evidence has been produced for melanophore-concentrating nerves in *Mustelus*. In a few teleosts, catecholamines cause dispersal rather than the normal concentration of melanophore pigments, and it is not clear how this will affect the chromatophore innervation. There is very little information concerning the innervation of other chromatophore types (xanthophores and erythrophores), although it is well known that they may, for example, disperse under conditions which induce melanophore concentration. Virtually nothing is known of the control of internal melanophores. Perhaps the most striking feature of chromatophore control in fish is that the animals can not only match the shade of the background but can also match its

pattern, an ability which is highly developed in various flatfish. The presence of such responses indicates that the sympathetic system can be used to pick out single or, at most, small groups of chromatophores as required, showing a fineness of control which is not usually attributed to the sympathetic system.

There is no clear evidence for the existence of a "spinal parasympathetic" autonomic outflow in fish. But Friedman has shown that stimulation of posterior spinal nerves in skates causes erection of the claspers by both muscular contraction and vasodilatation and elicits secretion from the clasper gland. As Nicol has remarked, these phenomena resemble effects of stimulation of the sacral parasympathetic system in mammals.

INDEX

A

B

C

D

E

F

G

H

I

L

T

U